Lecture Notes in Computer S

Commenced Publication in 1973
Founding and Former Series Editors:
Gerhard Goos, Juris Hartmanis, and Jan van Leeuwen

Editorial Board

Pietro Liò   Eiko Yoneki
Jon Crowcroft   Dinesh C. Verma (Eds.)

Bio-Inspired Computing and Communication

First Workshop on Bio-Inspired Design of Networks, BIOWIRE 2007
Cambridge, UK, April 2-5, 2007
Revised Selected Papers

 Springer

Volume Editors

Pietro Liò
Eiko Yoneki
Jon Crowcroft
University of Cambridge
The Computer Laboratory
William Gates Building, 15 J.J. Thomson Avenue, Cambridge CB3 0FD, UK
E-mail: {pietro.lio, eiko.yoneki, jon.crowcroft}@cl.cam.ac.uk

Dinesh C. Verma
IBM T.J. Watson Research Center
P.O. Box 704, Yorktown Heights, NY 10598, USA
E-mail: dverma@us.ibm.com

Library of Congress Control Number: 2008940650

CR Subject Classification (1998): F.1, F.2, C.2, I.6, D.1-2, H.2

LNCS Sublibrary: SL 1 – Theoretical Computer Science and General Issues

ISSN   0302-9743
ISBN-10   3-540-92190-7 Springer Berlin Heidelberg New York
ISBN-13   978-3-540-92190-5 Springer Berlin Heidelberg New York

springer.com

© Springer-Verlag Berlin Heidelberg 2008

Typesetting: Camera-ready by author, data conversion by Scientific Publishing Services, Chennai, India
Printed on acid-free paper   SPIN: 12568628   06/3180   5 4 3 2 1 0

Preface

This volume contains the papers from BIOWIRE 2007, the first in a series of workshops on the bio-inspired design of networks, and additional papers contributed from the research area of bio-inspired computing and communication. The workshop took place at the University of Cambridge during April 2–5, 2007 with sponsorship from the US/UK International Technology Alliance in Network and Information Sciences. Its objective was to present, discuss and explore the recent developments in the field of bio-inspired design of networks, with particular regard to wireless networks and the self-organizing properties of biological networks. The workshop was organized by Jon Crowcroft (University of Cambridge), Don Towsley (University of Massachusetts), Dinesh Verma (IBM T.J. Watson Research Center), Vasilis Pappas (IBM T.J. Watson Research Center), Ananthram Swami (ARL), Tom McCutcheon (DSTL) and Pietro Liò (University of Cambridge).

The program for BIOWIRE 2007 included 54 speakers covering a diverse range of topics, categorized as follows:

1. Self-organized communication networks in insects
2. Neuronal communications
3. Bio-computing
4. Epidemiology
5. Network theory
6. Wireless and sensorial networks
7. Brain: models of sensorial integration

The BIOWIRE workshop focuses on achieving a common ground for knowledge sharing among scientists with expertise in investigating the *application domain* (e.g., biological, wireless, data communication and transportation networks) and scientists with relevant expertise in the *methodology domain* (e.g., mathematics and statistical physics of networks). The aim of the workshop is to bring researchers together, to further our insights into bio-inspired computing and communications, and to investigate collectively the challenges that remain.

As an outcome from the workshop, we decided to edit the book *Bio-Inspired Computing and Communication* that includes some of the research papers presented at the workshop and additional papers contributed from the research area of bio-inspired design of networks. We received many papers of high quality, and we selected 35 papers in the following categories. Jon Crowcroft provides further insight into our selection in the introductory section.

1. Biological networks (6 papers)
2. Network epidemics (4 papers)
3. Complex networks (7 papers)
4. Bio-inspired network model (4 papers)
5. Network protocol in wireless communication (5 papers)

6. Data management (2 papers)
7. Distributed computing (5 papers)
8. Security (2 papers)

We would like to express our deep appreciation to the authors for submitting articles, and for sharing the results of their research work with the rest of the community. Finally, we would like to thank Springer for their excellent cooperation and our sponsoring organization.

August 2008

Pietro Liò
Eiko Yoneki
Jon Crowcroft
Dinesh C. Verma

Table of Contents

Complex Networks

Bio-Inspired Network Model

Network Protocol in Wireless Communication

Data Management

Distributed Computing

Security

Bio-Inspired Computing and Communication

Jon Crowcroft

University of Cambridge Computer Laboratory
Cambridge CB3 0FD, United Kingdom
Jon.Crowcroft@cl.cam.ac.uk

This volume of papers was put together by the Editors after a successful workshop that was held in the University of Cambridge in April 2007, partly sponsored by the UK-US ITA project. The goal of the workshop (and hence the purpose of this publication) was to bring together ideas from the natural world and problems in the area of wireless networks.

Wireless networking is more prevalent than any other kind of artificial communications system on Earth at the time of writing. There are over three billion digital cellular telephones in the world, and billions of analogue and digital TV receivers, as well as numerous specialized wireless networks in civilian and military life.

Wireless network system design is a very complex discipline, as so many factors interact. Radio propagation alone presents a huge challenge, as it is dependent on frequency, terrain, atmospheric conditions and so forth. Antennae design, transmission power control, coding and modulation, protocol design for sharing spectrum, for sharing the medium, for coping with interference, for cooperation in transmission and coding and modulation are all inter-related. Routing over multiple hops, using multiple radios simultaneously, managing interference, and disruption/delay/disconnection tolerance are all computationally difficult problems to solve. Increasingly, many of the tasks can be solved in software, making radio systems much more flexible, and requiring the systems designer to confront more choices than in the past limited by technology and cost constraints.

Networks exist in the natural world too, in many many forms, whether biological or otherwise. Such systems have typically evolved to fit certain ecological niches; however, they often exhibit characteristics that contain solutions to problems that we would like to solve in artificial networks that we are designing and building. This we seek to define as a sub-discipline of communications science, which is the bio-inspired design of networks, and (for the purposes of this work in particular), wireless networks.

Natural systems are often very large in terms of the number of nodes and edges between nodes (if we think of neural or cell-signalling networks, for example). While the networks they form may not be especially efficient in resource utilization on any particular axis (indeed they are often profligate with resources), they often manage to operate over a far wider range of parameter values, not just in terms of numbers of nodes or simple scale of the system, but more crucially in terms of dynamics too. These abilities to scale to Moles' worth of nodes and to shift operating modes (phase changes) to cope with diversity, with changing topologies or attacks, or varying degree distributions or (speaking metaphorically) link speeds, are highly attractive to future wireless network designers.

P. Liò et al. (Eds.): BIOWIRE 2007, LNCS 5151, pp. 1–8, 2008.

We attempted to group the papers in this volume by topic; however, it is clear that there are several dimensions along which one can classify each work, and so we encourage readers to take with a pinch of salt the simple taxonomy we use in the introduction here. One might wish to sort the papers into sections on: topology/graphs from nature; or control and routing, or phase transition, or emergence, and the so-called self* properties, or perhaps adaptation (not just autonomic, but also through different operating regimes), e.g., space vs. time optimization in file systems; or sloppiness and optimization; or sources of inspiration whether physical, chemical, biological, including cellular neural, ecological, botanical; and so on. However, this would be reductionist.

In any case, we give a list of the papers with some brief comments on the way that we (as networking researchers) take messages home from the ideas presented from these other, sometimes distant, subject disciplines.

1 Biological Networks

1.1 A Complex Network Approach to the Determination of Functional Groups in the Neural System of *C. elegans*, by Arenas et al.

This paper is very useful for introducing the tools and techniques that can be learned from the biological disciplines. *C. elegans* possesses one of the simplest, and most completely understood structures (and morphogenesis) of any creature to date.

In this paper, a technique to comprehend structure modularity is applied to understanding the clustering of components in the neural system of the worm. This technique has direct applications in diverse areas such as social networking, software fault diagnosis and threat analysis in artificial communications networks.

1.2 Modelling Gene Regulatory Networks, by Gelenbe

In this paper, the role of regulation in genetic systems is used to explore the potential application of the ideas for control (e.g., traffic engineering and routing) in communications systems.

1.3 The Role of Simplifying Models in Neuroscience: Modelling Structure and Function, by Kronhaus et al.

This paper brings together ideas from both of the previous papers, and includes an understanding of the evolution over time of structure and control as well as clustering and modularity. One key take-home message from this paper is that while neural systems can be very complex, it is sometimes possible to break them down into components.

1.4 An Artificial Chemistry for Networking, by Meyer et al.

This paper contains ideas from a very different subject discipline, and illustrates that we can learn ideas of control via analogies with diffusion in chemical reactions. Similar bi-sociative creative ideas have crossed over before from statistical mechanics. We will see similar ideas in another paper on molecular communication later in this volume. Key ideas are that control may be simple and stable even though it is decentralized.

1.5 Biomimicry: Further Insights from Ant Colonies, by Ratnieks

The large-scale behavior of insect populations in terms of organizing has been successfully applied to the problem of decentralized routing. Given that communication to control routing must flow over the same network as users' traffic itself, and that in any reasonable scale network there is significant latency in transmitting information, any solution for routing should be decentralized simply so that reactions to changes are localized and not based on stale information, and also that there is no dependency on remote, potentially unavailable servers to compute routes. Decentralizing down to the level of every individual node (as happens in insect colonies) and using continuous functions to organize feedback to control behavior seems like a promising combination of techniques.

1.6 Network-Related Challenges and Insights from Neuroscience, by Peck et al.

This paper gives some crucial insights into the devils in the details of neural networks looking at the problem of calibration (correct detection of a signal and not merging or splitting of separate signals), a further look at structure and sub-networking, and finally revisiting self-organization.

2 Network Epidemics

2.1 Networks in Epidemiology, by Eames et al.

Natural epidemics involve the spread of diseases via some vector (water, air, touch, parasite etc.). Understanding the epidemic processes leads to understanding mechanisms that can be used for content distribution, or indeed, for prevention of dissemination of unwanted content (computer viruses and worms). The design space is nicely mapped out via the epidemic equations the simplest of which involves only susceptibility, infectiousness and recovery.

The earliest work in computer networks in this area was the paper on distributed database update for Grapevine by Alan Demers and others at Xerox PARC, 20 years back. There are many applications in today's Internet and cellular phone networks.

2.2 Epidemiology and Wireless Communication: Tight Analogy or Loose Metaphor?, by Eubank et al.

This paper goes into some detail about the use of tools and techniques from epidemiology in communication networks by use (very roughly) of the analogy of wireless communication as a vector.

The actual communications architecture can be designed by taking account of choice of functions for signalling and data communications via packets, not just the control of communications.

2.3   Epidemic Spreading of Computer Worms in Wireless Networks, by Nekovee

This paper directly models artificial diseases and their propagation, detection/prevention in wireless networks. This topic comes up later (but more structurally) when looking at the overall topic of security (and defence).

2.4   Wireless Epidemic Spread in Dynamic Human Networks, by Yoneki et al.

This paper looks at the structure of networks formed by humans (social, mobile, encounter based networks), and proposes the idea of using epidemics for content distribution (a la Gossip) between smart devices carried by members of society.

3   Complex Networks

3.1   Stochastic Spreading Processes on a Network Model Based on Regular Graphs, by Fallert et al.

In this paper, we move to the next level of network complexity (beyond the topology and control) where the topic of phase transitions is introduced. One of the ways in which large systems cope with a very wide range of operating environments is to switch between different response functions by phase changes. As mentioned above, this idea is far more powerful than classical autonomic control (feedback) having much more than one operating regime. The idea has not seen much application yet in artificial systems, beyond (say) simple two-region schemes.

3.2   Weighted and Directed Network on Travelling Patterns, by Miguens et al.

This paper models demand dynamics and again looks beyond the simple network and control dynamics we have used to date in communications (or transport) system design. In cellular networks (mobile phones) and in future vehicular control and information networks, input from models like this will be essential.

3.3   Communication Networks in Insect Societies, by Nicolis

Re-capitulating the social approach of large insect populations, we now start to look at phase transitions, as well as clustering and other patterns. The watchword here is emergence. Useful properties may be present in the dynamics rather than in a static behavior of a system in a stable environment.

3.4   The Topological Fortress of Termites, by Perna et al.

Termites build complex physical architectures within which they live. There are predators on termites, and the physical systems are "designed" to provide defensible structures, with interesting topology in terms of connectedness. This is the spatial analogy which is explored also in a later paper on defence as well as in covert networks (spies, terrorism, etc.).

3.5 Evolutionary and Temporal Dynamics of Transcriptional Regulatory Networks, by Babu

In yet another area of genetics, as discussed by Gelenbe earlier, we have regulatory systems that have very useful properties in terms of emergent behavior, and phase change/dynamic adaption to a changing "environment".

3.6 Phase Patterns of Coupled Oscillators with Application to Wireless Communication, by Diaz-Guilera et al.

In this paper, we look at more direct modelling of more regular systems where the structure and phase transition are perhaps easier to model (quantitatively as well as qualitatively) and have application in the area of decentralized control. This idea also connects with the organization of timers in sensor networks, as we see next.

3.7 Self-organizing De-synchronization and TDMA on Wireless Sensor Networks, by Degesys et al.

This paper looks at the problem of battling synchronization. In some networks, this is a desired property, but in many sensor networks, the goal is to prolong the battery life of the devices while maintaining a reasonable probability of timely reports of readings and/or status from each device. Thus the goal is to prevent synchronization by suitable randomization in time through a self-organizing algorithm for allocation of TDMA time-slots. This was also introduced in the Internet when it was discovered that routing control messages between different devices tended to synchronize, leading to traffic and route computational spikes in load.

4 Bio-Inspired Network Model

4.1 Bio-Inspired Multi-agent Urban Data Harvesting in Vehicular Sensing Platforms, by Lee et al.

In this paper, a direct mapping of chemotaxis to the problem of gathering info (e.g., in sensor networks on moving vehicles) is reported. Again, we see that not only biology and ecology can inform us but also chemistry.

4.2 Bio-Inspired Approaches for Autonomic Pervasive Computing Systems, by Miorandi et al.

This paper is, rather like this whole volume, a survey of techniques.

4.3   Biologically Inspired Self-healing Routing with Preferred Path Selection, by Szymanski et al.

As with ants following pheromones, we can design decentralized routing systems that use tropisms as control, such systems can be resilient to node outage (someone stepping on an ant) and self-heal. Self-* properties are of interest in ad hoc wireless networks, and in peer-to-peer systems. Indeed, the routing in the Internet had, as an original design goal, a strong requirement to be self-healing.

4.4   Biologically Inspired Approaches in Networks; The Bio-networking Architecture and the Molecular Communication, by Suda et al.

Here we take inspiration from cell signalling and communication between molecules as a way to build networks. This paper reports on what is part of a growing field of synthetic biology.

5   Network Protocol in Wireless Communication

5.1   User-Centric Mobility Models for Opportunistic Networking, by Boldrini et al.

In this work, it is observed that people are not ants.

5.2   Wavelet-Domain Statistics of Packet Switching Networks Near Traffic Congestion, by Lio et al.

This paper reports on the notions of traffic modelling using mathematics usually used to describe images or signals with self-similarity. Such signals may be generated via coupled oscillators, and exhibit interesting phase shifting characteristics too.

5.3   A Circulatory System Approach for Wireless Sensor Networks, by Pappas et al.

This paper jumps to a completely different area of the natural world, that of blood circulation. Again, there are clear applications in routing, and (even before that) for topology discovery. (See also ants!)

5.4   Epcast: Controlled Dissemination in Human-Based Wireless Networks by Means of Epidemic Spreading Models, by Scellato et al.

Again, we look at the problem of content distribution between devices carried by people, but now using directly observed patterns.

5.5   Maintaining Spatial-Temporal Knowledge Through Human Interaction, by Lenando et al.

This paper is more introspective, looking at understanding the information gathered about human social contact. (See Passarella in contrast.)

6   Data Management

6.1   Beta Random Projection, by Lu et al.

In this paper, we are working bottom up, from information theory, to devise structures for finding items in a large structured worked, by informational proximity (hamming distance), despite entropy. Indexing data so that inaccurate (partially, or poorly specified) queries can be satisfied is a common problem in genomic and in search in peer-to-peer systems, and also in intrusion detection systems were an attack may be only partially pre-known.

6.2   Biologically Inspired Classifier, by Patti et al.

Here, clustering is driven more directly by knowledge of the dataset.

7   Distributed Computing

7.1   Human Heuristics for Autonomous Agents, by Bagnoli et al.

This paper presents techniques for agent organization and control for decentralized programs, i.e., not just for traffic. The approach is modelled after neural nets, so one should look to Gelenbe, but the result is more Eckert/von Neumann.

7.2   Designing Biological Computers: Systemic Computation and Sensor Networks, by Bentley

This paper presents a completely novel architecture for programming and control based on the biological paradigm, and is a shift away from any conventional von-Neumann (even distributed agent) based computing. This is also one of the topics for the recent UK Challenges in computer science.

7.3   A Rule System for Network-Centric Operation in Massively Distributed Systems, by Dressler et al.

A key goal here is to understand levels of scale that are cellular (10^10 nodes), and far beyond those of today's artificial networks.

7.4   Field-Based Coordination for Pervasive Computing Applications, by Mamei et al.

Again, the inspiration is ants. This time, the stigmergic concept is developed to propose it for decentralized network control.

7.5   Coalition Games and Resource Allocation in Ad-Hoc Networks, by Gibbens et al.

Here, the inspiration is from games in the most general sense, and their use as an organizing principle for decentralized control.

8   Security

8.1   Bio-Inspired Topology Maintenance Protocols for Secure Wireless Sensor Networks, by Gabrielli et al.

Natural systems are messy and inefficient, as we have said above several times, but they may operate in many environments Here, we see a proposal to design defensible wireless networks that use heterogeneity directly.

8.2   The Topology of Covert Conflict, by Nagaraja et al.

In this paper, we return once more to the graphs and node degree distribution within natural networks (whether social or signalling) and how these structures are reflected in artificial networks. This leads to a very general model for such structures, and the ability to understand wherein lies the key weakness in terms of attack or defence.

A Complex Network Approach
to the Determination of Functional Groups
in the Neural System of C. Elegans

Alex Arenas, Alberto Fernández, and Sergio Gómez

Universitat Rovira i Virgili, Departament d'Enginyeria Informàtica i Matemàtiques,
Avinguda dels Països Catalans 26, 43007 Tarragona, Spain
alexandre.arenas@urv.cat
http://deim.urv.cat/~aarenas/

Abstract. The structure of real complex networks is often modular, with sets of nodes more connected between them than to the rest of the network. These communities are usually reflecting a topology-functionality interplay, whose discovery is basic for the understanding of the operation of the networks. Thus, much attention has been driven to the determination of the modular structure of complex networks. Recently it has been shown that this modular organization appears at several scales of description, which may be found by a synchronization process on top of these networks. Here we make use of it for a tentative uncovering of functional groups in the neural system of the nematode C. elegans.

1 Introduction

Complex networks are graphs representative of the intricate connections between elements in many natural and artificial systems [1,2], whose description in terms of statistical properties has been largely developed in the curse for a universal classification of them. However, when the networks are locally analyzed some characteristics that become partially hidden in the statistical description emerge. The most relevant perhaps is the discovery in many of them of *community structure*, meaning the existence of densely (or strongly) connected groups of nodes, with sparse (or weak) connections between these groups [3].

The study of the community structure helps to elucidate the organization of the networks and, eventually, could be related to the functionality of groups of nodes [4]. The most successful solutions to the community detection problem, in terms of accuracy and computational cost required, are those based in the optimization of a quality function called *modularity* and proposed in [5], that allows for the comparison of different partitioning of the network. Given a network partitioned into communities, being C_i the community to which node i is assigned, the mathematical definition of modularity [6] is expressed in terms of the weighted adjacency matrix w_{ij}, that represents the value of the weight of the link between nodes i and j (0 if no link exists), as

$$Q = \frac{1}{2w} \sum_i \sum_j \left(w_{ij} - \frac{w_i w_j}{2w} \right) \delta(C_i, C_j) \ , \tag{1}$$

P. Liò et al. (Eds.): BIOWIRE 2007, LNCS 5151, pp. 9–18, 2008.
© Springer-Verlag Berlin Heidelberg 2008

where the strength of node i is $w_i = \sum_j w_{ij}$, the total strength of the network is $2w = \sum_i w_i$, and the Kronecker delta function $\delta(C_i, C_j)$ takes the value 1 if node i and j are into the same community, 0 otherwise.

The modularity of a given partition is then, up to a multiplicative constant, the probability of having edges falling within groups in the network minus the expected probability in an equivalent (null case) network with the same number of nodes, and edges placed at random preserving the strengths of the nodes. The larger the modularity the best the partitioning is, because more deviates from the null case. Note that the optimization of the modularity cannot be performed by exhaustive search since the number of different partitions are equal to the Bell or exponential numbers [7], which grow at least exponentially in the number of nodes N. Indeed, optimization of modularity is a NP-hard (Non-deterministic Polynomial-time hard) problem [8]. Several authors have attacked the problem proposing different optimization heuristics [9,10,11,12,13,14].

Maximizing modularity one obtains the "best" partition of the network into communities. This partition represents an intermediate topological scale of organization, or *mesoscale*, that in many cases has been shown to coincide with known information about subdivisions in the network [5,15]. However, recently it has been pointed out that the optimization of the modularity has a characteristic scale related to the number of links in the network, that delimits the resolution beyond which no separation into smaller groups can be obtained when optimizing modularity, although these smaller partitions, and then different levels of description, are plausible to exist from direct observation [16]. The problem seems to be that modularity, as it has been prescribed, does not have access to these other levels of description. The reason for this is that the topological scale at which we have access by maximizing modularity has a limit.

Here we propose the use of a synchronization process [17,18,19,20,21] between nodes in the network, for the determination of the mesoscales in complex networks. In particular we show its applicability to the determination of several scales of organization in the synaptic connectivity of the neuronal system of the nematode *C. elegans* from actual data compiled from [22] and arranged by [23].

2   Determination of the Mesoscales

The main idea we propose here to detect the mesoscales is to assimilate all nodes with identical oscillators following the coupling proposed by Kuramoto [24]. The Kuramoto model consists of a population of N coupled phase oscillators where the phase of the ith unit, denoted by $\theta_i(t)$, evolve in time according to the following dynamics

$$\frac{d\theta_i}{dt} = \omega_i + \sum_j K_{ij} sin(\theta_j - \theta_i) \quad i = 1, ..., N \tag{2}$$

where ω_i stands for the natural frequency of the oscillator and K_{ij} describes the coupling between units. The original model studied by Kuramoto assumed mean-field interactions $K_{ij} = K$, $\forall i, j$. If the oscillators are identical ($\omega_i = \omega$, $\forall i$) there is only one attractor of the dynamics: the fully synchronized regime where $\theta_i = \theta$, $\forall i$.

The temporal mesoscales of the dynamics of synchronization (of phase oscillators) near the synchronization attractor are governed by the solutions of the linear dynamics:

$$\frac{d\theta_i}{dt} = -k \sum_j L_{ij}\theta_j \quad i = 1, ..., N \tag{3}$$

where k is a constant, θ_j are the phases of the nodes and L_{ij} the Laplacian matrix of the network, defined as $L_{ij} = w_i \delta_{ij} - w_{ij}$, where w_i is the strength of node i, δ_{ij} is the Kronecker delta and w_{ij} is the element of the weighted adjacency matrix.

To identify patterns of synchronization corresponding to temporal mesoscales, we use a discretization of the matrix $\rho_{ij} = \langle cos(\theta_i - \theta_j) \rangle$ where $\langle \cdots \rangle$ stands for the average over different realizations of the initial conditions. In all cases presented here we have averaged 10^5 realizations [17]. The solution of Eq.3 in terms of the normal modes $\varphi_i(t)$ reads

$$\varphi_i(t) = \sum_j B_{ij}\theta_j = \varphi_i(0)e^{-\lambda_i t} \quad i = 1, ..., N \tag{4}$$

where λ_i are the eigenvalues of the Laplacian matrix, and B is the matrix of its eigenvectors. The different intermediate scales are separated according to gaps in the mode decay times defined by the difference between the inverse of consecutive (ordered) eigenvalues $1/\lambda_i - 1/\lambda_{i+1}$. Note that the smallest (different from 0) eigenvalue of the Laplacian matrix determines the time scale for the whole network to synchronize.

By using this method it is possible to reveal the structural mesoscales following the synchronization process, from the beginning of the process (t=0) when oscillators are desynchronized, to the end of the process when the whole system is at the synchronization fixed point. The mesoscales are defined as the communities of synchronization by taking snapshots of the evolution at discrete times and finding the synchronization patterns as described above.

3   Analysis of the Mesoscales

The different structural patterns observed along the synchronization process can be represented as a set of matrices at different times. For a comprehensive representation of the whole mesoscale that allows for the extraction of information, we propose to represent each matrix after processing it as follows: (i) for each pair of nodes we compute the degree of synchronization; (ii) the matrix is reordered from left to right by the size of the connected components with larger synchronization values. The darker colors in the scale represent groups of nodes that are more synchronized.

In the following subsections we give the details of the mesoscales determination for a toy model, in order to clarify all the steps involved.

A

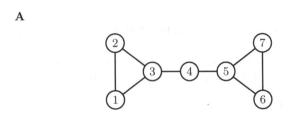

B

patterns found		% length
$\{1,2,3,4\}$ $\{5,6,7\}$		58.25%
$\{1,2,3\}$ $\{4\}$ $\{5,6,7\}$		36.52%
$\{1,2\}$ $\{3\}$ $\{4\}$ $\{5\}$ $\{6,7\}$		5.23%

C

	1	2	3	4	5	6	7
1	1.00	1.00	0.95	0.58	0.00	0.00	0.00
2	1.00	1.00	0.95	0.58	0.00	0.00	0.00
3	0.95	0.95	1.00	0.58	0.00	0.00	0.00
4	0.58	0.58	0.58	1.00	0.00	0.00	0.00
5	0.00	0.00	0.00	0.00	1.00	0.95	0.95
6	0.00	0.00	0.00	0.00	0.95	1.00	1.00
7	0.00	0.00	0.00	0.00	0.95	1.00	1.00

D

Fig. 1. (A) Sample network for the determination of its mesoscales. **(B)** Lengths corresponding to each optimal configuration. **(C)** Mesoscales table, formed by the lengths of pairs of nodes in the same community, normalized by the total length. **(D)** Mesoscales matrix (the contrast has been adjusted to enhance the visibility of the four different length levels present in the mesoscales table).

3.1   Mesoscales Matrix

Let us consider the undirected graph in Fig. 1A, with all weights equal to 1. We study the mesoscales using a discretization of the synchronization process. Any graphical representation of the whole temporal mesoscale should take into account, for every pair of nodes, the proportion of mesoscales at which they belong to the same community. Each mesoscale has a natural *length* (see Fig. 1B) defined by the range of values of time (in logarithmic scale) at which the patterns are represented. Thus, the length proportion for a pair of nodes is the sum of the lengths corresponding to mesoscales in which they belong to the same community, normalized by the total length (see Fig. 1C). The graphical representation of this table, which we call *mesoscales matrix*, is shown in Fig. 1D.

3.2   Filtered Mesoscales Matrix

The previous example is quite simple since the mesoscales obtained are hierarchical and their representation following the hierarchical order is convenient

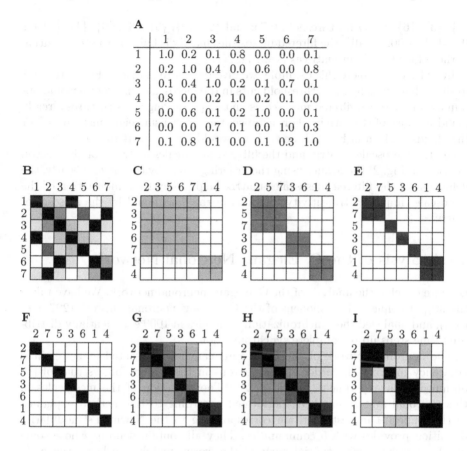

Fig. 2. Determination of the filtered mesoscales matrix. **(A)** Sample mesoscales table. **(B)** Corresponding mesoscales matrix. **(C)** Connected components of the mesoscales matrix at the threshold of 0.25. **(D)** Threshold of 0.50. **(E)** Threshold of 0.75. **(F)** Threshold of 1.00. **(G)** Filtered mesoscales matrix (4 levels). **(H)** Filtered mesoscales matrix (8 levels). **(I)** Mesoscales matrix using the ordering defined by the filtered mesoscales matrix.

to extract information. However, let us suppose that, after averaging, we have obtained the mesoscales table in Fig. 2A, whose mesoscales matrix is shown in Fig. 2B. We define the *filtered mesoscales matrix* which is obtained by the application of several thresholds to the mesoscales matrix, i.e. the lengths below the threshold are discarded, and the connected components of the graph defined by the remaining lengths are found. Figures 2C–F show the results after the application of thresholds 0.25, 0.50, 0.75 and 1.00. The first threshold divides the network in two connected components, which ordered by size are: $\{2,3,5,6,7\}$, $\{1,4\}$. This partition gives the reference for the rest of the process. The following connected components, ordered by size within each one of the groups found in the previous threshold, are: $\{2,5,7\}$, $\{3,6\}$, $\{1,4\}$ for threshold 0.50; $\{2,7\}$,

$\{5\}$, $\{3\}$, $\{6\}$, $\{1, 4\}$ for threshold 0.75; and $\{2\}$, $\{7\}$, $\{5\}$, $\{3\}$, $\{6\}$, $\{1\}$, $\{4\}$ for threshold 1.00. Finally, the filtered mesoscales matrix is built by the composition of these four threshold matrices (see Fig. 2G).

In order to complete this example, we give two more matrices. First, we want to show that by using more threshold cuts in the mesoscales matrix we would obtain a more detailed filtered mesoscales matrix preserving the structures already found because of transitivity. For instance, the result using eight instead of four thresholds is given in Fig. 2H. Second, we would like to assert the difference between the mesoscales matrix and the filtered mesoscales matrix. For this reason we show in Fig. 2I the former using the ordering found by the latter. Clearly, the definition of the filtered mesoscales matrix helps to extract information of the mesoscales imposing transitivity relations in the data found by the mesoscales matrix.

4   Analysis of the C. Elegans Neuronal Network

Here we develop the analysis of the C. elegans neuronal network. We have taken the largest connected component of the C. elegans neuronal network (297 neurons), and analyzed the synchronization dynamics in 1000 exponentially distributed time steps up to complete synchronization.

The neuronal network of the C. elegans can be represented as a weighted adjacency matrix. The order of the neurons in the matrix follows that of [1], obtained from experimental data in [22]. The detection of the mesoscales in this neuronal system has been performed according to the method explained in this paper. The best partition corresponding to the Newman's modularity definition provides with 5 communities. They all contain neurons whose soma can be correlated with spatial parts of the worm, mainly the head, the body and the tail (see Fig. 3). This coarse graining provides then with a large scale structural level in this system. We use the Newman's partition as a reference for the substructures found by the method, i.e. Newman's scale corresponds to the threshold equal to 0 in the mesoscales matrix.

Any trial of classification by the functional role of neurons in the C. elegans is extremely delicate because of the multi functional aspects they have. Many neurons participate in different synaptic pathways resulting in different functionalities. To extract information from the results obtained, we use the filtered mesoscales matrix as explained in the previous section. By fixing a threshold in the length value, we are able to unravel sub-structural scales that could correspond to groups of neurons involved in different functionalities. The most interesting information is that provided at a large value of the threshold, because in this case the substructures found contain small groups of neurons whose activity is most likely associated to a specific action. With this information at hand, and the wide description of each neuron found at the public database of C. elegans [22], we propose a tentative classification of some groups of neurons by functionality.

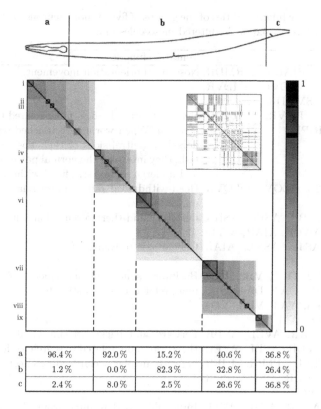

	a				
a	96.4%	92.0%	15.2%	40.6%	36.8%
b	1.2%	0.0%	82.3%	32.8%	26.4%
c	2.4%	8.0%	2.5%	26.6%	36.8%

Fig. 3. Mesoscales matrix (inside) and filtered mesoscales matrix of the C. elegans neuronal network

We have studied the filtered mesoscales matrix at a threshold value of 0.6. Fixing our attention at this level of description, we present a tentative functional classification for the groups of five or more neurons (see Fig. 3). We have used the information presented in [25] and [22] for each neuron position and individual functionality, as a guide for the classification of specific actions. Our purpose, after identification of individual functionalities, has been to assign a specific action to the whole group of neurons.

The results of the analysis of the filtered mesoscales matrix for the *C. elegans* neuronal connectivity show that: i) the substructures that prevail at different topological scales are most of them in agreement with the location of the soma of neurons along the body of the worm, and ii) the functionality of the different substructures found by the method are correlated with specific actions of the worm which allows for a tentative classification of functional groups. The classification obtained (see Table 1) does not pretend to be exact but to provide biologists with a useful information for future research.

Table 1. Tentative functionalities of the groups of five or more neurons of the C. elegans at a threshold level of 0.6 in the filtered mesoscales matrix

Neurons	Tentative function
i RIAL, RIAR, RMDR, RMDVR, SMDVL, SMDVR, RMDDL, SMDDR	Nose/head orientation movement.
ii IL1DR, IL1VR, IL2DR, IL2VR, RIPR	Head-withdrawal reflex, more related to dorsal relaxation. When worms are touched on either the dorsal or ventral sides of their nose with an eyelash, they interrupt the normal pattern of foraging and undergo an aversive head-withdrawal reflex.
iii IL2L, IL2R, OLQVL, OLQVR, RIH	Head-withdrawal reflex, more related to ventral relaxation.
iv ADLR, AIBR, ASEL, ASHR, AWCL, AWCR, AIAR, AIYL	Olfactory and thermo sensation reflex.
v ASGL, ASJL, ASKL, AIAL, PVQL	Chemotaxis to lysine reflex.
vi DB1, DB2, DD1, VB2, VD2, AS3, DA2, DA3, DA4, DA5, DB3, DB4, VA3, VD3, VD4, VD5, VD6, WM	Backward/sinusoidal movement of the worm, more related to touch stimulus.
vii AVAL, AVAR, AVBL, AVBR, AVDL, AVDR, AVEL, AVER, DA1, FLPL, FLPR, RIFR, PVDL, PVDR, PVPR, PQR, PVCL, PVCR	Forward and backward/sinusoidal movement of the worm, more related to search for food in starving case, involve social feeding effect.
viii AVHL, AVHR, AVJL, AVFL, AVFR	Impossible to determine from the experimental data available. There is not any specific function known for any of these neurons.
ix AVKL, AVKR, PDEL, PDER, PVM, DVA, WN	The functionality of this group could be related to a relaxation state similar to a sleep state, with reduced motor activity, decreased sensory threshold, characteristic posture and easy reversibility, basically mediated by PDs neurons.

5   Conclusions

In this paper we have introduced a method to uncover information from the several scales of description found in many real complex systems. The result is a *mesoscales matrix* whose representation provides with a structural map of the topology of the network. The mesoscale matrix for the sample network presented reveals how nodes form groups at different scales. Nevertheless, the symmetries of the network play in favor of this clear visualization. In real complex networks, where these symmetries are usually absent, a filtering process is needed to reveal the same information.

Hence, we have also designed what we call the *filtered mesoscales matrix*, consisting in to: (i) fix a mesoscale (a level color) and remove from the mesoscales matrix the elements under this level (lighter colors), (ii) calculate the connected components of the remaining elements (groups), and (iii) reorder the matrix from left to right in decreasing size within the groups obtained at previous levels. This process is iterated starting from the lowest mesoscale to the highest one, accumulating the results of previous stages. This way, without losing any information from the original mesoscales matrix, we achieve a clearer representation of the structural map.

We have applied the complete method to unravel the mesoscales of the neuronal connectivity of the nematode *C. elegans*. The whole nervous system of the nematode can be represented as a weighted adjacency matrix. We have calculated the corresponding filtered mesoscales matrix. The results of the analysis of the filtered mesoscales matrix for the *C. elegans* show some interesting correlations between synchronization patterns with the location of the soma of neurons, and with the functionalities in the worm. These results could help biologists to design specific targeted experiments based on the classification of neurons according to their roles at different topological scales.

Acknowledgments. This work has been supported by Spanish Ministry of Science and Technology Grant FIS2006-13321-C02-02.

References

1. Strogatz, S.H.: Exploring complex networks. Nature 410, 268–276 (2001)
2. Boccaletti, S., Latora, V., Moreno, Y., Chavez, M., Hwang, D.-U.: Complex networks: structure and dynamics. Phys. Rep. 424, 175–308 (2006)
3. Girvan, M., Newman, M.E.J.: Community structure in social and biological networks. Proc. Natl. Acad. Sci. USA 99, 7821–7826 (2002)
4. Guimerà, R., Amaral, L.A.N.: Functional cartography of metabolic networks. Nature 433, 895–900 (2005a)
5. Newman, M.E.J., Girvan, M.: Finding and evaluating community structure in networks. Phys. Rev. E 69, 026113 (2004)
6. Newman, M.E.J.: Analysis of weighted networks. Phys. Rev. E 70, 056131 (2004a)
7. Bell, E.T.: Exponential Numbers. Amer. Math. Monthly 41, 411–419 (1934)
8. Brandes, U., Delling, D., Gaertler, M., Goerke, R., Hoefer, M., Nikoloski, Z., Wagner, D.: Maximizing Modularity is hard. arXiv:physics/0608255 (2006)
9. Clauset, A., Newman, M.E.J., Moore, C.: Finding community structure in very large networks. Phys. Rev. E 70, 066111 (2004)
10. Duch, J., Arenas, A.: Community identification using Extremal Optimization. Phys. Rev. E 72, 027104 (2005)
11. Guimerà, R., Amaral, L.A.N.: Cartography of complex networks: modules and universal roles. J. Stat. Mech., P02001 (2005b)
12. Newman, M.E.J.: Fast algorithm for detecting community structure in networks. Phys. Rev. E 69, 066133 (2004b)
13. Newman, M.E.J.: Modularity and community structure in networks. Proc. Natl. Acad. Sci. USA 103, 8577–8582 (2006)
14. Pujol, J.M., Béjar, J., Delgado, J.: Clustering Algorithm for Determining Community Structure in Large Networks. Phys. Rev. E 74, 016107 (2006)

15. Danon, L., Díaz-Guilera, A., Duch, J., Arenas, A.: Community analysis in social networks. J. Stat. Mech., P09008 (2005)
16. Fortunato, S., Barthélemy, M.: Resolution limit in community detection. Proc. Natl. Acad. Sci. USA 104, 36–41 (2007)
17. Arenas, A., Díaz-Guilera, A., Perez-Vicente, C.J.: Synchronization reveals topological scales in complex networks. Phys. Rev. Lett. 96, 114102 (2006)
18. Arenas, A., Díaz-Guilera, A., Perez-Vicente, C.J.: Synchronization processes in complex networks Physica D 224, 27–34 (2006)
19. Gómez-Gardeñes, J., Moreno, Y., Arenas, A.: Paths to synchronization on complex networks. Phys. Rev. Lett. 98, 034101 (2007)
20. Gómez-Gardeñes, J., Moreno, Y., Arenas, A.: Synchronizability determined by coupling strengths and topology on Complex Networks. Phys. Rev. E 75, 066106 (2007)
21. Boccaletti, S., Ivanchenko, M., Latora, V., Pluchino, A., Rapisarda, A.: Detecting complex network modularity by dynamical clustering. Phys. Rev. E 75, 045102(R) (2007)
22. White, J.G., Southgate, E., Thompson, J.N., Brenner, S.: The structure of the nervous system of the nematode caenorhabditis elegans. Phil. Trans. Royal Soc. London. Series B 314, 1–340 (1986)
23. Achacoso, T.B., Yamamoto, W.S.: AY's Neuroanatomy of C. Elegans for Computation. CRC Press, Boca Raton (1992)
24. Kuramoto, Y.: Self-entrainment of a population of coupled nonlinear oscillators. Lect. Notes in Physics 30, 420–422 (1975)
25. Durbin, R.M.: Studies on the Development and Organisation of the Nervous System of Caenorhabditis elegans. PhD Thesis, University of Cambridge (1987)

Modelling Gene Regulatory Networks

Erol Gelenbe*

Intelligent Systems and Networks Group
Department of Electrical and Electronic Engineering Department
Imperial College
London SW7 2BT UK
e.gelenbe@imperial.ac.uk

Abstract. We consider methods to compute analytical solutions for the probabilities of activation in gene regulatory networks with positive and negative feedback loops, similar to those introduced by René Thomas, and show how discrete state-space and continuous time probability models called can be used to compute their steady-state behaviour. The inclusion of logical dependencies in stochastic regulatory networks is the developed in detail.

Keywords: Gene regulatory networks, G-networks, Stochastic models, Equilibrium solutions.

1 Introduction

René Thomas states that[1] "Most biological regulatory systems involve complex networks of interactions. Theoretical modelling, together with simulations and computational approaches, provides a useful framework for integrating data and gaining insights into the dynamical and functional properties of such networks. In this perspective, a major aim of [his] research is to contribute to the understanding of how regulatory mechanisms at various scales (e.g. molecular, cellular and intercellular) act synergistically or competitively to achieve degrees of regulation not attainable by one mechanism alone. Key issues are the variety of attractors possible for a network, the nature of transition states and transition dynamics, and the role of the network in emergent behaviour. These issues are examined in terms of systems of differential equations, automata networks and probabilistic models." This ambitious research programme has been developed by its author over a quarter of a century, yielding elegant insights and biological applications [1,2,3,4,6], and inspiring the work of others [5,15,16,17,18]. Computer scientists on the other hand, have shown how such models can be enriched

---

* The author is grateful for discussions with Professor Gilles Bernot and Dr Luca Cardelli, both of which provided personal motivation to the author to pursue this work. The opportunity offered to us by Gilles Bernot, Jean-Paul Comet and Franck Quessette to give the opening lecture at the Wotkshop on "Réseaux d'Interactions: Analyse, Modélisation et Simualtion (RIAMS)" in Lyon on 29th November 2006 was a great source of encouragement.

[1] See http://www.ulb.ac.be/cenoliw3/theoretical.html

P. Liò et al. (Eds.): BIOWIRE 2007, LNCS 5151, pp. 19–32, 2008.

with formal methods inspired from computational logic [15,17,16,18] so as to be able to use computer based tools that allow the detailed study of network transformation sequences within a semantically consistent framework.

In a regulatory network, although the effect of each individual agent on other agents can be explicitly indicated, the resulting emergent behaviour of the network is difficult to apprehend. One approach would be to describe the temporal dynamics of the network and to characterise it by solving an appropriate equational system so that the successive values of the state of each agent, in time, can be computed.

Another approach is to seek the stationary (steady-state) value of the state of each agent, and to use this as the predictor of which agents are *in fine* active, and which have been deactivated, as a result of the complex interactions between agents. This is the approach that we propose here.

Thus the purpose of this paper is to develop an approach to model the behaviour of regulatory networks with positive and negative feedback loops so as to compute the probability of activation of the agents in the presence of complex interactions. We also show how this framework can be used to include the effect of Boolean dependencies between agents when, for instance, the state of some agent is determined by a Boolean function of the state of other agents. Further work about the approach that we propose can be found in [19], and the more fundamental problem of modeling chemical reactions which are relevant to biochemistry can be found in [22]. Other recent papers on biological signaling include [20,21].

In Section 2 we develop a probability model for regulatory networks, and discuss its stationary solution in Section 3. In Section 3.1 we develop a simple example to illustrate show how the model can be used to predict the probability of activation or deactivation of agents in regulatory networks. Then in Section 4 and Section 5 we develop probabilistic models which include Boolean dependencies between agents, in addition to the activation/inactivation type dependencies, which include the standard normal forms of Boolean Algebra.

2   A Probability Model for Regulatory Networks

We adopt the abstract model of a regulatory network that is proposed through the work in [15,18], consisting of a set V of n nodes, and a set of directed and weighted arcs between the nodes, so that the model becomes a directed graph with weighted arcs.

The nodes represent biological objects, such as genes, proteins, other active biochemical substances, or organisms such as cells, viruses, while the arcs carry weights in the form $(+, x)$ or $(-, x)$ where:

- The $+$ sign denotes a positive effect (facilitating or exciting), while the $-$ sign denotes a negative effect (inhibiting), and the variable x is a non-negative real number and represents a possible threshold.

- Thus an arc from node a to b with a weight $w_{ab}(+, x)$ means that if (say) gene a has a concentration of at least x quantity, then it will facilitate the "expression" of gene b. Here by "expression" in the case of genes we would mean the synthesis of the proteins which are coded by gene b.

Note that the model we have just described does not have an explicit representation of time. The time factor could be introduced by supposing that each transition or activation takes unit time. However, weights which also represent activation rates could also be introduced in the model. For instance, a weight could have the form $w_{ab}(-, x, z)$ where z is the rate (or inverse of the average time) at which the interaction from a to b occurs. In that case, one has implicitly introduced a continuous time model.

G-networks [7,8] are stochastic dynamical models with an unbounded discrete state-space, which operate in continuous continuous time. In this section we will describe a special instance of G-networks which is adapted to the needs of modeling regulatory networks. The model will be composed of:

- *Agents*, which are the primary objects of interest; they represent genes or other active biochemical or living objects whose levels of activity we wish to represent, and
- *Gates* which represent the interactions between agents; gates are either *binary* in nature (i.e. they describe the effect of agent i on agent j), or they are *ternary* and describe the joint effect of two agents on a third agent, or they are multi-valued, representing the impact of a set of agents on a given agent.

In fact by chaining agents with ternary gates, we obtain joint effects of multiple agents on a single agent. We will now set up the probability model for regulatory networks, and discuss its analytical solution [7,8,9]. The probability model is defined via the following quantities defined for $i, j \in \{1, \dots, N\}$. They are:

- The $K_i(t) \geq 0$ are integer valued random variables which represent the concentration or quantity of the agents i at time $t \geq 0$.
- $\Lambda_i \geq 0$ is a real number representing the rate at which agent i is being replenished from some external source; similarly $\lambda_i \geq 0$ is the rate at which agent i is being depleted, provided that agent i is present in some concentration.
- The $r_i \geq 0$ are real numbers representing the activity rates of each agent i, provided again that the agent is present in some non-zero amount. In precise terms, λ_i, λ_i, r_i are the parameters of exponential distributions, and Λ_i, λ_i are the arrival rates of independent Poisson processes of signals which, respectively, increase or decrease the level of the variable $K_i(t)$. Similarly r_i is the average time between successive interactions of agent i with other agents.
- and $P^+(i, j)$, $P^-(i, j)$, $Q(i, j, l)$ are the probabilities, respectively, that agent i acts on j in a facilitating (excitatory) mode, or an inhibitory mode, or that (i, j) together act on l in a facilitating mode. We will assume that $P^+(i, j).P^-(i, j) = 0$, so that at most one of the two excitatory or inhibitory effects can occur from agent i to j. Finally for any i,

$$d_i + \sum_{j=1}^{n} [P^+(i, j) + P^-(i, j) + \sum_{l=1}^{n} Q(i, j, l)] = 1. \tag{1}$$

- where d_i is the probability that agent i does not act on any other agents. In many cases we will have either $d_i = 0$ (when the agent only has a role as an activator or inhibitor of other agents) or $d_i = 1$, when the agent does not act on other agents at all, for instance if it is the end product of a series of other interactions.
- It will be more convenient, and more compatible with our introductory remarks about the graph model for regulatory networks and the weights, to replace the probabilities by weights in the following manner:

$$w^+(i,j) = r_i P^+(i,j) \tag{2}$$

$$w^-(i,j) = r_i P^-(i,j) \tag{3}$$

$$w(i,j,l) = r_i Q(i,j,l). \tag{4}$$

To include the effect of the activation threshold x_i, we assume that we are given a vector $x = (x_1, \ldots, x_n)$ of non-negative integers so that agent i is only activated when $K_i(t) \geq x_i$.

The dynamics of the G-network can now be represented by a system of Chapman-Kolmogorov differential and difference equations that govern the random process $K(t) = [K_1(t), \ldots, K_n(t)]$, $t \geq 0$. This process represents the number of units, or the concentration, of the n different types of agents.

Denote by $k = [k_1, \ldots, n]$ an n-vector of non-negative integers, and let $P(k,t) = Prob[K(t) = k]$ be the probability that $K(t)$ takes that particular value k. In order to write the C-K equations, define e_i to be the n vector all of whose elements are zero $except$ for the $i - th$ element whose value is $+1$. The dynamic behaviour of the G-network is then given by:

$$\frac{dP(k,t)}{dt} = \sum_{i=1}^{n} [\, P(k + e_i, t)(\lambda_i + r_i d_i) \tag{5}$$

$$+ \Lambda_i P(k - e_i, t) 1[k_i > 0] - P(k,t)[\Lambda_i + 1[k_i > 0](\lambda_i + r_i)]$$

$$+ \sum_{j=1}^{n} [\, 1[k_i + 1 \geq x_i](P(k + e_i - e_j, t) 1[k_j > 0] w^+(i,j) \tag{6}$$

$$+ P(k + e_i + e_j, t) w^-(i,j)$$

$$+ \sum_{l=1}^{n} P(k + e_i + e_j - e_l, t) 1[k_i + 1 \geq x_i] 1[k_j + e_j \geq x_j]$$

$$\times (\, w(i,j,l) + w(j,i,l) \,) \,] \,]$$

where all the terms $P(y,t)$ in the right or left hand side of the equation are zero if any of the elements of the vector y are negative. Notice that the effect of the activation thresholds x are explicitly included in (6).

2.1 Discussion

The probability model that we have presented provides a level of detail for the interaction of agents which goes beyond just activation, since it also includes the *level or degree* of activation or concentration of agents through the quantities $K_i(t)$. Thus:

- As indicated earlier, $K_i(t)$ represents the *activation level*, amount or concentration level of the agent i. The equations (6) describe the case where any agent i is only activated if $K_i(t) \geq x_i \geq 1$.
- Through the parameters Λ_i, the natural replenishment of agent i, for instance via some biochemical reaction, or via infiltration from an external medium, is being represented. The parameters λ_i in turn represent a deletion of agent i. Both Λ_i and λ_i are specific to a single agent and do not represent inter-agent interactions.
- The parameters r_i represent the overall rate at which the agent i interacts with other agents; a higher r_i represents the fact that agent i is more active. At the same time, r_i is the depletion of agent i as a result of the agents' interaction with other agents, or via removal of the agent from the medium being considered through the rates $r_i d_i$. Note that $r_i = \sum_{j=1}^{n} [w^+(i,j) + w^-(i,j) + \sum_{l=1}^{n} w(i,j,l)]$.
- The parameters $w^+(i,j)$ and $w^-(i,j)$ represent the replenishment or depletion of agent j, or the excitation or inhibition effect, as a result of agent i.
- Finally, the parameters $w(i,j,l)$ represent the excitation/activation of agent l through the effect of i and j, or the rate of increase of the amount of l through the effect of i,j.

Thus the probability model we have described will represent both the inter-agent relations with respect to activations, and/or the amounts or concentrations of the agents and the manner in which this affects their interactions and activation.

3 Exact Solution for the Probability Model When All $x_i = 1$

Consider the equations (6) in which we have set $x_i = 1$, $i = 1, \ldots, n$. In other words, as long as there is at least one agent of type i, agent i is activated. In this case, the model we have presented is a special case of the "G-network with triggered customer movement" which we have introduced previously in the context of queueing theory [8]. WE will provide a proof in the Appendix so that this paper may be self-contained.

Consider now the manner in which the system behaves in the long run, represented by its steady-state probability distribution $P(k) = \lim_{t \to \infty} P(k,t)$, and introduce the term:

$$q_i = min[1, \frac{\Lambda_i + \sum_{j=1}^{n} q_j w^+(j,i) + \sum_{j,l=1,\ l \neq j}^{n} q_j q_l w(j,l,i)}{r_i + \lambda_i + \sum_{j=1}^{n} q_j w^-(j,i) + \sum_{j,l=1\ l \neq j}^{n} q_l w(l,i,j)}], \quad (7)$$

for $i = 1, \ldots, n$, which represents the probability that agent i is activated in steady-state.

Theorem 1. Consider the case where $x_i = 1$, $i = 1, \ldots, n$. For any subset $I \subset \{1, \ldots, n\}$ such that $q_m < 1$ for each $m \in I$, and $I = \{m_1, \ldots m_{|I|}\}$:

$$P(K_m = k_m) = q_m^{k_m}(1 - q_m), \ and \tag{8}$$

$$P(K_{m_1}, \ldots, K_{m_{|I|}} = k_{m_1}, \ldots, k_{m_{|I|}}) \tag{9}$$

$$= \Pi_{i=1}^{|I|} q_{m_i}^{k_{m_i}}(1 - q_{m_i})$$

The proof of this theorem, stated in a slightly different manner, can be found in [8]. Notice that (7) is a system of non-linear equations; thus we need to determine the conditions under which these equations have a solution, and also to determine whether they have a *unique* solution. Fortunately this was also proved in [8]:

Theorem 2. If all $x_i = 1$, then the solution of (6) with $x = (1, \ldots, 1)$ as provided by (7), (9), (10) exists and is unique.

3.1 A Simple Example

In this section we develop a simple example to illustrate the use of the approach we have introduced. In this example, three type of agents interact. The agents or entities (C, V, A) interact via facilitation/excitation, inhibition. Also joint facilitation of an agent by two others is possible and is represented by the "ternary" interaction operator $w(i, j, l)$ where an agent of some type i can influence an agent of type j to activate an agent of type l.

Agent C in isolation. In the system we consider, we would like to observe whether the agent C is activated. When it exists in isolation, with a replenishment rate Λ_c and a depletion rate r_c, using (7) we have:

$$P(K_c > 0) = \rho_c = \frac{\Lambda_c}{r_c}. \tag{10}$$

if $\Lambda_c < r_c$, while if $\Lambda_c \geq r_c$ then $P(K_c > 0) = 1$ and Agent C is constantly activated; in particular, if $r_c = 0$ there is no natural depletion of agent C.

The effect of Agent V. If Agent V is introduced into the system at some rate Λ_v, and V has an inhibitory effect on C represented by $w(v, c, v)$. Thus not only does V deplete C but it also re-activates itself V in the process, so that it is both depleting C and maintaining its own importance. We suppose that agent V is not subject to some other natural form of removal from the medium, except through its effect on agent C. Thus $r_v = w(v, c, v)$. As a result when V is present we now have:

$$q_c = \frac{\Lambda_c}{r_c + w(v, c, v)q_v} = \frac{\Lambda_c}{r_c + \Lambda_v}, \tag{11}$$

$$q_v = \frac{\Lambda_v + q_v q_c w(v, c, v)}{w(v, c, v)}, \tag{12}$$

so that

$$\frac{\Lambda_c}{r_c + \Lambda_v} \leq q_c \leq \frac{\Lambda_c}{r_c + w(v,c,v)} < \frac{\Lambda_c}{r_c}. \tag{13}$$

In particular, when $r_c = 0$, we see that the introduction of Agent V results in having

$$P(K_c > 0) = \frac{\Lambda_c}{\Lambda_c + \Lambda_v} < 1, \tag{14}$$

instead of $P(K_c > 0) = 1$. In fact, if $\Lambda_v > \Lambda_c$, then $P(K_c > 0) < 0.5$ which may be unacceptably low. As a result, we now take the following step.

Introducing Agent A. Now in order to limit the effect of V we introduce an agent A which has an inhibitory effect on V so that, still assuming that $r_c = 0$, we have:

$$q_a = \frac{\Lambda_a}{w^-(a,v)}, \tag{15}$$

$$q_v = \frac{\Lambda_v + q_v q_c w(v,c,v)}{w(v,c,v) + q_a w^-(a,v)}, \tag{16}$$

$$q_c = \frac{\Lambda_c}{q_v w(v,c,v)} \tag{17}$$

Conclusion. From the above equations, if Agent A is introduced in sufficient concentration or at sufficient rate so that:

$$\Lambda_a > \frac{w(v,c,v)\Lambda_v}{\Lambda_c} \tag{18}$$

then $P(K_c > 0) = 1$ and Agent C remains constantly activated despite the presence of Agent V.

3.2   A Heuristic Expression When the $x_i \geq 1$

When we have to deal with the conditioning of activation or inhibition between agents based on the "level of activity" or quantity present of some agents, which we have represented with values of $x_i > 1$, we currently do not have known a closed form solution for the steady-state probability distribution $P(x)$ resulting from equations (6). Thus we propose a heuristic solution inspired by the previous result, which is consistent with the exact solution (7), (9), (10). However we cannot prove that this heuristic is correct in exact terms and its value can only be determined from practical use.

Heuristic Solution 1. The approximate heuristic solution for equations (6) when the $x_i > 1$, where:

$$g_i \approx min[1, \frac{\Lambda_i + \sum_{j=1}^n g_j^{x_j} r_j P^+(j,i) + \sum_{j,l=1}^n g_j^{x_j} g_l^{x_l}(r_j + r_l)Q(j,l,i)}{r_i + \lambda_i + \sum_{j,l=1}^n q_l^{x_l} r_l Q(l,i,j)]}]$$

for $i = 1, \ldots, n$ represents the approximate probability that agent i is activated, and is given for $k_m \in I$, $k_{m_i} \in I$ where I is the set of indices of the agents for which $g_i < 1$, $P(K_m = k_m) \approx g_m^{k_m}(1 - g_m)$, $P(K_{m_1}, \ldots, K_{m_{|I|}} = k_{m_1}, \ldots, k_{m_{|I|}}) \approx \Pi_{m_i \in I} g_{m_i}^{k_{m_i}}(1 - g_{m_i})$. Note that we have used the terms of the form $q_j^{x_j}$ because if the expression were exact, then $P[K_m \geq x_m] = g_m^{x_m}$.

4   Logical Dependency of an Agent on Several Others: A First Approach

We will now continue exploiting the exact solution provided in (10). In the previous section we have covered the action of some agent i on an agent j, as well as the joint action of agents (i, j) on some third agent l. In this section we would like to consider how a set of agents $[a_1, a_2, \ldots, a_\alpha]$ and $[b_1, b_2, \ldots, b_\beta]$ can jointly act on some agent l. More specifically we would like to represent a logical equation of the form

$$Agent_l = [\wedge_{s=1}^{\alpha}(Agent_{a_s})] \bigwedge [\wedge_{s=1}^{\beta}(\neg Agent_{b_s})], \tag{19}$$

where the $Agent_j \in \{0, 1\}$ are binary variables indicating whether the agent l is activated (1) or inactive (0), and the notations:

- $\wedge$ denotes the logical "and",
- $\vee$ denotes the logical "or",
- and $\neg$ denotes the negation or complement of the logical variable that follows it.

We will also exploit the identity:

$$[\wedge_{s=1}^{\beta}(\neg Agent_{b_s})] = \neg[\vee_{s=1}^{\beta}(Agent_{b_s})] \tag{20}$$

The approach we take is via the steady-state probability distribution of the integer representing the internal state of the agent $P(K_{A_j}) = \lim_{t \to \infty} P(K_{A_j}(t))$ with:

$$A_j = 1 \Leftrightarrow K_j > 0, \tag{21}$$

so that we compute $q_j = P(K_j > 0)$. In order to do this we:

4.1   Construction

- Introduce a set of "dummy agents" $A_1, A_2, \ldots, A_\alpha$ that act as intermediaries between the set of agents a.
- a_1 acts upon A_1, (a_2, A_1) act upon A_2 and so on. Finally $(a_\alpha, A_{\alpha-1})$ act upon A_α, and A_α acts upon agent l in an excitatory manner with $w^+(A_\alpha, l) = 1$.
- Furthermore we set $\Lambda_{A_s} = \lambda_{A_s} = 0$, $r_{A_s} = 1$ for $1 \leq s \leq \alpha$, $w(a_1, a_2, A_2) = 1$ and $w(a_s, A_{s-1}, A_s) = 1$ for $s = 3, \ldots, \alpha$.

- We also introduce dummy agents B_1, ... , B_β so that (b_1) acts upon B_1 in an excitatory manner with $w^+(b_1, B_1) = 1$, (b_2) acts upon B_2 similarly, and so on, and b_β acts upon B_β in an excitatory manner with B_β with $w^+(b_\beta, B_\beta) = 1$.
- Then each B_s acts upon agent l in an *inhibitory* manner with $w^-(B_s, l) = \gamma$, $1 \le s \le \beta$.
- We set $\Lambda_{B_s} = \lambda_{B_s} = 0$, $r_{B_s} = 1$ for $1 \le s \le \beta$.

Using (7), we immediately obtain:

$$q_l = min[1, \frac{\Lambda_l + \sum_{j=1}^n q_j w^+(j, l) + q_{A_\alpha} w^+(A_\alpha, l)}{r_l + \lambda_l + \sum_{j=1}^n q_j w^-(j, l) + \sum_{s=1}^\beta \gamma q_{B_s}}, \tag{22}$$

$$q_{A_\alpha} = q_{a_1} \cdots q_{a_\alpha},$$

$$q_{B_s} = q_{b_s}, \quad s = 1, \ldots \beta,$$

so that we have:

Proposition 3. The logical equation between agents

$$Agent_l = [\wedge_{s=1}^\alpha (Agent_{a_s})] \wedge [\wedge_{s=1}^\beta (\neg Agent_{b_s})], \tag{23}$$

is obtained by setting $\Lambda_l = \lambda_l = r_l = 0$ with $w^+(j, l) = w^-(j, l) = 0$ for all $j \notin [1, \ldots, n] \cup a \cup b$, and using only the connections between agents provided by *Construction*. As a result we have:

$$q_l = min[1, \frac{\prod_{s=1}^\alpha q_{a_s}}{\sum_{s=1}^\beta \gamma q_{b_s}}] \tag{24}$$

and we use a *decision threshold H* above which any q_{a_s}, q_{b_s} must be in order to interpret A_{a_s} or $A_{b_s} = 1$, and they must be under the value L in order to interpret any of them as A_{a_s} or $A_{b_s} = 0$. We will therefore need to relate H and L for the thresholds used to evaluate the A_{a_s}, A_{b_s} on the right-hand-side of (24) to θ that is used for evaluating A_l on the left-hand-side of (24).

4.2 Selecting the Thresholds

From the previous discussion, and (24), we have the following requirements for θ:

$$\frac{H^\alpha}{\beta\gamma L} \ge \theta$$

$$\frac{LH^{(\alpha-1)}}{\beta\gamma L} \le \theta$$

$$\frac{H^\alpha}{\gamma H + (\beta-1)\gamma L} \le \theta$$

resulting in the following simple choice of the *threshold θ*:

$$\frac{H^{(\alpha-1)}}{\gamma} < \theta \le \frac{H^\alpha}{\gamma\beta L} \tag{25}$$

which also requires that $H > \beta L$.

5   Boolean Dependencies between Agents

The *conjunctive (CNF) and disjunctive (DNF) normal forms* are standard representations for Boolean functions. Each of them is universal in the sense that it allows the representation of any Boolean function. Consider a set of binary literals $A_j \in [0,1]$, $j \in [1, \dots, n]$, and consider a term $T_i = X_{i1} \lor \dots \lor X_{in}$ where X_{ij} is either A_j or it is $\neg A_j$.

The Boolean function $F : [0,1]^n \to [0,1]$ is in DNF if it is written as:

$$F = \wedge_{i=1}^m T_i, \tag{26}$$

while it is in CNF when it is written as

$$F = \vee_{i=1}^m \tau_i. \tag{27}$$

where the $\tau_i = X_{i1} \wedge \dots \wedge X_{in}$ with X_{ij} being either A_j or $\neg A_j$, and they too are disjoint. Clearly we can transform an expression in CNF into DNF and vice-versa using (20).

In this section we see how the expression (24) can be used to derive the sate probability for a logical expression in CNF. Let us consider the CNF given in (27) and assume that each literal A_j corresponds to a distinct agent. We will associate an agent A_F with the function F with $q_F = \lim_{t \to \infty} P(K_F(t) > 0)$. Similarly we associate dummy agents A_{T_i} with the terms T_i, and exploit Section 4.1 that was previously presented.

From the above discussion we see that once we have a way of representing the logical expression (19) using a G-network, it is quite direct to obtain the G-network counterpart for an expression in CNF. The approach requires an exact representation of (19) which includes the negated terms $\neg Agent_{b_s}$, and is based on constructing agents whose probabilities of being active are $[1 - q_{b_s}]$ where $q_{b_s} = \lim_{t \to \infty} P[K_{b_s}(t) > 0]$, $Agent_{b_s} = 1$ at time t if and only if $K_{b_s}(t) > 0$.

Turning to expression (7), we see that for any q_i the term $\rho_i = [1 - q_i]$ is:

$$\rho_i =$$

$$\frac{r_i + \lambda_i - \Lambda_i + \sum_{j=1}^n q_j [w^-(j,i) - w^+(j,i)] + \sum_{j,l=1,\ l \neq j}^n q_l [w(l,i,j) - q_j w(l,j,i)]}{r_i + \lambda_i + \sum_{j=1}^n q_j w^-(j,i) + \sum_{j,l=1,\ l \neq j}^n q_l w(l,i,j)]}$$

Note that we would like to have an agent, say $Agent_{c_i}$, whose state is the complement of agent $Agent_i$ so that ρ_i is the stationary distribution that $Agent_{c_i}$ is activated. Thus we require that the parameters in the expression for ρ_i have the following properties:

If $Agent_i$ has – in the same network – a complementary agent A_{c_i}, then:

$$\rho_i = \frac{L_i + \sum_{j=1}^n q_j \Omega^+(j,i) + \sum_{j,l=1,\ l \neq j}^n q_j q_l \Omega(l,j,i)}{R_i + l_i + \sum_{j=1}^n q_j \Omega^-(j,i) + \sum_{j,l=1,\ l \neq j}^n q_l \Omega(l,i,j)}, \tag{28}$$

with

$$(I)\ \ L_i = r_i + \lambda_i \geq \Lambda_i \tag{29}$$

$$(II)\ \ \Omega^+(j,i) = w^-(j,i) - w^+(j,i) \geq 0\ for\ j \neq i$$
$$(III)\ \ \Omega^-(j,i) = w^-(j,i)\ \ for\ j \neq i$$
$$(IV)\ \ w(l,i,j) > 0 \Rightarrow w(l,j,i) = 0\ for\ l,j \neq i$$
$$(V)\ \ w(l,i,j) = 0 \Rightarrow w(l,j,i) = 0\ for\ l,j \neq i$$
$$(VI)\ \ r_i + \lambda_i - \Lambda_i \geq 0$$

Notice that for $Agent_i$ for which we wish to construct a complementary agent within the same network we arrive at the following constraints:

$$(VII) \Rightarrow w^-(j,i) \geq w^+(j,i)\ for\ j \neq i$$
$$(VIII) : (IV)\&(V) \Rightarrow w(l,i,j) = 0\ for\ l,j \neq i$$
$$(IX) L_i = r_i + \lambda_i - \Lambda_i \geq 0$$
$$(X) R_i + l_i = l_i + \sum_{j=1, j \neq i}^{n} [\Omega^+(i,j) + \Omega^-(i,j) + \sum_{l=1, l \neq i,j}^{n} \Omega(i,j,l)] = r_i + \lambda_i$$

Note that (VII) is easy to satisfy, while (VIII) is compatible with the development in Section 4.1, since it implies that all the agents A_{a_s} may initiate joint actions of the form $(a_s, A_s) \to A_{s+1}$, but that actions of the form $(X, a_s) \to Y$ are not allowed.

Finally we have the following constraints for the A_{c_i}:

$$(II) \Rightarrow \Omega^+(j,i) = w^-(j,i) - w^+(j,i)\ for\ j \neq i \tag{30}$$
$$(IV)\ \&\ (V)\ \ \ \Rightarrow \Omega(l,i,j) = \Omega(l,j,i) = 0\ for\ l,j \neq i$$

5.1 Expressions in Conjunctive Normal Form

Consider the Boolean function $F : [0,1]^n \to [0,1]$ in CNF if it is written as:

$$F = \vee_{i=1}^m \tau_i. \tag{31}$$

where the $\tau_i = X_{i1} \wedge \ldots \wedge X_{in}$ with X_{ij} being either A_j or $\neg A_j$.

Define the set of indices $\Upsilon_i = \{j : X_{ij} = A_j\}$ and $\Phi_i = \{j : X_{ij} = \neg A_j\}$. Using (28) and the *Construction* in Section 4.1, we obtain q_F, the probability of activation of Agent A_F, as

$$q_F = min[1, \sum_{i=1}^m \prod_{j \in \Upsilon_i} q_j \prod_{j \in \Phi_i} \rho_j \tag{32}$$

The expression for the state probability of an agent whose state depends on others' state according to a Boolean function in DNF can also be constructed in a similar manner.

6 Discussion

The analysis we have presented can be used to compute the probability that in steady-state each agent in a regulatory network is activated or deactivated, as well as the joint probability of the state of all of the agents.

These results do not allow us to compute the successive states that a network of agents will enter into in the course of time.

Suppose that we consider that a regulatory network acts as the control system of a "biochemical nano-factory", with the rate of production of certain compounds being determined by the probability that certain sets of agents are activated. Then our analysis would enable the computation of the rate of production of these compounds over a period of time.

When the system being considered has some form of cyclic behaviour, with agents being successively activated and deactivated through their interactions, then our analysis can provide two things (a) the proportion of time that each agent or each combination of agents is in some combination of active or inactive states, and (b) by fixing the state of some of the agents in the right-hand-side of expressions of the form (24) or (32), we can predict the state of some of the agents when other agents' states are known.

Thus our approach does not replace a discrete event simulation of a regulatory network based on the full semantics of agent interactions, but does offer a means to evaluate and predict the overall behaviour of the network over a long period of time.

As a final illustrative examples, consider the following toy regulatory network[2] composed of four agents, call them $\{A_0, \ldots A_3\}$ connected cyclically so that the $i - th$ agent inhibits agents $(i + 1)mod3$ and agent $(i + 2)mod3$, and there are no other dependencies. Assume that agents have just two states (on and off).

The timing in this simple model may be either deterministic, where each agent changes state in exactly unit time, or random (e.g. exponentially distributed) of average value 1 for all agents, or each agent can have a different timing behaviour. Thus the resulting behaviour of this synchronous or asynchronous system can be quite different depending on what is assumed about the time between state transitions of the agents. Another important assumption about such a network concerns the state the agents will enter when they are quiescent, i.e. when they are left to themselves. Clearly, if the agents left to themselves all enter the 0 (off) state, then the model has little interest since all agents will remain in that state, assuming that they start there. On the other hand, if we assume that they spontaneously enter the 1 or "on" state when they are not acted upon by another agent, then more interesting behaviours can result. Also, the meaning of these interconnections can be interpreted in at least two different ways, for any $i = 1, \ldots, 3$:

– Interpretation (1): $A_i = \neg A_{(i-1)mod3} \wedge A_{(i-2)mod3}$.
– Interpretation (2): Both agents $A_{(i-1)mod3}$ and $A_{(i-2)mod3}$ inhibit the activation of agent A_i.

Assume now that all agents start in the same initial state, that all state transition times are exponentially distributed with average value 1, and that when they are quiescent (i.e. free of inputs from other agents) they all reset themselves to the value 1 ("on"). For both interpretations the probabilistic state of all agents

---

[2] We thank Dr Luca Cardelli for suggesting this example.

will be identical, and their stationary distribution $q = \lim_{t \to \infty} P[A_i(t) = 1]$ is given by:

- Interpretation (1): Using (32) we write $q = (1 - q)(1 - q)]$ so that $q = 0.382$.
- Interpretation (2): Using (7) we have $q = \frac{1}{1+2q}$ so that $q = 0.5$.

Under Interpretation (2) we see that the agents will all spend half of the time being "on" and the other half being "off", and all 16 states represented by the vector of four binary variables, will be equally likely with probability 1/16 in steady state. With Interpretation (1) they spend more time in the "off" state than in the "on" state; in fact in this case the state $(0, 0, 0, 0)$ is 6.854 times more likely to occur than the state $(1, 1, 1, 1)$. Thus we see that the manner in which the interactions between agents are precisely defined has significant impact on the analysis that our modeling approach can offer.

References

1. Thomas, R.: Boolean formalisation of genetic control circuits. J. Theor. Biol. 42, 565–583 (1973)
2. Thomas, R., Gathoye, A.M., Lambert, L.: A complex control circuit: regulation of immunity in temperate bacteriophages. Eur. J. Biochem. 71, 211–227 (1976)
3. Thomas Thomas, R.: On the relation between the logical structure of systems and their ability to generate multiple steady states or sustained oscillations. Springer Series in Synergetics, vol. 9, pp. 180–193 (1981)
4. Thomas, R.: Logical description, analysis and synthesis of biological and other networks comprising feedback loops. Adv. Chem. Phys. 55, 247–282 (1983)
5. Kaufman, M., Andris, F., Leo, O.: A Logical Analysis of T Cell Activation and Anergy. Proc. Natl. Acad. Sci. USA 96, 3894–3899 (1999)
6. Thomas, R., Kaufman, M.: Multistationarity, the Basis of Cell Differentiation and Memory. I. Structural Conditions of Multistationarity and Other Non-Trivial Behaviour, and II. Logical Analysis of Regulatory Networks in Terms of Feedback Circuits, Chaos 11, 170–195 (2001)
7. Gelenbe, E.: Queueing networks with negative and positive customers. Journal of Applied Probability 28, 656–663 (1991)
8. Gelenbe, E.: G-networks with instantaneous customer movement. Journal of Applied Probability 30(3), 742–748 (1993)
9. Gelenbe, E.: G-Networks with signals and batch removal. Probability in the Engineering and Informational Sciences 7, 335–342 (1993)
10. Fourneau, J.M., Gelenbe, E., Suros, R.: G-networks with multiple classes of positive and negative customers. Theoretical Computer Science 155, 141–156 (1996)
11. Gelenbe, E., Fourneau, J.M.: G-Networks with resets. Performance Evaluation 49, 179–191 (2002)
12. Fourneau, J.M., Gelenbe, E.: Flow equivalence and stochastic equivalence in G-Networks. Computational Management Science 1(2), 179–192 (2004)
13. Gelenbe, E.: Learning in the recurrent random neural network. Neural Computation 5(1), 154–164 (1993)
14. Gelenbe, E., Pujolle, G.: Introduction to Networks of Queues, 2nd edn. J. Wiley & Sons, Chichester (1998)

15. Bernot, G., Comet, J.-P., Richard, A., Guespin, J.: Application of formal methods to biological regulatory methods: extending Thomas' asynchronous logical approach with temporal logic. J. Theoretical Biology 229(3), 339–347 (2004)
16. Bernot, G., Guespin-Michel, J., Comet, J.-P., Amar, P., Zemirline, A., Delaplace, F., Ballet, P., Richard, A.: Modeling, observability and experiment: a case study. In: Proc. Dieppe School on Modeling and Simulation of Biological Processes in the Context of Genomics, pp. 49–55. Frontier Group Pub. (2003) ISBN: 2-84704-036
17. de Jong, H., Geiselmann, J., Hernandez, C., Page, M.: Genetic analyzer: qualitative simulation of genetic regulatory networks. Bioinformatics 19(3), 336–344 (2003)
18. Comet, J.-P.: De la bio-informatique textuelle à une approche formelle de la biologie des systèmes, Habilitation Thesis (Thèse d'Habilitation à diriger des Recherches), Université d'Evry-Val-d'Essonne, France, November 21 (2006)
19. Gelenbe, E.: Steady-state solution of probabilistic gene regulatory networks. Physical Review E 76(1), 031903 (2007); Virtual Journal of Biological Physics Research (September 15, 2007)
20. Gelenbe, E.: Network of interacting synthetic molecules in equilibrium. Proc. Royal Society A (Mathematical and Physical Sciences) (accepted for publication)
21. Gelenbe, E., Timotheou, S.: Synchronized interactions in spiked random networks. The Computer Journal (accepted for publication)
22. Gelenbe, E.: Network of interacting synthetic molecules in equilibrium. Proc. Royal Society A (Mathematical and Physical Sciences) (accepted for publication)

The Role of Simplifying Models in Neuroscience: Modelling Structure and Function

Dina M. Kronhaus[1] and Stephen J. Eglen[2]

[1] Computer Laboratory, Cambridge University
dk323@cam.ac.uk
[2] Cambridge Computational Biology Institute
sje30@cam.ac.uk

Abstract. The adult human brain has around 10^{11} neurons and 10^{15} connections between these neurons, thus forming an incredibly complex network. In this article, we first describe two complementary approaches to modelling brain function, namely simplifying and realistic models. We then demonstrate, by way of two examples, the utility of building simplifying neural models. In the first example, we consider the development of neuronal positioning. In the second example, we investigate the stability of a cortical network under control and perturbed conditions.

Keywords: simplifying neural models, retinal mosaics, cortical compensation, anterior cingulate.

1 Background

This article introduces, by way of examples, the utility of theoretical modelling in understanding aspects of neural circuitry. Perhaps one of the best examples of theoretical modelling aiding our understanding in neuroscience is the seminal work of Hodgkin and Huxley concerning the ionic basis of the action potential (reviewed by Hodgkin, 1958). Since their work over fifty years ago, many researchers have built upon this framework, developing more complex models to account for new experimental findings. This work is an example of a "realistic" model such that the computational model tries to account for all known relevant details of a particular system (Sejnowski et al., 1988). The model is then evaluated by comparing model output with experimental results.

One crucial issue with this kind of modelling is deciding just how much known experimental detail to include in a model. As we now have a lot of information about the details of individual neurons, should we include all those details when we model a neuron? The answer may, perhaps, be yes if our model investigates the behaviour of just one neuron. However, if, for example, we wish to study the collective dynamics of a large population of neurons, modelling each neuron in detail may not be possible, as the system may become too large to be carefully studied: for example, the model may require the specification of too many parameters, or require prohibitively long computation resources. Furthermore,

P. Liò et al. (Eds.): BIOWIRE 2007, LNCS 5151, pp. 33–44, 2008.

it may be the case that for a particular study, certain details, although known, may be irrelevant for the model in question.

For example, the Hodgkin-Huxley equations capture the detailed time evolution of an action potential by modelling ion channels and voltage-dependent conductances. However, if the detailed shape of an action potential is not important, but rather we are interested in the *timing* of the action potential, then a simpler integrate and fire neuron may be sufficient for our purposes. In this case, we can refer to the integrate and fire neuron as a "simplifying model" (Sejnowski et al., 1988). It clearly does not attempt to model all the details underlying action potentials, but can be used as the basis for building larger models to study e.g. network dynamics. This distinction between realistic models and simplifying models is obviously not unique to neuroscience modelling. For example, in ecology, the corresponding terms are "mechanistic" versus "phenomenological" models (Nathan and Muller-Landau, 2000). Mechanistic models attempt to simulate mechanisms underlying key behaviours, whereas phenomenological models are more concerned with replicating the behaviours, irrespective of the actual mechanisms that may generate those behaviours.

A valid criticism against simplifying models is often that they are "biologically implausible" and, hence, do not tell us how the brain solves the task. That often is indeed the case. However, the model may still tell us interesting things about other aspects of the problem. For instance, the back-propagation learning algorithm was used to train a multiple layer perceptron to investigate how children might learn to pronounce words (Sejnowski and Rosenberg, 1987). Specifically, the network was trained to associate letters within a word to the corresponding phonetic representation. This is a difficult task to achieve (at least in English) since the context of the letter is important: consider the pronunciation of the letter *i* within the words *bite* and *bit*. The back-propagation learning algorithm was used to adapt the connection strengths within the network to learn this association. This learning algorithm is so-named because of the way that during learning, error signals "back-propagate" from the output layer back to the input layer. This back-propagation of error is unlikely to occur in neural systems, and hence this learning algorithm is validly regarded as biologically implausible. However, if one regards this learning algorithm as simply a way of training a network, we can still explore the properties of the network, and compare them with human performance. For example, during training there is a stage-like progression of behaviour, seen in young children, where an early babbling-phase can be distinguished from the later, more-refined, performance. Analysis of the structure of the internal representation formed by the networks suggests that the network autonomously learns to distinguish vowels from consonants (Sejnowski and Rosenberg, 1987). Therefore, even though the learning rule may be biologically implausible, the model as a whole gives useful insights into how the brain may acquire the ability to pronounce words.

In this article, we review two recent applications of simplifying models to two different problems in neuroscience. The first model investigates an aspect of structural formation of neural circuitry, illustrating how modelling can help us study early developmental events. The second application considers how models can help us understand functional aspects of adult circuitry.

2   Structural Development of Neuronal Positioning

The human central nervous system is an incredibly complex structure: billions of neurons connect to each other to form complex networks. A key challenge in neuroscience is to understand how such complex networks are generated during development. There are too many neurons and connections for the network to be genetically encoded in the form of "wiring diagrams" that specify which neurons connect to which other neurons. Instead, the nervous system is likely to take advantage of principles of self-organisation to create such circuitry.

One key step in the generation of neural circuitry is for neurons to be appropriately positioned within their target tissue. Once neurons have been generated, they need typically to migrate to their destination layer and move to a particular location, respecting the position of neighbouring neurons. This process is most strikingly observed in the retina (the light-sensing neural structure at the back of the eye). Figure 1 shows an example of the position of all neurons of a particular class within a rectangular field of view. From this figure it is evident that there is some spatial ordering within this population of neurons: cells are not too close to each other, nor are they too far apart from each other. This semi-regular structure is called a "retinal mosaic" due to the way that the cell bodies (and their surrounding dendritic trees, not shown here) "tile" the surface of the retina. This neuronal arrangement serves to ensure that there are cells located throughout the retinal layer (rather than leaving "holes" in the surface), and may help the subsequent wiring of neurons within different retinal layers.

Many developmental mechanisms have been implicated in the formation of retinal mosaics, reviewed by Cook and Chalupa (2000). These vary from relatively early events of neuronal differentiation through to lateral interactions mediated by dendritic interactions. In addition to many experimental approaches to understanding the formation of retinal mosaics, there has been considerable interest in using theoretical modelling approaches to investigate developmental constraints in mosaic formation. To quantitatively compare real and simulated mosaics, here we use the simplest, and most popular, measure of spatial arrangement, the regularity index (RI) (Wässle and Riemann, 1978). The RI is simply the mean of the nearest-neighbour distances divided by the standard deviation of those distances (Figure 1). Informally, the higher the RI, the more regular the spatial distribution of neurons; values less than two typically indicate a random arrangement of neurons.

2.1   The d_{min} Model

The basic concept underlying the d_{min} model is that each neuron has a circular exclusion zone surrounding the cell body, which prevents neighbouring cells from coming too close to it. We take a phenomenological approaching by assuming

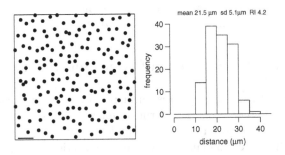

Fig. 1. Example retinal mosaic and quantification of its spatial organisation. Left: positioning of rat cholinergic amacrine cells (data from Lucia Galli-Resta). Cells drawn assuming 10 μm diameter; scale-bar (bottom left): 50 μm. Right: distribution of nearest-neighbour distances of the neurons on the left. The regularity index of this population, 4.2, indicates a mildly regular arrangement of neurons.

that biological mechanisms (which we do not explicitly model) can somehow enforce the exclusion zone. To model a retinal mosaic, we create a region **A** of the same size as the real mosaic being modelled. Initially this region is empty; neurons are added to **A** using a serial algorithm, positioning neurons one by one into the array until the number of neurons in the model region **A** matches the real mosaic. To position a neuron into **A** we follow the following steps:

1. Generate a trial neuron position and exclusion zone (x, y, d). The position of the neuron (x, y) is determined by uniform sampling of **A**. The effective diameter of the exclusion zone, d_{min} is drawn from a truncated Normal distribution with fixed mean and standard deviation (μ, σ). The Normal distribution is truncated at some lower bound d_{low} so that d cannot be smaller than e.g. the typical cell body diameter.
2. Find the distance of the trial neuron to all other neurons that have previously been accepted into **A**. The smallest of those distances is labelled d.
3. If $d < d_{min}$, the trial neuron is too close to an existing neuron. The trial neuron is thus rejected. Otherwise, if $d \geq d_{min}$, the trial neuron is accepted.

To model a given retinal mosaic, the only parameters required by this model are the mean and standard deviation of the exclusion zone. These parameters can be determined by trial and error, or by systematic searching over a range of suitable values. Furthermore, the model can also be used to generate randomly distributed neurons within a layer, subject only to the constraint that their cell bodies do not overlap, by setting the mean and s.d. of the d_{min} model to match the mean and s.d. of observed cell body diameters. Figure 2 compares a real mosaic against two d_{min} simulations: one where the exclusion zone has been selected to generate patterns similar to those observed, and one where the exclusion zone simply reflects non-overlap of cell bodies. Although in this case the regularity index of the simulated mosaic (6.3) is higher than the regularity index of the retinal mosaic (4.7), as will be discussed later, multiple simulations

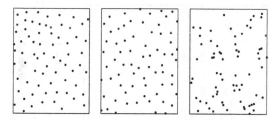

Fig. 2. Comparison of a retinal mosaic with two d_{min} simulations. Left: positions of off-centre beta retinal ganglion cells from cat; data taken from (Wässle et al., 1981). Cell bodies drawn to size, assuming 15 μm diameter. Regularity index: 4.7. Middle: matching d_{min} simulation ($\mu = 130$ μm; $\sigma = 25$ μm). Regularity index: 6.3. Right: d_{min} simulation with exclusion zone set to only reflect non-overlap of cell bodies ($\mu = 15$ μm; $\sigma = 0.01$ μm). Regularity index: 2.4. (Each sample field 1090 $\times$ 750 μm).

generate a range of RIs similar to the observed value. By contrast, when the exclusion zone merely prevents overlap of cell bodies (Figure 2 right), the pattern is typically quite disorganised, and both visually and quantitatively distinct from the retinal mosaic: many areas of the sample field are devoid of neurons.

So far, all the d_{min} model shows us is that some local-acting mechanism is sufficient for generating patterns similar to those observed experimentally. It therefore does not constrain the underlying biological mechanism that may generate such a local exclusion zone (Galli-Resta et al., 1997). This is a limitation of this simplifying model, and we must rely on either experimental results or more detailed theoretical models to inform us. (In this case, both experimental and theoretical evidence suggest the exclusion zone could be the product of lateral migration mediated by dendritic interactions (Eglen et al., 2000).) However, we can now use this d_{min} model to allow us to ask other questions, outlined next.

2.2 Bivariate Patterning of Retinal Mosaics

Figure 2 shows a sample of off-centre beta retinal ganglion cells (RGCs); these cells respond to the offset of light stimulation. By contrast, there is a complementary group of neurons, the on-centre beta RGCs that respond to the onset of light stimulation. The cell bodies of these neurons occupy the same layer of the retina, generating a bivariate mosaic pattern, as shown in Figure 3, left (data from Wässle et al., 1981). A striking feature of this pattern is that neighbouring neurons are usually of the opposite type, which has led to the question of whether there are interactions between the two cell types during development that generate such a pattern (Eglen and Willshaw, 2002).

To address this question, the d_{min} model can be extended to generate bivariate patterns. The null hypothesis of the model is that there are no functional interactions between neurons of opposite type. In this case, we generate N_1 type 1 neurons (here the on-centre neurons), and N_2 type 2 neurons (off-centre). When positioning a neuron in the array, e.g. a type 1 neuron, the type 2 exclusion

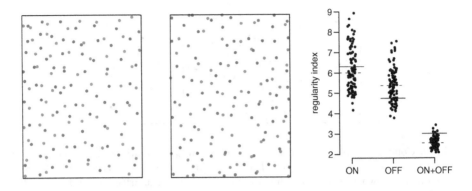

Fig. 3. Bivariate d$_{min}$ simulation. Left: bivariate pattern of on- and off-centre beta retinal ganglion cells (on-centre coloured green and off-centre coloured red). Sample field as in Figure 2. Middle: typical output from bivariate simulation. Parameters used: ($\mu_1 = 116$ μm, $\sigma_1 = 20$ μm, $\mu_2 = 130$ μm, $\sigma_2 = 25$ μm, $d_{12} = 9$ μm). Right: quantitative comparison of the regularity index of the retinal mosaic (horizontal red lines) with each of 99 simulations (black dots; dotted black line indicates median). In each case, the regularity index of the real mosaic is within the range generated by the simulations.

zone is ignored, and the trial neuron is rejected if it falls within the exclusion zone of an existing type 1 neuron. Each exclusion zone is again described by a Normal distribution with given (μ, σ). However, even under the assumption that there are no functional interactions, cell bodies of two neurons of opposite type still cannot overlap, and so a trial neuron is also rejected if the distance to the nearest neuron of opposite type is less than some small value, d_{12}, which typically is around 9–15 μm, matching cell body diameter.

Typical results from this bivariate d$_{min}$ simulation are shown in Figure 3. The regularity index of either solely the on-centre cells, the off-centre cells, or both types of cells fall within the range observed from 99 runs of the simulation, leading us to accept the null hypothesis that there are no functional interactions between neurons of opposite type, contrary to results from earlier modelling work (Eglen and Willshaw, 2002). A fuller treatment of this problem, using a more general style of model for simulating point patterns (the pairwise interaction point process model) is given elsewhere (Eglen et al., 2005).

3   Investigating Compensation in Cortical Networks

In our second example, we consider a higher level problem: inferring the putative activity dynamics from connectivity maps that describe effective connections between a network of cortical areas performing a cognitive task (Kronhaus and Willshaw, 2006). Again, we take the approach of using a simplified model that allows us to investigate subsequent ideas. In this case, we study the stability of network performance in the presence of global and local perturbations, as well as compensation in altered networks.

3.1   Cortical Interactions Implicated in the Delayed Match to Sample Task

Positron emission tomography (PET) was used to image brain activation during performance of a recognition task, the Delayed Match to Sample (DMTS) task. In this task, subjects were asked to identify a stimulus that was presented earlier in the experiment (Haxby et al., 1995). Here, two conditions were considered. In the first condition, perceptual matching, there were no distracting stimuli between the two presentations of the stimulus to be identified, and just one second delay between the two stimuli. In the second condition, long-delay, four distracting stimuli appeared over an interval of 21 seconds. The stimuli used in this study were pictures of male and female faces. The imaging data under the two conditions were analysed, generating activity maps for each condition. Structural equation models were then applied to estimate (effective) connectivity between key brain regions thought to be involved in these tasks (McIntosh et al., 1996). The path coefficients estimated by McIntosh et al. (1996) were used to guide the generation of the networks studied here (Figure 4, Table 1).

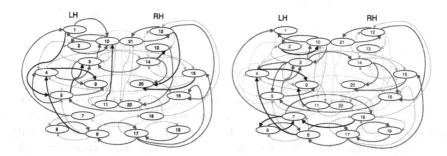

Fig. 4. Human perceptual matching (left) and long-delay (right) networks inferred from PET imaging data. Areas in each hemisphere are identified using the index described in Table 1, which can be used to identify the corresponding Brodmann area. The areas are arranged topographically, respecting their order within the brain. The colour and width of arrows indicates the value and type (excitatory/inhibitory) of connection: thick red (+0.65), thin red (+0.35), thin blue (−0.35) and thick blue (−0.65). Connectivity data derived from McIntosh et al. (1996).

3.2   Modelling Cortical Dynamics

We assume that N brain regions are being modelled. (For the McIntosh data, $N = 22$, eleven of the same brain regions from each hemisphere.) Each brain region i is summarised simply by a real value a_i denoting the overall activity of that area; the vector $\mathbf{a}(t)$ represents the activity of all brain regions at given time t. Cortical interactions (for each of the two conditions: perceptual matching and long-delay) were encoded into an $N \times N$ connectivity matrix $\mathbf{W}$, with

Table 1. Brodmann Areas used in this study. Left and Right hemisphere (LH/RH) index denote the region number assigned to this area in the model.

Lobe	Brodmann Area	Region	LH index	RH index
Frontal	BA46	Middle frontal	1	12
	BA10	Inferior frontal	2	13
	BA47	Ventral inferior frontal	3	14
Temporal	BA21	Middle temporal gyrus	4	15
	BA37	Inferior temporal gyrus	5	16
Visual	BA18v	Fusiform	6	17
	BA19d	Cuneus	7	18
	BA17/18	Cuneus	8	19
Limbic	GH	Hippocampus	9	20
	BA24	Anterior cingulate gyrus	10	21
	BA23	Posterior cingulate gyrus	11	22

positive/negative values denoting excitatory/inhibitory connections, respectively. Given a pattern of activity at time t, the activity at time $t + 1$ is given by:

$$\mathbf{a}(t + 1) = f(\mathbf{W}\mathbf{a}(t)) \tag{1}$$

$$\text{where} \quad f(x) = \frac{1}{1 + \exp(-k(x - \theta))} \tag{2}$$

The sigmoidal function $f(\cdot)$ is applied elementwise to $\mathbf{a}$ to ensure that the activity of each unit stays bounded within [0,1] (typically k=10, θ=0.5.) Given an initial pattern of activity at time 0 (described in the results below), network activity was typically updated for 100 iterations by which time the network dynamics usually converged to a stable pattern.

3.3   Characteristic Behaviour

The basic network behaviour was examined by initialising the network with a small amount of activity, and seeing how the activity propagated. In particular, one brain region (i) was selected to be initially active, whilst the remaining brain regions were silent. Figure 5 shows typical results of the spread and stabilisation of brain activity. Brain activity was normally initialised in BA18 (left hemisphere; i=6); this area was chosen as it is part of the visual cortex, and so likely to be activated by visual stimulation. For the perceptual matching network, transient visual activity led to sustained activity in several left hemisphere regions; some activity propagated transiently to the right hemisphere, but this did not persist. By contrast, activation of only BA18 did not lead to persistent activity in the long-delay network (data not shown). Instead, to generate persistent activity, initial input was needed in both BA18 and another area, such as BA37 (Figure 5 right). Visual comparison of the activity patterns in the perceptual matching and long-delay networks shows that although persistent activity

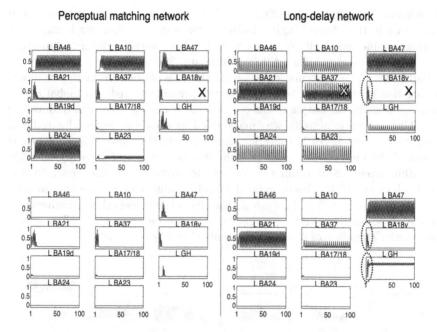

Fig. 5. Characteristic activation in the perceptual matching (left) and long-delay (right) networks. Each graph shows the activity of one brain region for 100 time steps. The legend above each graph denotes the brain region being plotted and its hemisphere (L/R). X within a graph (e.g. L BA18v on left) denotes the activation of this area was set to 1 at time 0; remaining areas were set to 0. This figure adapted from (Kronhaus and Willshaw, 2006), with permission from Oxford University Press.

is generated by both networks, in the long-delay network more activity persists in the right hemisphere. This result could not be predicted by visual comparison of the two networks in Figure 4.

As in Section 2, so far, this model can be regarded as a simplifying model; clearly simplifying the neural activity within each brain region down to a single number is not biologically plausible. However, we believe that we can use the characteristic activity patterns (e.g. from Figure 5) as a signature of activity that we can use as a reference when comparing these networks under different perturbations to represent various experimental conditions, as shown next.

3.4 Compensation for Localised Dysfunction

In this section we show how our network approach can be applied to investigate how cortical networks might adapt to either global or local impairment in a clinical population of interest, namely depressed patients. One key experimental finding suggests that in depressed patients there is an decrease in cortical excitability throughout the entire brain (Shajahan et al., 1999). We have therefore investigated how characteristic network behaviour is shaped by a reduction in cortical excitation. Surprisingly, characteristic network behaviour is unaffected by global reduction in all excitatory weight connections (inhibitory connections were held constant) by at least 20% (Kronhaus, 2004).

Furthermore, we systematically tested the role of each brain region in sustaining this characteristic behaviour in the presence of reduced global excitation. Specifically, all output connections from a chosen area were reduced by 10%. Characteristic activity was maintained in each case except for reducing connectivity from L BA24 (anterior cingulate). In this case, this local perturbation of the anterior cingulate had the dramatic effect of silencing all activity within around 20 iterations. Since network architecture was constrained by McIntosh et al. (1996), this prominent role of the anterior cingulate was an unexpected, emergent, finding. This novel result from the model echoed findings from the experimental literature reporting reduced neuronal and glial cell density in the anterior cingulate of depressed patients (Drevets et al., 1997).

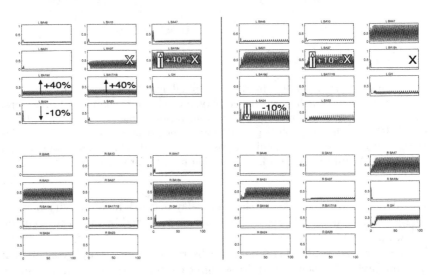

Fig. 6. Compensation in the long-delay network. Left: recovery by mostly visual areas; Right: compensation by BA37. Arrows indicate the direction in which an area's outgoing connections were modified by the given percentage. X indicates the areas which were initially stimulated. This figure adapted from (Kronhaus and Willshaw, 2006), with permission from Oxford University Press.

How might the brain compensate for such changes in network connectivity? One possibility is that activity in other areas may increase to compensate for these changes. To test this idea, we focused on alterations to the long-delay network, i.e. overall reduction of all excitatory connections by 20%, together with reducing all outgoing connections from L BA24 by a further 10%; in such a network, activity does not persist. However, it is possible to recover sustained activity, as shown in Figure 6, if we increase the outgoing connections from other areas. In the first case (Figure 6 left), an increase of outgoing connections from three visually-related areas led to a recovery in activity in both hemispheres. This activity is far from the characteristic pattern (Figure 5) observed in the long-delay network, as many areas that were previously active (e.g. both L and R BA47) were inactive. Further, selective increase of outgoing connections from just one area that plays a prominent role in generating the characteristic activity pattern (10% increase to L BA37) produced sustained activity in many areas, similar to that observed in the default network. Comparing Figure 6 right with Figure 5 right, activation patterns are broadly similar, although L BA46 and L BA10 are noticeably different. Thus, our model predicts that network dynamics in brains where both global and local activity are compromised (e.g. depressed patients) may be restored by intensifying activity elsewhere in the network.

4   Concluding Remarks

This paper has reviewed two recent examples of how simplified models of brain structure and function can be useful to investigate problems in neuroscience. Even though aspects of each model (e.g. the exclusion zone in the d_{min} model, and the characterisation of brain activity within a region by a single scalar value) do not have a direct biological interpretation, we believe that making such simplifying assumptions in models allows us to use these components within a bigger framework to test particular hypotheses and predict the outcome under novel situations. Furthermore, even though both models exclude fine-grained neurobiological details, they are consistent with the neural underpinnings. Thus, changes in efficacy (excitation or inhibition) are not simply ascribed to excitatory or inhibitory neurotransmitters. Instead, we argue that *dynamic* changes in the network are more appropriate to emulate behavioural phenomena such as those observed in clinical conditions such as depression.

In the context of this Volume, it is interesting to speculate on how these neuroscience models may inspire the next generation of wireless networks. One key feature in both our neuroscience models is robustness, in both design and function. Rather than hard wiring a neural circuit, developmental processes adapt to the local environment and create networks that are robust to environmental differences. Moreover, once the network is compromised, different features of network wiring (e.g. excitatory and inhibitory connections) allow the network to self-adjust, restoring function. These principles allow the nervous system to develop and function under a wide range of conditions. We suggest that such principles may also be of benefit to artificial systems, such as wireless networks.

Acknowledgements. Dina Kronhaus is supported by the Heller Research Fellowship, St Catharine's College, Cambridge. Thanks to Dr Lucia Galli-Resta, Prof. Heinz Wässle and Prof. Randy McIntosh for providing experimental data.

References

Hodgkin, A.L.: The Croonian lecture: Ionic movements and electrical activity in giant nerve fibres. Proc. R. Soc. Lond. B 148, 1–37 (1958)

Sejnowski, T.J., Koch, C., Churchland, P.S.: Computational neuroscience. Science 241, 1299–1306 (1988)

Nathan, R., Muller-Landau, H.C.: Spatial patterns of seed dispersal, their determinants and consequences for recruitment. Trends Ecol. Evol. 15, 278–285 (2000)

Sejnowski, T.J., Rosenberg, C.R.: Parallel networks that learn to pronounce English text. Complex Systems 1, 145–168 (1987)

Cook, J.E., Chalupa, L.M.: Retinal mosaics: new insights into an old concept. Trends Neurosci. 23, 26–34 (2000)

Wässle, H., Riemann, H.J.: The mosaic of nerve cells in the mammalian retina. Proc. R. Soc. Lond. B 200, 441–461 (1978)

Wässle, H., Boycott, B.B., Illing, R.B.: Morphology and mosaic of on-beta and off-beta cells in the cat retina and some functional considerations. Proc. R. Soc. Lond. B 212, 177–195 (1981)

Galli-Resta, L., Resta, G., Tan, S.-S., Reese, B.E.: Mosaics of Islet-1-expressing amacrine cells assembled by short-range cellular interactions. J. Neurosci. 17, 7831–7838 (1997)

Eglen, S.J., van Ooyen, A., Willshaw, D.J.: Lateral cell movement driven by dendritic interactions is sufficient to form retinal mosaics. Network: Comput. Neural Syst. 11, 103–118 (2000)

Eglen, S.J., Willshaw, D.J.: Influence of cell fate mechanisms upon retinal mosaic formation: a modelling study. Development 129, 5399–5408 (2002)

Eglen, S.J., Diggle, P.J., Troy, J.B.: Homotypic constraints dominate positioning of on- and off-centre beta retinal ganglion cells. Vis. Neurosci. 22, 859–871 (2005)

Kronhaus, D.M., Willshaw, D.J.: The cingulate as a catalyst region for global dysfunction: a dynamical modelling paradigm. Cereb. Cortex 16, 1212–1224 (2006)

Haxby, J.V., Ungerleider, L.G., Horwitz, B., Rapoport, S.I., Grady, C.L.: Hemispheric differences in neural systems for face working memory: a PET rCBF study. Hum. Brain Mapp. 3, 68–82 (1995)

McIntosh, A.R., Grady, C.L., Haxby, J.V., Ungerleider, L.G., Horwitz, B.: Changes in limbic and prefontal functional interactions in a working memory task for faces. Cereb. Cortex 6, 571–584 (1996)

Shajahan, P.M., Glabus, M.F., Gooding, P.A., Shah, P.J., Ebmeier, K.P.: Reduced cortical excitability in depression. Impaired post-exercise motor facilitation with transcranial magnetic stimulation. Br. J. Psychiatry 174, 449–454 (1999)

Kronhaus, D.M.: Neuroinformatics approaches to understanding affective disorders. PhD thesis, University of Edinburgh (2004)

Drevets, W.C., Price, J.L., Simpson Jr., J.R., Todd, R.D., Reich, T., Vannier, M., Raichle, M.E.: Subgenual prefontal cortex abnormalities in mood disorders. Nature 386, 824–827 (1997)

An Artificial Chemistry for Networking

Thomas Meyer, Lidia Yamamoto, and Christian Tschudin

Computer Science Department, University of Basel
Bernoullistrasse 16, CH–4056 Basel, Switzerland
{th.meyer,lidia.yamamoto,christian.tschudin}@unibas.ch

Abstract. Chemical computing models have been proposed since the 1980ies for expressing concurrent computations in elegant ways for shared memory systems. In this paper we look at the distributed case of network protocol execution for which we developed an online artificial chemistry. In this chemistry, data packets become molecules which can interact with each other, yielding computation networks comparable to biological metabolisms. Using this execution support, we show how to compute an average over arbitrary networking topologies and relate it to traditional forms of implementing load balancing. Our long-term interest lies in the robust implementation, operation and evolution of network protocols, for which artificial chemistries provide a promising basis.

Keywords: artificial chemistry, network protocols, distributed algorithms, load balancing, Fraglets.

1 Introduction

Chemical computing models have been proposed since the 1980ies for expressing concurrent computations in a natural way [1,2,3,4,5]. In parallel, artificial chemistries [6,7,8] were constructed to model chemical phenomena related to life and its origins. Some of these chemistries express and evolve computer programs [7,9], potentially representing new models of computation. However the vast majority of these artificial chemistries have remained at the level of simulations, where the actual pace of the chemical system with respect to the real time and among different devices is not an issue.

In this contribution we examine the potential of such chemically-inspired computation models for networking. On one hand, we extend our Fraglet language [10] to a full artificial chemistry setting. On the other hand, we extend the artificial chemistry concept to a distributed system, in two steps: The first step is to extend it to an online environment where time synchronization becomes essential for responding to external events with the expected impact. The second step is to take the network topology into account, extending the centralized analysis methods to a distributed system.

Finally, we show a case study on a distributed equilibrium algorithm that is able to find the equilibrium concentrations of a target molecule among a set of nodes in an arbitrary topology. The algorithm is applied to a load balancing problem in which tasks should be equally distributed among all nodes.

P. Liò et al. (Eds.): BIOWIRE 2007, LNCS 5151, pp. 45–57, 2008.
© Springer-Verlag Berlin Heidelberg 2008

Many sophisticated approaches to load balancing exist [11,12,13,14,15]. A classification can be found in [12]. More recently, a chemotaxis-inspired load balancing approach was proposed [15]. Although also chemically-inspired, the approach in [15] did not rely on a general purpose artificial chemistry. In our system, in contrast, the load balancing algorithm implicitly emerges as an effect of the chemical reaction network that is constructed and distributed among nodes. We do not claim that the resulting algorithm is superior to the current state of the art in load balancing. However, it offers a different perspective upon problem solving in distributed systems: instead of pre-programming the system numerically to achieve a desired stable state, an equilibrium can be reached autonomously via the exchange of virtual molecules among nodes.

This paper is structured as follows: Section 2 provides some basic background information on artificial chemistries and chemical computing, that will be needed for the rest of the paper. Section 3 briefly describes the Fraglet language and the instructions used in the load balancing case study, which is then presented in Sect. 4. Section 3 also defines Fraglets as an artificial chemistry, then proposes extensions of artificial chemistries to online and distributed systems. The equilibrium study in Sect. 4 also relies on these definitions and extensions.

2   Artificial Chemistries and Chemical Computing

Chemical computing [4,5,8] includes real (wet computation with real molecules) and artificial models inspired by chemistry but executing top of conventional computer architectures. This paper focuses on the latter only, and their extension to networked environments.

In [8] chemical computing models are classified within *Artificial Chemistry*, the subfield of Artificial Life devoted to the dynamics of chemical phenomena related to life and organizations in general.

The term *artificial chemistry* also refers to the specific chemical model used. In this sense, an artificial chemistry [8] is defined by a triple (S, R, A), where S is the set of molecules, R is the set of reaction rules, and A is an algorithm that determines how the rules are applied to the molecules. For example, in [7] the set S contains expressions from λ-calculus, the set R contains conditions under which two molecules from S may react and the way reactions take place: the reactants remain in the reactor, the corresponding products are inserted, and two other molecules are chosen at random for decay. The algorithm A just picks two molecules at random and performs the reaction or not depending on the conditions in R. Such a simplified chemistry can nevertheless show the spontaneous emergence of self-sustaining organizations out of an initial "soup" of random molecules. However, the algorithm A becomes computationally expensive if the probability of two random molecules reacting with each other is small.

For simulating real chemistries in an efficient way, variants of the Gillespie algorithm [16] are widespread. This algorithm simulates the stochastic dynamics of a real-world well-stirred chemical reactor tank. For each time step iteration, it calculates: *(i)* the next reaction to occur, taking into account for each reaction rule, the collision probabilities of their reactants; *(ii)* the virtual time τ when the chosen reaction is expected to occur. The complexity of this algorithm is $\mathcal{O}(|R|)$ for each iteration step: it is of course more complex than just choosing two molecules at random at every iteration step, but on the other hand, it only selects those molecules that do react in fact, and does not spend cycles on inert molecules. Therefore it is more efficient when only a small subset of those molecules in the reactor may actually react.

3 Organizing Interacting Packets as Chemical Reactions

In computer networking, the most frequently executed action on data packets is the rewriting of header fields. For example, on each leg of a packet's route through a sequence of Ethernets, the packet must obtain a new destination field to reach the next hop. Fraglets are a special form of data packets, based on the same principle: By rewriting a packet's header fields, we can implement distributed computations like communication protocols or a load balancing algorithm.

In this section, we first describe the Fraglet communication environment before in Sect. 3.2 we show how Fraglets form an artificial chemistry. Relating artificial time with real time and by interconnecting the artificial chemical reactors, we obtain a distributed artificial chemistry that is able to implement network functionality in ways beyond classic forwarding tasks, as we show in Sect. 3.3 and 3.4, respectively.

3.1 Fraglets

Formally, a string rewriting system is a pair (Σ, P) where Σ is a finite alphabet of symbols and P is a set of production rules. A production rule is a string substitution pattern that operates on words $w \in \Sigma^*$. The Fraglet language is an instance of a string rewriting system in which substitution patterns are limited to those which, on their left side, only depend on the first symbol of a word. For example, the rule

[exch S T U TAIL] → [S U T TAIL]

when applied to the word [exch a b c d] will result in [a c b d] – that is, two symbols are swapped. The exch acted as a prefix command for the rest of the word whereas the new leftmost symbol 'a' serves as a continuation pointer for further processing of the result.

This type of string rewriting systems, where the leftmost symbol identifies the rule to apply, is also called a tag system. Unlike Post's original tag system [17], which operates on one initial word and asks about the system's expansion, we place ourselves in a multiset context where the production rules are applied to all words in a multiset. In Fraglets, we interconnect several multisets such that they form a network of packet processing nodes. Thus a Fraglet system is a tuple (Σ, P, N, E) where N is a set of nodes $n_1, \ldots, n_k$, each containing a multiset of words over Σ to be transformed according to the rules P, and the nodes being interconnected according to edges $(n_i, n_j) \in E$. An excerpt from the set of production rules for Fraglets is shown in Table 1.

Table 1. Selected production rules of a Fraglet system. `S`, `T` and `U` are placeholders for symbols $\in \Sigma$, `TAIL` stands for a potentially empty word $w \in \Sigma^*$.

Op	input output
exch	`[exch S T U TAIL]` $\rightarrow$ `[S U T TAIL]`
node	n_i`[node S]` $\rightarrow$ `[S` n_i`]` *(get node's name)*
send	n_i`[send` n_j `TAIL]` $\rightarrow$ n_j`[TAIL]` *(if $(n_i, n_j) \in E$, ϵ otherwise)*
	n_i`[send any TAIL]` $\rightarrow$ n_j`[TAIL]` *($\exists j : (n_i, n_j) \in E$; anycast)*
	n_i`[send all TAIL]` $\rightarrow$ n_j`[TAIL]` *($\forall j : (n_i, n_j) \in E$; broadcast)*
split	`[split PART1 * PART2]` $\rightarrow$ `[PART1]`+`[PART2]`
sum	`[sum S` i_1 i_2 `TAIL]` $\rightarrow$ `[S` i_1+i_2 `TAIL]` *(do. for mult etc)*
match	`[match S TAIL1]`+`[S TAIL2]` $\rightarrow$ `[TAIL1 TAIL2]`
matchp	`[matchp S TAIL1]`+`[S TAIL2]` $\rightarrow$ `[matchp S TAIL1]`+`[TAIL1 TAIL2]`
mmatchp	`[mmatchp n S`$_1$ `...` `S`$_n$ `TAIL`$_0$`]` $\rightarrow$ `[mmatchp n S`$_1$ `...` `S`$_n$ `TAIL`$_0$`]`
	`+[S`$_1$ `TAIL`$_1$`]+...+[S`$_n$ `TAIL`$_n$`]` `[TAIL`$_0$ `TAIL`$_1$ `...` `TAIL`$_n$`]`

As an example, the fraglet `[split a b * c d e]` will result in two fraglets `[a b]` and `[c d e]`. The `match` rule lets two fraglets react together which share a common symbol at the second and first position, respectively:

$$\texttt{[match a b c] + [a x y z]} \rightarrow \texttt{[b c x y z]}$$

and the result is the concatenation of the two tails. The special persistent form of `match` is called `matchp` and permits to define "catalytic" processing rules that are not consumed during their reaction.

As a final example the following execution trace shows how the `send` tag can be used to implement traditional packet forwarding:

`[send n2 send n3 send n4 my payload]`	*sender's fraglet executes at n_1*
`[send n3 send n4 my payload]`	*at n_2*
`[send n4 my payload]`	*at n_3*
`[my payload]`	*arrived at n_4*

This example demonstrates source routing where the sending node n_1 lets a packet work itself through a chain of nodes $n_2 \ldots n_4$.

3.2 Fraglets as an Artificial Chemistry

In accordance with the definition in [8] explained in Sect. 2, an artificial chemistry is characterized by the triple (S, R, A), which we now define for Fraglets in the following way:

The set of molecule species S corresponds to the set of all possible fraglets, i.e. all words $w \in \Sigma^*$. Thus $S \equiv \Sigma^*$. Similarly, the set of reaction rules R is equivalent to the set of production rules P: $R \equiv P$.

The algorithm A, which selects which molecules to process at each round, only takes the molecules' matching heads into account. This leads to a two-level hierarchy: the actual molecular species correspond to fraglets which are strings $w \in \Sigma^*$ of arbitrary length, each of which may occur several times in the multiset of a given node $n \in N$. The reactor algorithm only looks at their headers, defining the second level of hierarchy where all the molecules with the same matching head symbol are considered as the same reactant for the choice of the next reaction to perform. This makes the algorithm scalable in spite of a potentially large number of different fraglets.

The original reaction algorithm in Fraglets did not take into account the full dynamics of molecule concentrations as in [7,16]. This restricted its applicability to cases where linear dynamics would suffice. We have now extended the Fraglet interpreter with variants of the Gillespie algorithm [16] such that more complex dynamics can be expressed. This is essential for an analytically tractable control of molecule concentrations. The header matching scheme is preserved in any case, since the algorithm only operates at the second level of hierarchy, which inspects only the fraglet headers.

3.3   Online Artificial Chemistry

In this section we introduce the concept of an *online artificial chemistry* for an artificial chemistry that is embedded in a real world environment in which it has to react in a timely manner.

In chemistry, the amount of reactions that may happen in parallel is only limited by the amount of molecules present and their collisions. For instance, autocatalytic reactions may lead to exponential growth in substrate concentration. From an information processing perspective, this is equivalent to an exponential growth in processing capacity. This powerful property is largely exploited when computing with real molecules such as DNA computing.

Algorithms such as Gillespie's [16] emulate chemistries on top of classical von Neumann computers with fixed processing capacity. The elastic processing capacity of chemical systems is emulated with the help of a virtual time which is inversely proportional to the total sum of the products of concentrations of all potential reactants. Virtual time steps can be made arbitrarily small as the reactant concentrations grow. If such virtual time is used for making decisions such as in a robot or network, then mapping it to physical time in a coherent way is mandatory for the device to present consistent reaction times.

It follows that an online artificial chemical system has to map the calculated virtual time (τ) which would have elapsed until the next reaction in a real tank reactor, to a physical time (τ'), the time that will actually elapse. A simple algorithm is to assume a one-to-one mapping $\tau = \tau'$, and just sleep for $\tau - T$, where T is an estimation of the real time it takes to process the reaction in practice. This simple algorithm obviously requires $\tau \geqslant T$, meaning that there is an upper bound in the number of molecules in the reactor, above which the algorithm is unable to keep the pace of a real chemistry. On the other hand, there is also a lower bound below which the sleeping time becomes too large for the system to react to the real world and to the inflow of new molecules.

Finding this compromise is crucial to have artificial chemistries running online, and deserves further research.

3.4   Distributed Artificial Chemistry

In order to use an online artificial chemistry in computer networks, we must expand it into a model of distributed reaction systems. In this section we show the challenges of the distributed case, and that the resulting distributed artificial chemistry can analytically be treated like a local one.

First we define a network of artificial chemical reactors (nodes). The network topology, which interconnects the reaction vessels, describes a high-level structure, conceptually one layer above the reaction system inside a certain node. A single node still reflects a well-stirred reactor that does not consider spatial neighborhood of molecules.

In this sense, a network of reactors can be described as a undirected graph $G = \{N, E\}$, where $N = \{n_1, \ldots, n_k\}$ is the set of all nodes in the network. Each node $n_i \in N$ is an independent reactor, driven by an individual CPU. The edges $E = \{e_1, \ldots, e_l\}$ are bidirectional network links and connect neighbor nodes. Two nodes n_i and n_j are neighbors iff $\exists e = (n_i, n_j) \in E$. In this case we define $\text{adj}(i, j) = \text{adj}(j, i) = 1$, otherwise $\text{adj}(i, j) = \text{adj}(j, i) = 0$. Each node is able to emit molecules, which are sent along the path of a link to one of the neighbor nodes using unicast, broadcast, or anycast transmission primitives.

Since each node is simulated independently, a single node n_i is an individual artificial chemical reactor with its own (S_i, R_i, A_i), i.e. each node is defined by an individual set of molecules S_i, reaction rules R_i, and algorithm A_i. We write W_i for a molecule $W \in S_i$. The exchange of molecules between two nodes n_i and n_j can be treated like reactions that map the set of molecules S_i to S_j, for example $W_i \rightarrow W_j$. Thus, in addition to a local reaction network, each node also contains reaction rules that send molecules to other nodes. The overall reaction network is then defined by (S, R, A), where $S = \bigcup_{i \in N} S_i$ and $R = \bigcup_{i \in N} R_i$.

Consider the following network $G = \{N, E\}$, where $N = \{n_1, n_2\}$, and $E = \{(n_1, n_2)\}$. The reaction system over the overall set of molecules $S = \{W_i, W_j\}$, depicted in Fig. 1, achieves a balance in concentration of molecule W between the two nodes.

The corresponding program in Fraglets uses unicast transmission to let each node send one molecule W at the time to the other node.

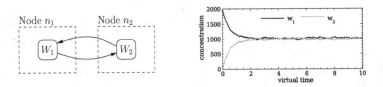

Fig. 1. Equilibrium reaction for a two node network topology

```
in node n1: [ matchp W send n2 W ]
            [ W ] 2000 (i.e., 2000 copies of [W])
in node n2: [ matchp W send n1 W ]
```

Even when starting with an unequal distribution of molecules W, the reaction system drifts into a state where both nodes contain the same amount of W. The stochastic simulation of this distributed reaction system is shown in the right side of Fig. 1.

There is one essential requirement for the algorithm A_i of a distributed reaction system: The virtual time evolution in all nodes participating the network must be the same. Generally, when a node sends molecules to another node, the sender generates a molecule stream of a certain rate with respect to the virtual time of its simulation algorithm. This rate is proportional to the concentration of the molecule. The molecule to be sent is then encoded as payload of a network packet, and is sent to the destination node, which is usually driven by another CPU. There the payload is converted back to a molecule, and is injected into the destination reaction vessel, which simulates its own virtual time evolution.

Therefore if the nodes of a network are allowed to be driven by different CPUs, it is important that the virtual time evolution in all nodes of the network is the same. Otherwise, a molecule stream may be received with an other rate than originally generated. If such a "synchronization" is achieved, for example by locally synchronizing the virtual time of each node to the physical time, then, the resulting distributed reaction system that is spawned across all nodes can analytically be treated like a local reaction system. This includes methods to analyze the topology of the network, stoichiometric analysis [18], metabolic control analysis [19], as well as results from the chemical organization theory [20].

4 A Chemical Protocol for Load Balancing

In this section we introduce a novel approach to balance work load in a network that exploits the dynamics of molecule reactions in an artificial chemistry. To this end we install flows of "job molecules" that seek to level out different concentration of job molecules on each node. The salient point is that load differences never need to be computed explicitly, as differences in packet rates are sufficient to steer the system into an equilibrium state.

Figure 2 depicts a typical network topology with columns representing the amount of jobs before and after load balancing. The left side shows the situation right after injecting a given amount of job molecules W into node n_1, whereas the right side delineates the job molecule distribution after the distributed reaction system reached its steady state.

We assume that all jobs are either independent and self-contained, or that we have job tokens which a node uses to request the actual job, that is: our system does not have to maintain a queueing policy for job molecules.

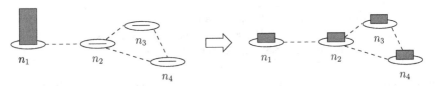

Fig. 2. Load balancing for a four node network topology. The columns represent the concentration of job molecules W in each node.

4.1   Distributed Equilibrium Algorithm

The basic idea of our algorithm is to *(i)* publish the local concentration of job molecules W to all neighbor nodes. Every node then *(ii)* requests job molecules W from overloaded neighbors in a stochastic manner. *(iii)* This promotes another reaction that carries job molecules from heavily to lightly loaded nodes.

We introduce three different families of molecule species: W_i molecules represent the work load, S_i are signalling molecules whose rates indicate to neighbors the current load level, while R_i molecules are requests from a node to obtain more work. Our algorithm can be expressed formally by the following abstract reaction system, where $X_i^{(j)}$ denotes molecule X, currently residing in node i, originally created and sent by node j.

$$W_i \longrightarrow W_i + \sum_{j \in N} \left((i,j) \cdot S_j^{(i)} \right) \qquad (i)$$

$$S_i^{(j)} \longrightarrow R_j^{(i)} \qquad (ii)$$

$$R_i^{(j)} + W_i \longrightarrow W_j \qquad (iii)$$

The numbering of the reactions corresponds to our list of principles introduced above. For example, reaction *(ii)* states that a received signaling molecule will be converted into a request molecule that is sent back to the originator of the signaling molecule.

The three reactions have a direct translation into the Fraglet language as shown below. Note that each of these three Fraglet rules must be present in all nodes of the network.

```
(i)   [ matchp W split W * split node N * match N send all S ]
(ii)  [ matchp S split node RCMD1 *
        split match RCMD1 exch RCMD2 W *
        split match RCMD2 RCMD R *
        match RCMD send ]
(iii) [ mmatchp 2 R W send ]
```

Before we explain in more details how these reactions work together, we introduce a graphical representation of the reaction network for a three-node string topology. As can be seen in Fig. 3, S molecules are deterministically transferred

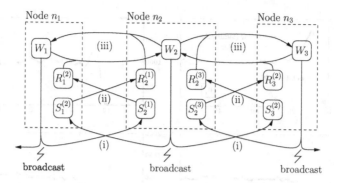

Fig. 3. Distributed reaction network

to a neighbor node and become a R molecule. Such a R molecule will stochastically bind to a W job molecule and drag it back to where the S molecule came from. The following paragraphs explain each of the three reactions in more detail:

(i) Concentration Tracking: Each node must be informed about the load of its neighbors. Therefore the first reaction's task is to broadcast signaling molecules S to all peer nodes. A signaling molecule contains the name of the source node, and is emitted with a rate that is proportional to the concentration of job molecules W. The opposite view is that each node receives a signal stream from its neighbors. The resulting concentration of signaling molecules S reflects the concentration of job molecules in the neighbor nodes.

(ii) Job Molecule Request: The second reaction stochastically picks and consumes one of the signaling molecules S_i. Since the Fraglet `matchp` reaction matches only the head symbol, this rule applies for all molecules $S_i^{(j)}$ received from any neighbor node j. This results in picking the signaling molecules of heavy loaded peer nodes more frequently, because the concentration of the signaling molecule reflects the concentration of job molecules W in the peer node, and due to the stochastic selection process. After picking a signaling molecule $S_i^{(j)}$, this reaction builds a request molecule $R^{(i)}$, tags it with the name of the local node, and sends it back to the node j, the originator of the signaling molecule. Request molecules are sent using unicast messages.

(iii) Job Molecule Transport: Request molecules promote the third reaction, that transmits a job molecule W_i to the neighbor node j that requested it.

The overall result of this reaction network is that if a node has a higher concentration of W molecules than any of its neighbors, it will emit more S molecules than the neighbors, thus be drained more aggressively through R molecules.

4.2 Results

The proposed three-stage mechanism regulates the exchange of job molecules W. Figure 4 shows the dynamic behavior of the concentration of molecule W in a four node network as depicted in Fig. 2. This result was obtained with a

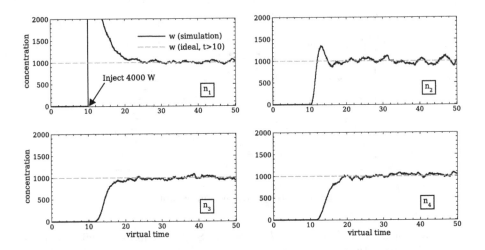

Fig. 4. Equilibrium of job molecules W for a four node topology without delay and packet loss. At time $t = 10$, a quantity of 4000 molecules of W is injected to node n_1.

Fraglet interpreter that simulated the behavior of the four nodes. We injected 4000 job molecules W at time $t = 10$ into node n_1. We ran the simulation of the above system such that all nodes are operating in non-saturated mode and that the network links are not afflicted with delay or packet loss.

The concentration of molecule W converges to the expected concentration of 1000 molecules in all nodes of the network. At $t = 13$ one can nicely see the onset of W molecules in node n_2: the rise of concentration is soon capped by nodes n_3 and n_4 which start to drain node n_2 from excess job molecules. The system quickly reaches the steady state where the concentrations fluctuate around an average value due to the stochastic notion of the reaction algorithm.

An important aspect of our approach is that we can verify the properties of chemical protocols by formal methods of flux base analysis [18]. In the case of the load balancing protocol, we studied the steady state of the system where the molecule concentrations do not change anymore. Solving the resulting equations for an arbitrary node shows that the concentration of W is equal to the average concentration of W in its neighbors. From this it follows that the concentration of job molecules must be equal in all nodes of the network for an arbitrary topology.

4.3   Discussion

In this paper we showed another, chemical way of looking at the problem of load balancing. In this section we first classify our algorithm and compare it to a similar chemotaxis-inspired method. Then we show why our initial approach, an even simpler, diffusion-based reaction network was not successful for an arbitrary network topology. Finally, we discuss the impact of imprecise virtual to physical time synchronization and network links with delay and packet loss.

Related Work. The chemical algorithm proposed in this paper can be classified according to [12] as a distributed dynamic load balancing algorithm: distributed, because the algorithm does not rely on central knowledge of the job distribution, and dynamic, because jobs may be dynamically produced and consumed during operation. Our algorithm is similar to the chemotaxis-inspired load balancing algorithm proposed in [15], which lets fast signals diffuse into the network. These signals then act as attractors to move jobs from overloaded nodes to nodes with available capacity. Similarly, in our algorithm, every node broadcasts signaling molecules to its neighbors. The role of signaling molecules is to promote the transfer of jobs. Unlike [15] we fully rely on the stochastic selection of reactions and molecules. For example, our algorithm never explicitly calculates the exact arithmetic difference of job molecules between nodes. Instead, the Gillespie algorithm more frequently picks those molecules that manifest in higher concentrations: The artificial chemical reaction vessel intrinsically balances the execution probabilities of interdependent reactions, and thus the resulting balance is an emergent property of the distributed reaction network.

Load Balancing with Less Than Three Reaction Types? The initial idea was to let job molecules immediately diffuse to the neighbors using anycast transmission. We hoped that a single reaction, which stochastically picks and alternatively sends a job molecule to a neighbor, would already yield a balance of work load. However, a formal analysis of the resulting reaction network, which strongly resembles the diffusion mechanism in physics, showed that the equilibrium can only be maintained for a fully meshed topology. Already a simple chain topology as in Fig. 3 results in an imbalance. At the end, we came up with the presented algorithm that obtains a work load balance for any network topology.

Real World Considerations. So far we have assumed that a CPU that executes the artificial chemical reactor is infinitely fast, that its clock is precise, and that the network is not afflicted with delay or packet loss. In the remaining paragraphs, we show what happens if we relax these constraints.

In reality, CPU clocks are subject to jitter and drift which affects the virtual to physical time mapping suggested in Sect. 3.3. Stochastic jitter is not harmful for chemical algorithms since they do not rely on deterministic execution of reactions. In contrast, clock drift between distributed reaction vessels lead to shifted reaction weights. When dilating the virtual time on a certain node the rate of incoming molecules increases while the rate of outgoing molecule decreases with respect to the virtual time. In this case the proposed algorithm establishes a distribution of job molecules proportional to the virtual time dilatation of the participating nodes. For example, if there is one node in which a reaction takes twice as long as in the other nodes, that node contains twice as much job molecules as the other nodes in steady state. Therefore one must assert that a node's load does not affect the speed of the artificial chemistry reactor, as otherwise our protocol will not achieve the desired result.

In addition to imperfect time mapping, in a real network the algorithm has to cope with packet loss and delay. In case of overloaded links, some of the signaling molecules S will be dropped. Consequently, signaling molecules are received at a lower rate, which leads to a lower concentration of S in the neighbor nodes. Hence, the second reaction reduces its activity, and generates less request molecules R, whereupon the algorithm gradually decreases the exchange of job molecules W, allowing the link to recover from the overload situation. Although this is a desirable behavior, it points out the need for another "job conservation" protocol that can handle molecule leaks.

Another problem is that in networks with non-negligible delay, the concentration of S follows the peer concentration of W with that delay. In the proposed algorithm we use the concentration of S for feedback control. A consequence of having delay in the feedback loop are oscillations. However by adapting the reaction constants accordingly, we are able to stabilize the algorithm for these situations.

5   Conclusions

In this paper we showed how Fraglets, a tag matching system to design network protocols, can be mapped to an artificial chemistry. By mapping Fraglets to an artificial chemistry, and by extending artificial chemistries to an online distributed environment, we obtained a system that enables us to implement "chemical protocols". We demonstrated such a protocol for load balancing, where the desired result emerges from the combination of the distributed reaction network and the stochastic execution algorithm.

"Chemical protocols" are interesting because they offer a new way of coupling network functionality: Instead of explicit numeric values (e.g. load differences) we use rate difference, which is much more elastic and receptive for cross talk from other network functions. Ultimately, we could organize network stacks as an ensemble of intertwined metabolic pathways, which we hope will be more robust and adaptive than the current static assembly of protocol modules.

Acknowledgments

This work has been partially supported by the Swiss National Science Foundation and the European Union, through SNF Project Self-Healing Protocols and FET Project BIONETS, respectively.

References

1. Banâtre, J.P., Métayer, D.L.: A new computational model and its discipline of programming, Technical Report RR0566, INRIA (1986)
2. Berry, G., Boudol, G.: The Chemical Abstract Machine. Theoretical Computer Science 96, 217–248 (1992)
3. Păun, G.: Computing with Membranes. Journal of Computer and System Sciences 61(1), 108–143 (2000)

4. Calude, C.S., Păun, G.: Computing with Cells and Atoms: An Introduction to Quantum, DNA and Membrane Computing. Taylor & Francis, Abington (2001)
5. Dittrich, P.: Chemical computing. In: Banâtre, J.-P., Fradet, P., Giavitto, J.-L., Michel, O. (eds.) UPP 2004. LNCS, vol. 3566, pp. 19–32. Springer, Heidelberg (2005)
6. Farmer, J.D., Kauffman, S.A., Packard, N.H.: Autocatalytic replication of polymers. Physica D 2(1-3), 50–67 (1986)
7. Fontana, W., Buss, L.W.: The Arrival of the Fittest: Toward a Theory of Biological Organization. Bulletin of Mathematical Biology 56, 1–64 (1994)
8. Dittrich, P., Ziegler, J., Banzhaf, W.: Artificial Chemistries – A Review. Artificial Life 7(3), 225–275 (2001)
9. Dittrich, P., Banzhaf, W.: Self-Evolution in a Constructive Binary String System. Artificial Life 4(2), 203–220 (1998)
10. Tschudin, C.: Fraglets – A Metabolistic Execution Model for Communication Protocols. In: Proc. 2nd Annual Symposium on Autonomous Intelligent Networks and Systems (AINS), Menlo Park, USA (2003)
11. Cybenko, G.: Dynamic load balancing for distributed memory multiprocessors. Journal of Parallel and Distributed Computing 7, 279–301 (1989)
12. Hosseini, S.H., Litow, B., Malkawi, M., McPherson, J., Vairavan, K.: Analysis of a graph coloring based distributed load balancing algorithm. Journal of Parallel and Distributed Computing 10, 160–166 (1990)
13. Xu, C.Z., Lau, F.C.M.: Analysis of the generalized dimension exchange method for dynamic load balancing. Journal of Parallel and Distributed Computing 16, 385–393 (1992)
14. Bahi, J., Couturier, R., Vernier, F.: Synchronous distributed load balancing on dynamic networks. Journal of Parallel and Distributed Computing 65, 1397–1405 (2005)
15. Canright, G., Deutsch, A., Urnes, T.: Chemotaxis-Inspired Load Balancing. In: Proceedings of the European Conference on Complex Systems (2005)
16. Gillespie, D.T.: Exact Stochastic Simulation of Coupled Chemical Reactions. Journal of Physical Chemistry 81(25), 2340–2361 (1977)
17. Post, E.: Formal Reductions of the Combinatorial Decision Problem. American Journal of Mathematics 65, 197–215 (1943)
18. Sauro, H.M., Ingalls, B.P.: Conservation analysis in biochemical networks: computational issues for software writers. Biophysical Chemistry 109, 1–15 (2004)
19. Hofmeyr, J.H.S.: Metabolic control analysis in a nutshell. In: Proceedings of the International Conference on Systems Biology, Pasadena, California, pp. 291–300 (2000)
20. Dittrich, P., di Fenizio, P.S.: Chemical organization theory: towards a theory of constructive dynamical systems. Bulletin of Mathematical Biology 69(4), 1199–1231 (2005)

Biomimicry: Further Insights from Ant Colonies?

Francis L.W. Ratnieks

Laboratory of Apiculture & Social Insects
Department of Biological & Environmental Science
University of Sussex
Falmer, Brighton BN1 9QG, UK
F.Ratnieks@Sussex.ac.uk

Abstract. Biomimicry means learning from nature. Well known examples include physical structures such as the Velcro fastener. But natural selection has also "engineered" mechanisms by which the components of adaptive biological systems are organized. For example, natural selection has caused the foragers in an ant colony to cooperate and communicate in order to increase the total foraging success of the colony. Ant colony optimization (ACO) is based on the pheromone trails by which many ant species communicate the locations of food in the environment around the nest. Computer algorithms based on ACO perform well in hard computational problems like the Traveling Salesman Problem. ACO algorithms normally use only a single attractive "pheromone". However, it seems that real ants use more. The Pharaoh's ant, *Monomorium pharaonis*, uses three different trail pheromones to provide short-term (volatile) and long-term attraction (non-volatile) and short-term (volatile) repellence so that foragers are directed to particular locations of the trail system where food can be collected. In addition, Pharaoh's ants also extract information from the geometry of the trail system and have division of labour among the forager workers, some of whom specialize in laying and detecting pheromone trails. ACO takes inspiration from ant colonies but does not need to faithfully model how ant colonies solve problems. For example, in ACO "pheromone" is applied retroactively once an "ant" has returned to the nest, which is something that can easily be implemented in a computer program but is obviously something that real ants cannot do. This raises the possibility that ACO might benefit from taking further inspiration from ant colonies. Presumably, real ants use multiple information sources and communication signals for a reason.

Keywords: Ant colony optimization, Pharaoh's ant, *Monomorium pharaonis*, honey bee, *Apis mellifera*, social insects, complex adaptive systems.

1 Learning from Nature

Learning from nature, or biomimicry, is common in engineering. Having spent the past 12 years living close to Chatsworth House in Derbyshire, England, one of my favourite examples concerns the giant water lily, *Victoria amazonica*, a spectacular plant with floating circular leaves up to 3m in diameter. A friendly rivalry developed

P. Liò et al. (Eds.): BIOWIRE 2007, LNCS 5151, pp. 58–66, 2008.
© Springer-Verlag Berlin Heidelberg 2008

between the Duke of Devonshire, the owner of Chatsworth, and the Duke of Northumberland, the owner of Syon House in London, to cause this plant to flower. Joseph Paxton, the head gardener at Chatsworth, succeeded in 1849 and presented one of the first blooms to Queen Victoria. The ribs and veins on the undersides of the giant leaves appeared to Paxton "like transverse girders and supports" and became his inspiration for designing a giant glasshouse at Chatsworth and later the Crystal Palace in London, at the time the largest building in the world. Less spectacular but no less ingenious is the Velcro fastener, which was invented in 1951 by George de Mestral, a Swiss engineer. De Mestral was inspired by the hooks by which burdock seeds (*Arctium* spp.) attach to animal fur for dispersal, and which also attached to his dog and his own clothing during summer walks in the Alps (information taken from Wikipedia).

The two above examples both concern physical structures. But learning from nature is not confined to physical structures. In *Biomimicry: Innovation Inspired by Nature*, Benyus [1] gives examples in areas as diverse as farming, energy supplies, healing, making things, storing information, and business. In *Biomimicry for Optimization, Control and Automation*, Passino [2] focuses on insights from the ways that biological systems, such as cells or organisms, control and regulate processes.

Life on Earth is more than 3 billion years old and has been subject to improvements "engineered" by natural selection throughout this period. Natural selection favours adaptations. These are normally features that cause organisms to be more successful in survival and reproduction [3]. All kinds of adaptations have been favoured, including physical structures such as the hooks on burdock seeds and the robust veins on giant lily leaves. But adaptations also include the internal mechanisms by which the different components of an organism function in a coordinated manner. In this way the organism functions effectively. That is, it stays alive, grows and reproduces. In short, natural selection has not only caused a wealth of adaptations involving materials and physical structures but also ways for controlling and organizing complex systems consisting of many component parts. The latter may be less immediately obvious, both in understanding how they work and in devising applications through biomimicry. But in a world in which humans rely more and more on systems consisting of many components, such as transport and communication networks, biology is a good place to look for ideas.

The purpose of this article is to showcase the mechanisms by which a particular type of adaptive biological system—the foraging system of an insect society—is controlled and coordinated as an example worthy of further study by engineers. The emphasis is on the Pharaoh's ant, *Monomorium pharaonis*, a species which we have been studying in my lab. I will also refer to the honey bee, *Apis mellifera*, which is the best studied of all social insects. Although computer scientists have already devised methods inspired by insect societies [4], such as Ant Colony Optimization (ACO) [5] and Swarm Intelligence [6,7], I think that more insights can be obtained. In particular, insect societies seem to be much richer in coordination and communication systems than seems to be minimally necessary. Presumably, there is a reason behind this, which may lead further insights and inspiration.

2 Insect Societies as Adaptive Biological Systems

The need to organize efficient communication and transport networks is shared by both insect and human societies. The colonies of many species of ants form networks of foraging trails. To forage efficiently a colony must send foragers to where the food is [8]. Because food locations change constantly this is not a trivial challenge. In fact, it is a biological example of a dynamic optimization problem. This is a type of problem increasingly of interest to engineers and computer scientists. It is a problem, for example, in organizing cell-phone networks given that both phone locations and traffic intensity are constantly changing.

Should insect colonies be of special interest? What about other groups consisting of multiple individuals? If you see a group of organisms, such as pigeons, of the same species this does not mean that they are working together to some common goal. They may well be aggregating to reduce predation ("selfish herding") or to find mates [9]. But the subset of the worker force in an insect society that collects food is designed by natural selection to cooperate and, thereby, to work together more effectively. This is because the food that the many forager workers collect all goes back to the same nest to feed the same larvae. If one worker ant or bee helps a worker from the same colony to collect more food then this cooperation will be favoured by natural selection. (Conversely, if one pigeon in a flock were to help another to collect more food, this would not normally be favoured by natural selection.) The most complex known animal communication signal, the waggle dance of the honey bee, is one way of doing this [10]. A forager bee that has returned to the nest will often make waggle dances to communicate to unemployed forager bees the direction and distance of the food source she has been visiting. Foragers are more likely to dance if they are working a highly profitable patch of flowers, thereby causing positive feedback to better foraging patches [11]. Ants don't have waggle dances but in many species they communicate the locations of food sources by pheromone trails or by leading recruits directly [11]. As such, the foraging system of an insect colony is a good place to look for insights into improving the performance of systems that rely on communication and cooperation among many components. Furthermore, social life in insects is a proven success. Both ants and honey bees are very successful organisms in terms of abundance and ecological importance [11].

Ant colonies solve problems via the self-organization of multiple agents [12]. A solution, such as an efficient foraging trail network, emerges from the actions of many agents—worker ants—each of which is individually ignorant of the overall network and is simply responding to local conditions. For example, if a forager ant reacts to finding food by depositing pheromone on its way back to the nest then this will result in positive feedback on trails that lead to food. Negative feedback on trails that lead to depleted food sources can be caused simply by the evaporation of attractive trail pheromone, or by the addition of a repellent trail pheromone [13]. In this way, the trail network from the entrance of the nest does not direct foragers to random locations in the surrounding environment. Rather, it directs them to the better feeding locations. This is analogous to the honey bee foraging system, where positive feedback to better feeding locations comes about because foragers that are working more profitable flower patches are more likely to make waggle dances [10].

3 Inspiration from Ants and Other Social Insects

One example of ant-inspired problem solving in computer science is ant colony optimization, or ACO [5]. Here, multiple-agent simulations incorporate an evaporating pheromone trail with successful agents laying more pheromone. In this way a wide range of potential solutions are explored. The better solutions are reinforced with pheromone, and the system converges on a good solution. Near optimum solutions to problems that cannot be solved analytically, such as the Travelling Salesman Problem, TSP, can be obtained [5] and the method also performs well in real world tasks [7].

ACO is inspired by ants but does not faithfully follow ant biology. But why should it? A computer algorithm need not be constrained in the same way as real ants. For example, in ACO "pheromone" is normally applied retroactively when an agent has "returned to the nest". By contrast, real ants can only lay pheromone as they walk. ACO also employs heuristics that real ants cannot use. For example, in the TSP, when two cities are equally attractive in terms of "pheromone", an agent may select the nearer. Ants may well have their own heuristics. In choosing between two branches at a trail bifurcation, more ants will likely take the branch that involves less angular deviation from their current path or direction.

4 The Foraging System of Pharaoh's Ants

The Pharaoh's ant, *Monomorium pharaonis*, is a good species for studying pheromone trails and foraging. The worker ants are only 2mm long. Short foraging distances between nest and food, such as 50cm, are both realistic and easy to set up experimentally. As a result, natural foraging can be studied in the laboratory. An ant colony is kept in a small wooden box or tube within a larger plastic box that acts as a foraging territory. Workers do not seem to be greatly guided by their own memory or landmarks when foraging, as occurs in some ants.

Most ant species are hard to breed in the lab. But in Pharaoh's ants nestmate males and young queens mate readily. In addition, colonies have multiple queens. To make two colonies, the ants and brood in a single colony are divided, making sure that each part has a few queens. This makes it simple, for example, to make up colonies of any desired size, and then to reuse or recombine the same ants to make up new colonies [14]. Pharaoh's ants are "unicolonial" meaning that colonies are not well defined. Ants from different nests can be combined without fighting. Pharaoh's ants are thought to originate from Africa, but have been spread worldwide by man. They are found in the UK, but only inside building where they can be pests.

It is easy to get a colony to establish a foraging trail. Laboratory colonies are normally fed water, sugar syrup, and dead insects *ad libitum*. Before an experiment they are deprived of sugar syrup for several days. Syrup is then provided in a small plastic tube with pin holes for the ants to drink from, or simply by placing a drop of syrup on a piece of plastic. The syrup is placed in a location suitable for an experiment, such as at the end of a trail apparatus that is placed in the foraging box or connected to it by a temporary bridge. The bridge can lead to an experimental trail system, such as a straight trail with a bifurcation [13,15,22] or the ants can be given access to a larger area and allowed to make their own trails. Trails can be made out of easily available materials including plastic [15] or photocopier paper [16]. Foragers normally discover the syrup within minutes. It then takes a colony of 1000-2000 workers approximately 20 minutes to establish a trail with approximately 100 ants per minute passing to and from the feeder [15].

Our research mainly investigates the behaviour of ants walking along pheromone trails at a behavioural level. For example, by determining the number of ants going left or right at a trail bifurcation or following a trail we can determine the ability of individual ants to detect the pheromones, and hence the decay rates of trails. It is also possible to study the pheromones directly, using chemical analyses of the trails [17], and to observe the trail laying behaviours themselves [18,19]. Pharaoh's ants have a sting that they extend to deposit pheromone. Marks left by this sting can be observed if the ants walk over smoked glass [16].

Our research has uncovered some unexpected and interesting properties of the pheromone trail system. Perhaps the main overall result is simply the richness of the system in terms of information and communication mechanisms. Trails are not marked by a single attractive trail pheromone, as is often assumed. Rather, the behavioural responses of workers to trails indicate that there are two attractive pheromones and one repellent pheromone. Information about the polarity of the trail (which direction leads back to the nest) is also extracted from the geometry of the trail system [16]. And individual foragers, although they look the same do not behave the same. Some specialize in trail laying [18] or trail detecting [20].

Why have multiple trail pheromones? Our working hypothesis [21,22] is that they have complementary functions (Fig. 1). One of the trail pheromones is attractive and volatile, decaying within approximately 20 minutes [15,22]. Having a trail that dissipates rapidly is presumably advantageous in that it reduces the duration over which foragers will be directed to a depleted feeding location. It seems to function like the "white line" on a road, providing something to follow. It also provides information as to which branch to take at a trail bifurcation [15]. One of the two other trail pheromones is also attractive, but is non-volatile lasting up to two days [20]. This probably acts as a "memory", allowing a colony to re-establish a trail to a previously rewarding feeding location. Many ant species forage only at a certain time of day, and some food sources are only profitable at certain times of day, such as flowers or aphids that secrete nectar or honeydew. This long-lived pheromone is detected by specialist "pathfinder" ants that walk more slowly and with their antennae in contact with the substrate [20]. The final pheromone is repellent and volatile, decaying in approximately 30 minutes. It seems to act as a "no entry" signal, and is deposited after a trail bifurcation on the branch that does not lead to food [13]. This is in contrast to the attractive volatile pheromone, which is deposited along a whole trail. The repellent signal is volatile and can be detected before an ant reaches the trail bifurcation [13]. At junctions, additional information is provided in the form of road signs (human roads) and no-entry pheromone (Pharaoh's ant trails). Ants detecting this signal are more likely to U-turn, and also zig-zag more, presumably in search of the alternative, rewarding, branch.

An additional source of information is provided by the geometry of the trail system [16]. The trails are analogous to the root system of a plant, bifurcating repeatedly. The two outward branches have an angle of approximately 60 degrees between them. As a result, the junction of the three trails is asymmetrical, and this asymmetry provides information concerning the polarity of the trail—which direction leads to the nest and which lead away from the nest. Ants walking through trail bifurcations often make U-turns if they are walking the wrong way. (For example, foragers with a full stomach who are walking in the direction that leads away from the nest, and *vice versa.*) By experimentally altering the angle between trail bifurcations, it is clear that it is the geometry of the trail bifurcation that provides the polarity information, not the trail pheromone itself. At an angle of 60 degrees, approximately 5 times as many correct U-turns were made as incorrect U-turns. The precise angle was not critical showing that the underlying mechanism is robust. Angles between 45 and 90 degrees all gave good results [16].

In addition, even though the forager ants all look the same they do not behave the same. Approximately 7% actively maintain the trail, making repeated U-turns and laying trail pheromone with their extended sting [18]. Pathfinders, the ants that walk with their antennae in contact with the ground in order to detect the non-volatile attractive trail pheromone comprise about 20% of the foragers [20].

5 Conclusions

Our research on the organization of Pharaoh's ant foraging systems shows that ant foraging trail networks are much more complex that previously thought, with individual specialization, multiple information sources, and multiple trail pheromones providing both positive and negative feedback [21]. As an evolutionary biologist, I am certain that this complexity is not accidental but exists to make the foraging system of a colony of Pharaoh's ants more effective at collecting food. In this way the workers will (indirectly) pass on more copies of their genes. At this stage of the research we have identified some of the mechanisms involved, but as yet how they work together (Fig. 1) is only a hypothesis.

The focus of much research in complex systems in biology is directed at the organismal level or below, particularly cells within multi-cellular organisms, genes within the genome, or molecules within cells. Ant colonies, as well as colonies of honey bees and other social insects, provide another level of adaptive organization for comparison. The study of insect societies has several potential advantages. First, the subunits are macroscopic and can be directly observed. Second, it is easy to manipulate the system. Third, the number of communication signals and feedback loops is small enough to make the system manageable to study and model but not so small as to make the system trivial. Fourth, insect societies can easily be studied in the field or lab. In addition, insect societies have evolved many times (at least 10), and are highly diverse (approximately 20,000 species), and so provide many independent solutions to the problem of organizing a complex system. Finally, many of the problems solved by insect societies are within networks [23], including the network provided by the trail system of an ant colony, and are "agent-based" and so have wide applicability in engineering.

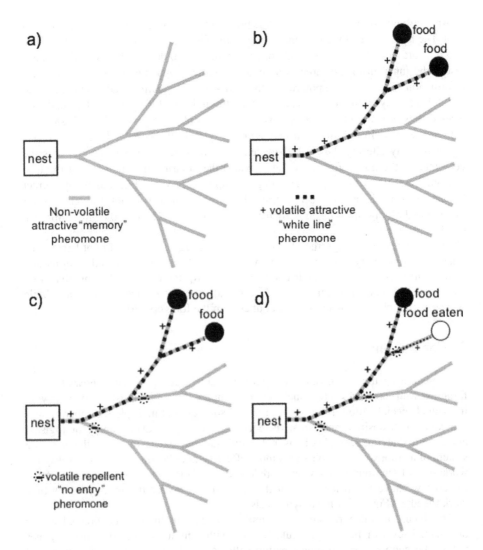

Fig. 1. Hypothesized complementary roles of multiple trail pheromones in Pharaoh's ants. a) The trail network is marked out with attractive non-volatile pheromone. This acts as a "memory" lasting several days of where ants have been walking and foraging, and where food may be found again. The memory can be "retrieved" by "pathfinder" ants that walk with their antennae in contact with the substrate. b) If food is found at a particular location the path is marked with the volatile "white line" attractive pheromone. This guides ants along the correct path and helps them chose the correct branch to take at trail bifurcations. c) At trail bifurcations the non-rewarding branch is marked with the "no-entry" pheromone. This helps ants chose the correct branch at a trail bifurcation in which only one branch leads to food. It is volatile and can be detected before an outgoing forager reaches the trail bifurcation. d) When the food at one location has been eaten, this branch becomes less attractive by the decay of the volatile "white line" pheromone, which decays in c. 20 minutes, and by laying "no entry" pheromone.

Do the trail networks of ants have anything to offer human engineers and computer scientists? The complexity we see in Pharaoh's ants may simply reflect the limitations that ants work under. Thus, the use of multiple trail pheromones may be because foraging ants make many mistakes when following a trail. For example, when a trail bifurcation is marked with attractive pheromone only approximately 75% of the ants choose the branch leading of food versus an unmarked branch [15]. An agent-based computer system could presumably make agents much more sensitive to differences in "pheromone concentration" at decision points, and this might obviate the need for multiple pheromones. On the other hand the realization that ant colonies use multiple trail pheromones might lead to improved ant-inspired problem-solving techniques. (Similarly, honey bees use at least 6 communication signals in organizing their foraging system [10,24,25,26]. ACO generally utilizes only a single evaporating attractive trail pheromone. What benefits might the implementation of multiple pheromones and the use of behaviourally specialized agents provide?

Acknowledgments. I thank the Leverhulme Trust for a Research Fellowship for the academic year 2006-7, which provided time to work with computer scientists and to attend the Biowire conference. I also thank the many PhD students, postdoctoral researchers, visiting scientists and colleagues who have worked with me in studying Pharaoh's ant trail systems over the past 12 years.

References

1. Benyus, J.M.: Biomimicry: Innovation Inspired by Nature. William Morrow (1997)
2. Passino, K.M.: Biomimicry for Optimization, Control and Automation. Springer, Heidelberg (2005)
3. Darwin, C.: On the Origin of Species by Means of Natural Selection. John Murray, London (1859)
4. Van Parunak, H.V.D.: Go to the Ant: Engineering Principles from Natural Agent Systems. Ann. Operations Res. 75, 69–101 (1997)
5. Dorigo, M., Stützle, T.: Ant Colony Optimization. Bradford Books, MIT Press, Cambridge (2004)
6. Bonabeau, E., Dorigo, M., Theraulaz, G.: Swarm Intelligence: from Natural to Artificial Systems. Oxford University Press, Oxford (1999)
7. Bonabeau, E., Theraulaz, G.: Swarm Smarts. Scientific American 282(3), 54–61 (2000)
8. Ratnieks, F.L.W.: Outsmarted by Ants. Nature 436, 465 (2005)
9. Alcock, J.: Animal Behavior: An Evolutionary Approach. Sinauer, Sunderland (2005)
10. Seeley, T.D.: The Wisdom of the Hive. Harvard University Press, Cambridge (1995)
11. Hölldobler, B., Wilson, E.O.: The Ants. Harvard University Press, Cambridge (1990)
12. Camazine, S., Deneubourg, J.-L., Franks, N.R., Sneyd, J., Theraulaz, G., Bonabeau, E.: Self-Organization in Biological Systems. Princeton University Press, Princeton (2001)
13. Robinson, E.J.H., Jackson, D.E., Holcombe, M., Ratnieks, F.L.W.: "No entry" Signal in Ant Foraging. Nature 438, 442 (2005)
14. Beekman, M., Sumpter, D.J., Ratnieks, F.L.W.: Phase Transition between Ordered and Disordered Foraging in Pharaoh's Ants. Proc. Nat. Acad. Sci. USA 98, 9703–9706 (2001)

15. Jeanson, R., Ratnieks, F.L.W., Deneubourg, J.-L.: Pheromone Trail Decay Rates on Different Substrates in the Pharaoh's ant, Monomorium pharaonis (L.). Physiol. Entomol. 28, 192–198 (2003)
16. Jackson, D.E., Holcombe, M., Ratnieks, F.L.W.: Geometry Gives Polarity to Ant Pheromone Trails. Nature 432, 907–909 (2004)
17. Jackson, D.E., Martin, S.J., Ratnieks, F.L.W., Holcombe, M.: Spatial and Temporal Variation in Pheromone Composition of Ant Foraging Trails. Behav. Ecol. 18, 444–450 (2007)
18. Hart, A., Jackson, D.E.: U-Turns on Ant Pheromone Trails. Current Biol. 16, R42–R43 (2006)
19. Jackson, D.E., Châline, N.: Modulation of Pheromone Trail Strength with Food Quality in Pharaoh's ant, Monomorium pharaonis. Anim. Behav. 73, 463–470 (2007)
20. Jackson, D.E., Martin, S.J., Holcombe, M., Ratnieks, F.L.W.: Longevity and Detection of Persistent Foraging Trails in Pharaoh's ants, Monomorium pharaonis (L.). Anim. Behav. 71, 351–359 (2006)
21. Jackson, D.E., Ratnieks, F.L.W.: Primer: Communication in Ants. Current Biol. 16(15), R570–R574 (2006)
22. Robinson, E.J.H., Jenner, E.A., Green, K.E., Holcombe, M., Ratnieks, F.L.W.: Decay Rate of Positive and Negative Pheromones in an Ant Foraging Trail Network. Insectes Soc. (in press)
23. Fewell, J.H.: Social Insect Networks. Science 301, 1867–1870 (2003)
24. Leoncini, I., et al.: Regulation of Behavioral Maturation by a Primer Pheromone Produced by Adult Worker Honey Bees. Proc. Natl. Acad. Sci. USA 101, 17559–17564 (2004)
25. Anderson, C., Ratnieks, F.L.W.: Worker Allocation in Insect Societies: Coordination of Nectar Foragers and Nectar Receivers in the Honey Bee. Behav. Ecol. Sociobiol. 46, 73–81 (1999)
26. Thom, C., Gilley, D.C., Hooper, J., Esch, H.E.: The Scent of the Waggle Dance. PLoS Biology 5, 1862–1867 (2007)

Network-Related Challenges and Insights from Neuroscience

Charles Peck[1,*], James Kozloski[1], Guillermo Cecchi[1], Sean Hill[1],
Felix Schürmann[2], Henry Markram[2], and Ravi Rao[1]

[1] IBM T. J. Watson Research Center, Yorktown Heights, NY 10598
cpeck@us.ibm.com
[2] Ecole Polytechnique Fédérale de Lausanne, 1015 Lausanne, Switzerland

Abstract. At nearly every spatio-temporal scale and level of integration,
the brain may be studied as a network of nearly unrivaled complexity. The
network perspective provides valuable insights into the structure and function of the brain. In turn, the structure and function of the brain provide
insights into the nature and capabilities of networks. As a consequence,
neuroscience provides a rich offering of network-related challenges and insights for those designing networks to solve complex problems. This paper
explores techniques for extracting and characterizing the networks of the
brain, classification of brain function based on networks derived from fMRI,
and specific challenges, such as the disambiguation of classification network
representations, and functional self-organization of cortical networks. This
exploration visits theory and data driven neural system modeling validated
respectively by capabilities and biological experiments, analysis of biological data, and theoretical analysis of static networks. Finally, techniques
that build upon the network perspective are presented.

Keywords: neuroscience, modeling, analysis, neural, network.

1 Introduction

For those seeking to exploit insights from biological networks for the understanding and design of network-based systems, neuroscience offers a compelling
subject of study for three key reasons. First, the nervous system, its elements,
and its activity can be characterized using network abstractions. Second, nervous systems possess capabilities often sought by designers of network-based
systems. Insights into how these capabilities are achieved may have engineering
value. Finally, the biological and physical constraints on a nervous system's elements and their interactions create new classes of challenges and problems, such
as maintaining stability, adapting to change, and eliminating representational
ambiguities, that must be solved enroute to satisfying the overarching requirements of nervous systems, such as producing coordinated, beneficial responses
to environmental threats and opportunities.

The network perspective can be applied at many levels of neuroanatomic
integration and abstraction, including:

---

* Corresponding author.

P. Liò et al. (Eds.): BIOWIRE 2007, LNCS 5151, pp. 67–78, 2008.
© Springer-Verlag Berlin Heidelberg 2008

- The brain modeled as a network of anatomically and functionally distinct structures, from small nuclei to major structures, such as the cerebellum and neocortex [5].
- Individual structures modeled as networks of neurons. For some structures it is not necessary to characterize every component and link of the network because highly stereotyped, repeated microcircuits, or network topologies applied to specific neuron types, have been demonstrated [17].
- Neurons modeled as electrophysiologically distinct, interacting compartments arranged according to dendritic and axonal morphologies [16].
- Compartments modeled as networks of interacting molecules, some with properties that change through these interactions.

Using various experimental modalities, such as functional magnetic resonance imaging and micro-electrode arrays, it is also possible to extract and analyze spatio-temporal networks of neural activity or brain function [7].

Similarly, emergent capabilities and the subnetwork challenges arising from biological and physical constraints occur at many levels of integration and abstraction. This interplay between network structure, biophysical constraints, emergent behavior, and functional requirements provides insights not exposed by any of these perspectives alone.

This paper attempts to illustrate the possibilities with a few examples. In particular, it explores network-related challenges and insights derived from all but the last of the abstraction levels itemized above. Sections 2–4 consider the cerebral cortex from biological and theoretical perspectives. Section 2 presents techniques to extract and characterize the neocortical microcircuit, the neurons within it, and their interactions. Section 3 explores a critical aspect of the perceptual "binding problem": how ambiguities arise in multi-layer classifier networks. It also explores how the dynamic properties of neurons may overcome this problem. Section 4 examines how the information maximization technique may be used to explain overcomplete representations in cortical maps. Section 5 presents a technique for extracting complex networks representing spatio-temporal activity patterns across the brain. It also draws conclusions about the topological features of these networks and how they can be used for classification and discrimination tasks. Conclusions are presented in Section 6.

2   Neocortical Column Calibration

The behavior of a network depends upon the characteristics of the network's elements and their relationships to each other. The Blue Brain Project (BBP) seeks to characterize the neocortical microcircuit of the rat somatosensory cortex in greater detail than has ever been attempted, translate this characterization into computational models, and use simulations to gain insights into neocortical function that are not possible through biological experimentation alone [26]. This section briefly describes the efforts of the BBP to characterize the neocortical microcircuit by using biological data to calibrate the modeled components and their structural and functional relationships to each other [14].

The Blue Brain Project calibration effort encompasses multiple interdependent components. To monitor and manage these interdependencies, the project has developed automated tools to score the resulting models for fit and completeness against each component of the calibration process. These scores are combined to create an overall measure of the precision and quality of fit for the resulting neocortical microcircuit model. Refinement of the model through these calibration steps is necessary to converge towards biological accuracy.

The calibration process examines the model across many levels: from ion channels in single cells to large-scale network phenomena. Each step of the calibration process includes a comparison of the model with biological experiments, a fitness analysis, and a score indicating the overall precision and quality of the fit. The calibration workflow for the neocortical column model checks the biological fitness of: 1) ion channel kinetics; 2) single cell electrical behavior; 3) dendritic integration properties including postsynaptic potential and backpropagating spike attenuation; 4) morphology repair; 5) monosynaptic properties including rise-time, amplitude, latency and short-term synaptic facilitation and depression; 6) polysynaptic loops including layer V pyramidal cell (L5PC)-Martinotti interactions; and 7) emergent phenomena including network oscillations and population responses to stimuli.

These calibration steps aim to apply experimental protocols to the model and provide a quantitative measure of the success of the model in recreating experimentally observed phenomena across multiple levels, from subcellular electrical properties to large-scale emergent phenomena. The calibration process provides a means to identify those areas where additional biological data is required and model aspects needing refinement to replicate the biological data more accurately. The calibration process is iterative and will be elaborated as new biological details become available.

Ion channel kinetics are modeled from voltage-clamp studies. These hold a cell or piece of membrane at various potentials for several intervals and measure the currents flowing between intracellular and extracellular space. The activation and inactivation conductance parameters and time constants for these voltage traces are fitted using Hodgkin-Huxley-style equations [32]. The calibration process for ion channel models entails applying the same stimulation protocol to the model ion channels and verifying the resulting current traces with experimental data. The difference between the model and experimental traces is computed and serves as a fitness score. In the event of a low score, the model must be further refined, possibly with additional experimental data.

A neuron's electrical properties are determined largely by the types, numbers, and proportions of ion channels and the way they are distributed throughout the neuron's morphology. The distributions of ion channels on the cell morphologies is determined by a genetic algorithm-based process to fit experimentally-observed firing properties [6].

The BBP protocol for calibrating the electrical properties of modeled neurons begins by stimulating real neurons with a sequence of computationally reproducible electrical patterns, called an eCode, and extracting key features, such as mean and resting potentials, spike height and width, first spike delay, firing rates, etc. A candidate computational model for the neuron is then stimulated in the same manner, the measured features of its response are compared to the actual, and a fitness score is computed.

The morphology of an individual neuron can be observed with a microscope after injecting the neuron with a dye. Because neuron morphologies are often damaged by the experimental preparation, Scholl analysis is applied to many instances of each neuron type to gather the expected numbers and properties of branches at each radius from the cell body and the missing portions of the neurons are repaired according to these statistics [1]. The morphology calibration step uses additional morphometric statistics to compare the repaired cells to the statistics of the population of the morphological class to which they belong. This ensures that repaired cells are consistent with biological classes of neurons.

Electrically and morphologically calibrated neurons are then used as building blocks for the calibration of the neocortical microcircuit. Completing the microcircuit calibration process requires determining the numbers of different neuron types and their relationships - both anatomically and functionally. Measurements of neuron types by cortical layer are used to both calibrate the numbers of neurons by type and complete the first step of calibrating their relationships: determining their spatial layout.

The precise location at which a neuron synapses on the dendritic and somatic morphology of another neuron influences the effect the presynaptic neuron has on the postsynaptic neuron's electrical response. It is therefore necessary to model these relationships accurately. While detailed information about synapse locations is generally not known, the number of synapses between two types of neurons is often known and can be used to validate specific configurations.

With 10,000 neurons in the neocortical column, each comprising hundreds to thousands of dendritic and axonal segments, finding configurations that yield the biologically observed touches is computationally daunting [19]. This problem has successfully been mapped to the 8,192 processor BlueGene/L supercomputer used throughout the Blue Brain Project. The task has been made easier and more effective by first fitting the connectivity within minicolumns and then fitting the connectivity between them. A minicolumn is a collection of neurons that migrated along the same radial glial cell during development.

With the anatomical relationships established, the calibration task shifts to functional relationships of increasingly higher order. This begins by calibrating the functional relationships in monosynaptic pathways. Experimental measurements have characterized the monosynaptic properties of pathways in the neocortex [25]. The key parameters of these relationships are those governing the post-synaptic response: the amplitude, rise time, latency, and the rate at which subsequent responses grow or decay.

Next, polysynaptic pathways are considered. In particular, a well studied circuit involving two layer five pyramidal cell (PC) and a Martinotti cell (MC) is used [36]. In this circuit, the first PC excites the two other cells and the MC inhibits both PCs. The first calibration step involves matching the effect of the first PC on the MC. The second involves matching the effect of the MC on each of the PCs. Once these effects are matched, the cascade of effects on the second PC from the first PC, both directly and through the MC, can be calibrated.

Finally, the emergent behavior of the entire column can be compared to biologically observed behavior. Comparisons to date have focused on cortically-generated slow oscillations, high-frequency stimulus-evoked oscillations and frequency dependent triggered recurrent activity. Analyzing the column's ability to reproduce these types of experimentally observed network behavior provides feedback on both the completeness of other calibration steps and the sufficiency of the levels of detail integrated into the model.

3   Dynamic Subnetworks Operating over Fixed Network Structures

Since McCulloch and Pitts spawned the field of artificial neural networks in 1943, researchers have sought to model neural systems and to recreate their capabilities using multi-layer networks of simple classifiers. In 1962, Rosenblatt identified a theoretical limitation of these networks [34]. This limitation, known as "Rosenblatt's Superposition Catastrophe" (RSC), applies when neural responses are binary and shared network resources are used to represent or classify multiple input patterns. In such networks, ambiguities can arise when multiple input patterns are presented simultaneously.

To illustrate how such ambiguities can arise, consider the following two layer network of neural classifiers. The first layer contains four neurons responding to features or attributes of a visual scene, without regard to location. The particular attributes are: 1) the color red, 2) the color green, 3) the presence of a circular contour, and 4) the presence of a triangular contour. Objects that may be present in the visual scene will have only one color and one shape. The second layer contains four neurons, each of which responds to a specific combination of attributes that may arise from objects in the visual scene: a red circle, a green circle, a red triangle, and a green triangle. If any single object with one of these attribute combinations is presented, the second layer will clearly respond properly. However, if two objects with mutually exclusive attributes are presented, such as a red circle and a green triangle, then all of the neurons in the first layer will respond. It is not possible to distinguish this response from that produced by from a green circle and a red triangle. This ambiguity, the RSC, erroneously causes all neurons in the second layer to respond.

The implications of the RSC are not only theoretical, but also biological. The problem is a central component of the "binding problem." The binding problem arises when reconciling the functional neuroanatomy of the cerebral cortex with the the properties of perception [24]. It has been observed that different attributes of a visual scene, like color, oriented edges, and motion, are represented by neural activations in distinct, distributed cortical areas [15]. Yet, visual perceptions are experienced in a unified way; the various attributes of the sensory space are "bound" together.

The root of the RSC is that the implicit relationship between an object's color and shape is lost when these two attributes are independently classified by the set of four classifiers in the first layer of the network example. Since this lost information is not propagated forward through the network, an ambiguity regarding the sources of attributes can exist at the second layer.

There are two solutions to this problem discussed in the literature [24]. The first is avoiding the ambiguities and superpositions with a combinatorially large network of classifiers. This solution is largely dismissed due the biological limitations on cortical networks.

The other solution is to augment the neuron amplitudes corresponding to classification with a second signal generated at the source of the classification signal, where all attributes are physically related to each other. Here, we'll refer to this second signal as relationship information. Biological observations suggest that synchronized neural activity reflects global properties of visual stimuli and that synchronization is correlated with recognition [38,9]. For these reasons, most researchers model the relationship information with the temporal properties of the classification signal, such as amplitude, frequency, and phase [24].

In [31], four requirements on relationship information were identified:

- *Uniqueness:* relationship information must be sufficiently distinct to avoid erroneous relationship interpretations,
- *Propagation:* relationship information must propagate forward and backward,
- *Aggregation:* classifiers must take the disparate, but sufficiently similar relationship values of its feedforward and feedback inputs and produce a unified relationship value; and
- *Selectivity:* relationship information must be used to modulate classifier responses to inputs.

If these requirements are satisfied, then the ambiguities in our example can be avoided. Let us assume that the relationship information is represented by the phase of signals propagating through the network. In our example, each pixel in the visual scene input participates in the classification of all attributes: color, shape, and, implicitly, location. For successful disambiguation, the process begins by assigning a unique phase to each pixel. This information is propagated forward to first layer classifiers. They produce their own phase based on the phases of signals that contributed most to their response. This phase information is propagated backward to its inputs, to synchronize them, and forward to the second layer classifiers. All classifiers respond solely to signals that are phase locked or nearly so. For this reason, second layer classifiers will not respond to combinations of attributes unless they are in phase and the ambiguity caused by multiple objects is avoided.

The effect of the relationship signal is to dynamically carve subnetworks of classifiers out of a large fixed network. Successful uses of this technique have been shown in the literature. For example, Rao et al. [33] constructed a network of oscillating elements that produces a sparse representation of objects presented to it. This network was able to demonstrate phase synchronization for single objects as well as superposed objects. This shows that the RSC can be overcome through the introduction of a relationship signal, such as phase.

4   Functional Self-organization of Cortical Maps

In 1986, Grinvald et al. pioneered the use of infrared imaging to reveal receptive fields of neurons in the primary visual cortex (V1) [11]. A visual receptive field corresponds to the region and pattern of visual space that evokes the strongest response in a neuron. In primates, the anatomical projections from the retinas of both eyes, through the thalamus, to the cortex are organized in an overlapping, retinotopic fashion; that is, retinal detectors responding to neighboring regions of the visual field project to neighboring neurons in the cortex, and detectors responding to distant areas project to distant neurons. Within this retinotopic framework, the receptive fields are organized into meandering, terminating, and forking stripes (or elongated "ocular dominance columns"), where all neurons within a stripe respond to information from one eye only.

Within these stripes, neurons are grouped into roughly cylindrical columns, similar to the columns investigated by the Blue Brain Project. The diameter of a column is approximately the width of the stripe that contains it. The columns within a stripe maintain retinotopic relationships to each other and may contain specialized groups of cells called "blobs," which respond to color information.

At a still finer scale, columns are made up of orientation-selective "mini-columns," in which all neurons respond best to bars of the same orientation. In 1991, Bonhoeffer and Grinvald showed that orientation-selective minicolumns are organized in characteristic topographic patterns [12]. Neurons preferentially responding to similarly oriented gratings tend to be close to each other in the plane of the cortex. As a grating is rotated from 0 to π, the pattern of preferentially responding neurons tends to radiate from the center of the column at progressively increasing or decreasing angles, like a pinwheel. While there are variations, iso-orientation bands form and typically begin and terminate at the pinwheel centers of adjacent columns.

This organization enables the cortex to produce a distinct response to every edge of every orientation located at any point in the visual space of either eye. As the neurons have overlapping inputs, it appears some form of cooperation is required in the organization of receptive fields to ensure the ability to respond distinctively to all inputs without undesirable redundancy. This raises an important question: "How does this organization arise in a network of neurons?"

Many ad hoc models of cortical learning processes have achieved key attributes of primary visual cortex receptive field organization [28]. There are limits to these models, however. For example, to be biologically plausible, models of the cortical network should derive their receptive fields from locally dense and distantly sparse interactions between network elements. This requirement is not satisfied by the many models that require interactions among all elements. Models with biologically realistic patterns of connectivity often use schedules to guide the learning of their receptive fields[2]. This can create new challenges to biological plausibility. For example, configuring multi-map systems based on these models creates biologically implausible learning schedule parameterization challenges. Multi-map systems are those where the outputs of one or more maps are used as inputs to others. Without a theoretical foundation, it is difficult to extend these models to solve this multi-map problem.

Information theoretic approaches, such as information maximization (Infomax) [22] [3], produce biologically realistic receptive fields and have a firm theoretical foundation. In related work, Linsker anticipated Grinvald's topographic observations by combining winner take all units and a wiring length minimization constraint [21]. Subsequently, Linsker showed that a globally optimal Infomax result could be achieved using information shared between neuron-like elements and self-contained, autonomous learning operations [23]. This technique is not hampered by biologically implausible learning schedules.

Until recently, the Infomax approach applied to these neuron-like elements could not yield topographic organization. The primary reason is that information maximization produces nearly statistically independent receptive fields and this removes the opportunity to use spatial correlations to generate topography.

This problem was recently overcome by Kozloski [18], who describes a multilevel network that first spatially smooths and rescales the output to successively smaller, higher level layers. Next, it eliminates redundancy in the smallest, topmost layer using an iterative Infomax technique. By successively rescaling and applying the changes to successively larger, lower-level layers, the system effectively reduces redundancy in the original output layer based on long-range statistics only. As learning proceeds, distant areas become statistically independent, while local dependencies increase. As long range statistics stabilize, shorter range statistics begin to dominate the learning process and local redundancy is eliminated. Consistent with the conventional interpretation of the multi-grid techniques from which this network was derived, the network first eliminates low spatial frequency redundancies and then redundancies of increasingly higher spatial frequencies. This continues until statistical dependencies in the output are largely, if not entirely, eliminated.

This system produces stable spatial organizations because as the receptive fields become statistically independent, the learning slows and the structure of the network helps to maintain the topography previously achieved. If one receptive field varies, it loses its statistical independence relative to other elements. This forces a cascade of learning until statistical independence and topography are approached once again.

There exists an important variation in the topographic organization achieved above and that observed by Grinvald. The organization observed by Grinvald, especially in the carnivore and primate (i.e., cat and monkey), is "overcomplete." Because there are more cortical outputs than thalamic inputs, the outputs must necessarily be algebraically redundant. This may be valuable for biology by not only representing independent components, but also explicit combinations of them. Mathematically, however, this is quite challenging. Which redundant components should be represented?

In [18], Kozloski achieves some degree of overcompleteness by composing his output layer from interwoven full rank Infomax subnetworks within the output layer. The elements of these subnetworks together create an overcomplete output. As above, they undergo spatial smoothing, and learn initially based on the collective statistics of distantly separated neighborhoods. However, in the overcomplete variation, these neighborhoods span multiple subnetworks. In this way, each subnetwork is co-embedded in the same topographic map, and components emerge that are linear combinations of components from other subnetworks and nearly statistically independent within each full rank subnetwork.

5   Classification of Brain Function from Spatio-temporal Activity Patterns

Until 15 years ago, electroencephalograms (EEGs) were the only practical, noninvasive technique to directly expose brain function. This changed in the early 1990's, when the functional Magnetic Resonance Imaging technique (fMRI) was developed [29,30,20]. This technique reveals locations and degrees of brain activity through localized variations in the magnetic susceptability of neural tissue. Neural activity causes these variations by altering the deoxyhemoglobin to oxyhemoglobin ratio in blood perfusing the tissue. This is known as the blood oxygen level dependent (BOLD) response.

Typically, fMRI data is analyzed with the General Linear Model (GLM) [8,37], which explicitly assumes that brain areas respond to stimuli (visual, tactile, etc.) or cause events (say, cognitive or motor) using linear mechanisms, and the activity of these brain areas are statistically independent. Even though the linear model has led to a number of remarkable findings, the above assumptions are extremely restrictive and certainly violated by the highly non-linear and interconnected nature of the brain. A different approach is required to formally capture and respect the brain's complex dynamics, neural connectivity, and functional interdependence.

One such approach is statistical network theory, pioneered by Erdös and further developed in the late 1990's [39]. It is a framework to model and analyze large-scale graphs based on their topological properties. The approach has been successfully applied to areas as diverse as social networks, the world wide web, gene networks, and linguistic structures [27].

Assuming that the topological characteristics of fMRI networks are more robust across subjects than the topographical mappings used in the GLM methodology, Eguiluz et al. applied statistical network theory to fMRI data [7]. They showed that networks defined by pair-wise correlations between functional voxels exhibit properties similar to other large-scale self-organized biological and technological networks, namely scale-free connectivity distributions and small-world topologies. These properties were also shown to be universal; that is, invariant across subjects and experimental conditions.

The demonstration that topological analysis of functional networks uncovers robust statistical regularities raised the possibility of classifying functional and dysfunctional brain states based on correlations with systematically-derived network motifs. The initial approach of Equiluz et al., however, does not provide enough information about the dynamical state of the brain to discriminate between sufficiently similar functional states. Consequently, Cecchi et al. [4] introduced a generalization of the correlation method to augment and refine the information for classification. In addition to the non-delayed or zero-lag correlations used by Eguiluz, the Cecchi method also computes delayed correlations. Significant delayed correlations are represented by directed links and significant zero-lag correlations are represented by undirected links. Furthermore, this network of directed and undirected links is simplified by eliminating redundant and ambiguous links, such as those generated by common source correlations and other correlation-based measures of functional causation.

The resulting hybrid networks were shown to effectively discriminate between subtly different brain states based on specific topological properties, such as their average mean geodesic path. The directed links are essential for this increase in discriminatory power, as they capture significant task-induced deviations of the functional dynamics from the default-mode state, which seems to be represented by the undirected (zero-lag) links [10]. Hybrid networks so constructed also provide a richer ensemble of topological patterns for identification and classification. Efforts are underway to implement motif analysis in massively parallel platforms.

6   Conclusions

It has been shown that the brain can be modeled using network abstractions at many levels of integration based on a variety of properties. The Blue Brain project models networks corresponding to neuron morphologies and the connections between specific classes of neurons. It calibrates various properties of these networks and analyzes emergent phenomena at many levels of network integration.

Rosenblatt's Superposition Catastrophe results from the properties of fixed classifier networks that share resources. Observed correlations of synchronized brain activity with global perceptual capabilities has inspired a solution to this problem using temporal modulation of classifier signals with relationship information that would otherwise be lost.

It has been shown that large-scale topographical organization of cortical maps can be achieved by competitive learning mediated through complex networks. If was further shown that overcomplete representations can be achieved through cooperative co-embedding of self-organizing networks.

The work of Eguiluz and Cecchi show that statistical network theory can be used to find robust properties of brain networks at the fMRI scale. Further, it was shown that these properties can be useful for classifying brain states.

Together, this body of work shows that the network perspective is useful for characterizing, analyzing and modeling the brain. Furthermore, it shows that understanding the brain through this lens can cast additional light onto the capabilities of networks and inspire solutions to network-related challenges.

References

1. Anwar, H., Riachi, I., Hill, S., Schürmann, F., Markram, H.: Capturing neuron morphological diversity. In: De Schutter, E. (ed.) Computational Neuroscience: Realistic Modeling for Experimentalists, 2nd edn. (to appear, 2008)
2. Bednar, J.A., Miikkulainen, R.: Neurocomputing 52-54, 473–480 (2003)
3. Bell, A.J., Sejnowski, T.J.: An information-maximisation approach to blind separation and blind deconvolution. Neural Computation 7, 1129–1159 (1995)
4. Cecchi, G.A., et al.: Identifying directed links in large scale functional networks: application to brain fMRI. BMC Cell Biology 8(suppl. 1), 5 (2007)
5. Doya, K., Kimura, H., Kawato, M.: Neural Mechanisms of Learning and Control. IEEE Control Systems Magazine, 42–54 (August 2001)
6. Druckmann, S., Banitt, Y., Gidon, A., Schürmann, F., Markram, H., Segev, I.: A novel multiple objective optimization framework for constraining conductance-based neuron models by experimental data. Frontiers in Neuroscience 1, 1 (2007)
7. Eguiluz, V.M., et al.: Scale-free functional brain networks. Physical Review Letters 94, 018102 (2005)
8. Friston, K.J., Holmes, A.P., Worsley, K.J., Poline, J.B., Frith, C.D., Frackowiak, R.S.J.: Statistical parametric maps in functional imaging: A general linear approach. Human Brain Mapping Volume 2(4), 189–210 (2004)
9. Gray, C.M., Singer, W.: Proc. Natl. Acad. Sci. USA 86, 1698–1702 (1989)
10. Greicius, M.D., et al.: Functional connectivity in the resting brain: a network analysis of the default mode hypothesis. PNAS 100, 253–258 (2003)
11. Grinvald, A., Lieke, E., Frostig, R.D., Gilbert, C.D., Wiesel, T.N.: Nature 324(6095), 361–364 (1986)
12. Bonhoeffer, T., Grinvald, A.: Nature 353, 429–431 (1991)
13. Hikosaka, O., et al.: Neurobiology of Learning and Memory 70, 127–149 (1998)
14. Hill, S.L., Chapochnikov, N., Druckmann, S., Gidon, A., Hay, E., Mace, A., Ramaswamy, R., Ranjan, R., Riachi, I., Schürmann, F., Srinivisan, K., Tränkler, T., Markram, H.: The Blue Brain Project: Calibrating a model of the neocortical column. Program 752.17. In: 2007 Neuroscience Meeting Planner, Society for Neuroscience, San Diego (2007)
15. Kandel, E.R., Schwartz, J.H., Jessup, T.M.: Principles of Neural Science, 4th edn. McGraw-Hill, New York (2000)
16. Koch, C., Segev, I. (eds.): Methods in Neuronal Modeling: From Ions to Networks, 2nd edn. MIT Press, Cambridge (1998)
17. Kozloski, J., Hamzei-Sichani, F., Yuste, R.: Stereotyped Position of Local Synaptic Targets in Neocortex. Science 293(5531), 868–872 (2001)
18. Kozloski, J., Cecchi, G.A., Peck, C.C., Rao, A.R.: Topographic Infomax in a Neural Multigrid. ISNN (2), 500–509 (2007)
19. Kozloski, J., Sfyrakis, K., Hill, S., Schürmann, F., Markram, H.: Identifying, tabulating, and analyzing contacts between branched neuron morphologies. IBM Journal special issue on Applications of Massively Parallel Systems (to appear, 2008)
20. Kwong, K.K., et al.: Dynamic magnetic resonance imaging of human brain activity during primary sensory stimulation. Proc. Natl. Acad. Sci. USA 89, 5675–5679 (1992)
21. Linsker, R.: From Basic Network Principles to Neural Architecture: Emergence of Orientation Columns. PNAS 83, 8779–8783 (1986)
22. Linsker, R.: Local synaptic learning rules suffice to maximise mutual information in a linear network. Neural Computation 4, 691–702 (1992)

23. Linsker, R.: A local learning rule that enables information maximization for arbitrary input distributions. Neural Computation 9, 1661–1665 (1997)
24. Von der Malsburg, C.: The what and why of binding: The modeler's perspective. Neuron, 95–104 (1999)
25. Markram, H., Lübke, J., Frotscher, M., Roth, A., Sakmann, B.: Physiology and anatomy of synaptic connections between thick tufted pyramidal neurones in the developing rat neocortex. J. Physiol. 500(Pt 2), 409–440 (1997)
26. Markram, H.: The Blue Brain Project. Nat. Rev. Neurosci. 7(2), 153–160 (2006)
27. Newman, E.J.: The structure and function of complex networks. SIAM Rev. 45, 167 (2003)
28. Obermayer, K., Sejnowski, T. (eds.): Self-Organizing Map Formation: Foundations of Neural Computation. MIT Press, Cambridge (2001)
29. Ogawa, S., Lee, T.-M., Nayak, A.S., Glynn, P.: Oxygenation-sensitive contrast in magnetic resonance image of rodent brain at high magnetic fields. Magn. Reson. Med. 14, 68–78 (1990)
30. Ogawa, S., Lee, T.M., Kay, A.R., Tank, D.W.: Brain magnetic resonance imaging with contrast dependent on blood oxygenation. Proc. Natl. Acad. Sci. USA 87, 9868–9872 (1990)
31. Peck, C.C., Kozloski, J., Cecchi, G., Rao, A.R.: A Biologically Motivated Classifier that Preserves Implicit Relationship Information in Layered Networks. In: Ribeiro, et al. (eds.) Adaptive and Natural Computing Algorithms: Proc. of the Int. Conf. in Coimbra, Portugal. Springer, Berlin (2005)
32. Ranjan, R., Druckmann, S., Gidon, A., Goodman, P., Hay, E., Hill, S.L., Ramaswamy, S., Schürmann, F., Markram, H.: The Blue Brain Project: Capturing parameters from genetically-prescribed ion channels for neuron modeling. In: Program 752.15. Neuroscience Meeting Planner, Society for Neuroscience, San Diego (2007) Online
33. Rao, A.R., Cecchi, G.A., Peck, C.C., Kozloski, J.R.: Unsupervised segmentation with dynamical units. IEEE Transactions on Neural Networks (January 2008)
34. Rosenblatt, F.: Principles of Neurodynamics: Perception and the Theory of Brain Mechanisms. Spartan Books, Washington (1962)
35. Seth, A.K., McKinstry, J.L., Edelman, G.M., Krichmar, J.L.: Cerebral Cortex May 13 Advanced access (2004)
36. Silberberg, G., Markram, H.: Disynaptic inhibition between neocortical pyramidal cells mediated by Martinotti cells. Neuron 53(5), 735–746 (2007)
37. SPM from the Wellcome Department of Cognitive Neurology, http://www.fil.ion.ucl.ac.uk/SPM
38. Varela, F., Lachaux, J.P., Rodriguez, E., Martinerie, J.: The brainweb: phase synchronization and large-scale integration. Nat. Rev. Neurosci. 2(4), 229–239 (2001)
39. Watts, D.J., Strogatz, S.J.: Collective dynamics of 'small-world' networks. Nature 393(6684), 440–442 (1998)

Networks in Epidemiology

Ken T.D. Eames[1] and Jonathan M. Read[2]

[1] DAMTP, Centre for Mathematical Sciences,
Wilberforce Road, Cambridge, CB3 0WA, UK
[2] Faculty of Veterinary Science, University of Liverpool, Leahurst Campus,
The Wirral, CH64 7TE, UK
ktde2@cam.ac.uk, jonread@liv.ac.uk

Abstract. We discuss the uses of networks as epidemiological tools to describe the interactions taking place within populations. The difficulties of accurate measurement of real-world social networks are discussed, along with modelling approaches designed to require only incomplete data. Properties of human contact networks such as clustering and variable strengths of interactions are seen to be important factors in the spread of an epidemic. We consider the evolution of a pathogen spreading through a dynamic network and show that the pattern of contacts within a host population determines the evolutionary pressures that a pathogen experiences.

Keywords: Infectious disease, social contact, epidemic, mathematical model, evolution.

1 Introduction

A wide range of mathematical models has been developed to study the spread of epidemics. When considering the spread of an infection through a population many factors are likely to be important, such as the transmissibility of the pathogen, the immune response of the host, the quality of health-care provision, and the behaviour of the host population. We will concentrate on the last of these and will discuss the use of networks as tools to examine interactions within a population. We will consider network approaches as applied to human populations but note that they are similarly applicable to animal and plant infections [1,2,3].

We begin by reviewing the basic principles of mathematical epidemiology, presenting briefly the standard modelling framework. We will then discuss the use of networks, and will cover some issues of more recent interest: data collection, modelling approaches, contact tracing and pathogen evolution.

2 Epidemiological Modelling

No simulation could accurately capture the multiple factors, their complexity and their variability, that combine to determine how and where an epidemic will spread. In order to gain insights into epidemic dynamics, to explain past events

P. Liò et al. (Eds.): BIOWIRE 2007, LNCS 5151, pp. 79–90, 2008.

or offer predictions of the future, practicality demands that simplifying approximations are made. It is the nature of models that all underlying assumptions are open to dispute, and in the case of epidemiological models this is particularly true. However, bearing in mind their necessary limitations, models have proved useful in the past and will be more and more relied upon in the future to direct public health planning and policy [2,4,5].

2.1   Basic Principles

The standard model of epidemic spread divides the population into a set of compartments according to infection status; a common formulation is the Susceptible-Infected-Recovered (SIR) model, in which individuals are assumed to be either susceptible to infection, infected and infectious, and recovered (if recovery does not result in long-term immunity, a Susceptible-Infected-Susceptible (SIS) model may be appropriate). The two key processes of infection and recovery are viewed as movements of individuals between compartments [1]. Infected individuals recover at a rate γ and, under the mass-action principle, the rate at which new infections arise is proportional to the product of the number of susceptibles and the number of infected individuals. The SIR model is described by the following system of differential equations:

$$\frac{dS}{dt} = -\beta SI \tag{1}$$

$$\frac{dI}{dt} = \beta SI - \gamma I \tag{2}$$

$$\frac{dR}{dt} = \gamma I \tag{3}$$

where S, I, and R are the number of susceptible, infected, and recovered individuals respectively and $N = S + I + R$ is the population size. β, the infection parameter, contains information about the behaviour of the hosts and the transmissibility of the pathogen. These models are easily formulated, simple to solve, and extremely adaptable, and have been applied to a wide range of contexts. We define the "basic reproductive ratio", R_0, to be the number of secondary cases generated by a single infected individual introduced into a susceptible population [1]. R_0 summarises the potential of a pathogen to spread (though not the time-scale at which transmission occurs). Here, $R_0 = \beta N/\gamma$. In a deterministic system an epidemic will occur if and only if $R_0 > 1$.

The model assumes a number of things: that infection status can be described by discrete compartments, that the numbers in each compartment can be represented as continuous variables, that the process is deterministic, that events occur at constant rates, and that interactions are mass action in nature. More complex modelling approaches have been developed to explore the effect of each of these simplifications and there are circumstances in which they must be reassessed. Here we will focus on the mixing assumption and particularly on the use of networks to represent interactions within a population.

3   Mixing Patterns and Networks

The mass action approximation is unlikely to be applicable to any directly transmitted infection. Therefore, several adaptations to the SIR model have been made
such as dividing the individuals into further categories (based on age, for instance)
and allowing different rates of interaction between those in different categories [1]
or assigning a spatial location to individuals or sub-populations and introducing
separation-dependent mixing [2]. Each is appropriate in the right context but neither solves the problem that, at some level, all individuals meet with all others
and each fails to take into account the fact that each individual in a population
has a specific set of contacts with whom he interacts [6].

Network approaches are inherently individual-based rather than population-
based. Individuals are represented as nodes and relevant interactions as links between nodes [7,8,9]. Precisely what constitutes a link will depend on the pathogen
of interest – for some, conversational contact will be sufficient for transmission
to occur, whereas for others links may correspond to sexual interactions [8,10].
Representing interactions as a network makes a number of points apparent. First,
each individual, in general, only interacts with a small subsection of the population. Second, pairs of individuals that interact tend to do so repeatedly. Thus,
while a social network will be dynamic, changing as new contacts form and old
associations disappear, a large number of links will remain in place day after day.
Networks are therefore often treated as static, showing all relevant interactions
over a timescale of interest (such as the duration of an epidemic).

For the same interaction rate, infection will spread more slowly on a network
than via mass action mixing; since contacts are repeated within a network infection becomes locally saturated – infected individuals continue to interact with
those whom they have already infected. Therefore, infectious contacts will be
wasted; this saturation is referred to as 'self-shading' in theoretical evolution
studies [11]. R_0 is also different on a network; in the mass action model increases
in the transmission rate or infectious period can increase R_0 indefinitely; in a
network R_0 is limited by the number of neighbours an individual has (his degree). Furthermore, since infection always occurs locally within a network, the
first few cases will interfere with each other since the first and second cases are
connected. Thus, local infection alters the epidemic threshold [12]. This effect is
further enhanced in a clustered network where connected individuals share other
contacts [13].

3.1   Measuring Networks

Network models are inherently more complex than mass action models and require more data to parameterise; in a mass action model all that is required
is a single parameter representing transmission whereas in a network model
ideally all interactions within a population would be known. Obtaining this information is a difficult and, often, impossible task [14,15]. Some information is
obtained through tracing chains of infection, but any network derived from infection tracing will only be a partial one, since it only gives information about

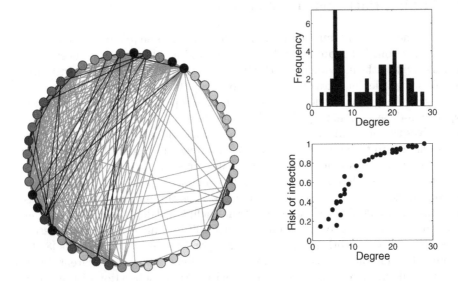

Fig. 1. a) Interaction network between 49 individuals each questioned on 14 separate days. Red lines show interactions where there was some physical contact, blue lines show conversational contact. The colour of a node represents its degree, with dark nodes being of higher degree. b) Degree distribution of this network. c) Risk of infection plotted against degree; a stochastic SIR epidemic is run repeatedly and the number of times each node becomes infected is counted.

those interactions that resulted in transmission; other contacts are not unveiled. Contact tracing provides a related source of network data: the contacts of infectious individuals are sought and tested/treated [8,16,17,18]. Contact tracing is commonly used for sexually transmitted diseases (STDs), where diagnosis is usually carried out by laboratory testing; information about all contacts is retained since it is not immediately apparent which require further intervention. Any infectious cases uncovered have their contacts sought, and the process continues. When records are well-maintained it is possible to connect individuals mentioned during contact tracing to form large networks of sexual interactions [8,17,18].

Some studies aim, by interviewing all individuals within a sub-population, to describe the entire network of social interactions within that subpopulation [6,14]. Such studies are intensive and do not necessarily give information that can be easily applied to an entire population, but are nevertheless often interesting and instructive. Fig 1a shows interactions between 49 members of the Mathematics and Biological Sciences departments at the University of Warwick: on 14 different days these individuals were asked to record all individuals with whom they had a contact that day, enabling a social network to be constructed. The heterogeneous degree distribution is shown in Fig 1b.

Often it will not be possible to obtain enough data to parameterise a network. In such cases, it may be appropriate to use an approximation of the true network structure. A number of different network idealisations have been studied: random networks, lattices, spatial, small-world and scale-free networks [10,11,19,20,21,22]. Some are based on convenience, others as means to explain particular features of social networks (e.g. small-world networks having high clustering but low diameter [22]; scale-free networks containing individuals with unusually high degree [19,20]). Alternatively, proxy measures are used to describe contact patterns: scientific collaboration graphs [23], the movement of banknotes [24], or travel patterns [25,26]. Of course, the network type employed within an epidemiological model should bear some resemblance to the network that it is intended to represent.

3.2 Modelling Epidemics on Networks

Stochastic epidemics on networks are in theory straightforward to simulate [21,25,26]; one must only keep track of each individual's infection status and the number of infected neighbours he has. However, stochastic simulations are not always convenient; they require a large amount of computational time and make exploration of an entire parameter space laborious. Furthermore, they require the entire network structure to be known or approximated. In cases where data is limited this may not be wise, in part because such models tend to conceal the limitations of their data [27].

An alternative approach is to use pair approximation methods [12,28,29]. These provide a deterministic model of spread through a network that lies somewhere between a full network model and the mass action approximation. As well as keeping track of the number of individuals in each state pair approximation methods keep track of the number of connected pairs of each type (e.g. susceptible-infected pairs); we denote the numbers of susceptible, infected, and recovered individuals by $[S], [I]$, and $[R]$ respectively, and let $[AB]$ represent the number of pairs within the network with one individual in state A and the other in state B (where A and B can be S, I, or R). Therefore

$$\frac{d[S]}{dt} = -\tau[SI] \tag{4}$$

where τ is the rate of transmission along a link. To iterate the model we must know how the numbers of each pair type change over time: infection can spread within a pair or can enter a pair from outside, thus, for example:

$$\frac{d[SI]}{dt} = -\tau[SI] - \gamma[SI] - \tau[ISI] + \tau[SSI] \tag{5}$$

To close the system at the level of pairs in a homogemeous network of degree k the moment closure approximation

$$[ABC] \approx \frac{(k-1)[AB][BC]}{k[B]} \tag{6}$$

is applied [12]. This model, and more complex extensions [13,29], allow the correlations that emerge between connected individuals to be accounted for without requiring the whole network to be known or modelled.

4   Use of Networks

Since each individual has a distinct location within the social network a range
of centrality measures can be used to assess an individual's infection risk. The
simplest, degree, is unsurprisingly correlated with risk of infection. For instance,
Fig. 1c shows the outcome of stochastic epidemics spreading though the network
shown in Fig 1a; highly connected nodes are more at risk of becoming infected.
Other measures – betweenness, centrality, etc – have also proved useful to determine which individual should be targeted for interventions [7,9,16,20,30,31,32,33].

Placing individuals within a network also allows the identification of core
groups – interconnected groups of people with a greater than average degree.
Such groups have been shown to act as reservoirs of infection and to drive infection dynamics within a population [17,31,34]. Even when most individuals are
unlikely to become infected or, if infected, to pass infection on, a small high-risk
subpopulation can allow infection to persist. Interventions that target the core
group often have the best chance of eradicating infection.

4.1   Contact Tracing Models

Contact tracing, described above, is an intervention that makes use of the network of interactions. The idea is to pursue infection up and down chains of
transmission, ideally catching up with it and snuffing out an epidemic. Network
models have been used to predict the effectiveness of contact tracing [35,36].
These demonstrate that, if tracing is carried out quickly and efficiently, it will
always be possible to prevent an epidemic; the difficulty in the real world is to
identify all risky contacts and to test and treat them before they spread infection.
Simple models demonstrate that if no more than one secondary case per index
case is untraced then rapid tracing can eradicate infection – tracing operates
similarly to vaccination within the neighbourhood of an infected individual.

Because contact tracing directs interventions to the neighbourhood of individuals known to be infected, it provides a well-targeted control measure. Since
tracing follows chains of transmission it naturally uncovers high-risk groups and
therefore results in the treatment of high degree nodes, thereby depriving a disease of those individuals most likely to pass it on.

5   Clustering

One important feature of networks is clustering: the existence of short loops [13].
In the simplest case we define a cluster as a group of three people who all know
each other, i.e. a triangle, and define a clustering coefficient, ϕ, as the probability
that two neighbours of an individual are themselves connected.

Some parts of a social network are expected to have high levels of clustering.
Households, schools, and workplaces, in which a large fraction of interactions
take place, are likely to be highly clustered locations, containing many groups
of people who all interact [4,25,26,37] – a household may well have a clustering
coefficient of 1. The network shown in Fig 1a, for example, has $\phi = 0.7$.

We see from the pair approximation model that clustering is important: the moment closure approximation, (6), assumes that the two individuals at the ends of a triple only interact via the central individual. However, if the triple forms a triangle then they also interact directly. This has prompted the development of an adapted triples approximation,

$$[ABC] \approx \frac{(k-1)[AB][BC]}{k[B]}\left((1-\phi) + \phi\frac{N[AC]}{k[A][C]}\right), \qquad (7)$$

that allows for the two possibilities [12,13,38].

In a highly clustered network connected individuals are likely to share other contacts, so there is a chance that some of the contacts of a secondary case will have been infected by the index case; this reduces the number of cases that the secondary case can generate. Clustering increases the amount of local saturation of infection and the number of "wasted" contacts, thereby slowing the spread of infection [13].

5.1 Pathogen Evolution

The structure of a network and the patterns of interactions within a population have an effect not only on the spread of a single epidemic but on the properties of the pathogen itself. The possibility that two infected individuals will both be attempting to infect the same susceptible individual has significant implications for pathogen evolution. Competition between strains on networks (and modified lattices) have been investigated using simulation models [11,21,38,39,40]. Much evolutionary work has focused on explaining observed differences of disease behaviour (in terms of transmission rate, infectious period and virulence) by invoking trade-offs between such properties [41]. However, differences may be explained by the host contact networks upon which pathogens transmit, mutate, survive or become extinct.

The impact of network structure on pathogen evolution can be modelled by allowing different strains to exist and infect hosts within a contact network, with the properties of a strain (such as transmission rate and infectious period) changing slightly on each transmission event, mimicking mutation. Different evolutionary outcomes arise in networks with different clustering coefficients [21,39] (Fig 2). When clustering is low, evolution leads eventually to long-lived strains with a relatively low transmission rate. However, in highly clustered networks direct competition between strains forces the transmission rate to evolve to a much higher value. The emerging structure of susceptible components within each network type re-enforce this evolutionary divergence. Direct competition between strains means that, to be successful, a strain must be the first to colonise new susceptible members of the network, even though this leads to burning through the entire network and epidemic fade out through lack of susceptibles, and confers an increased risk of extinction. Thus, the structural properties of the underlying contact network may be sufficient to select diseases with differing transmission characteristics.

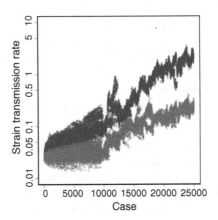

Fig. 2. Example evolution of simulated pathogen transmission rate on clustered (red, $\phi \approx 0.5$) and unclustered (blue, $\phi \approx 0$) networks. Both networks have identical degree, primary strain, size, and demographic parameters. The initial epidemic in each realisation is the first 10,000 cases, thereafter infection is imported at a constant probabilistic rate, where imported strain properties reflect the history of strain characteristics for that simulation.

6   Weighted Networks

Standard network models assume that all contacts within a network are equal: infection spreads at the same rate along all links. In the absence of data to the contrary, this is a reasonable simplifying assumption. However, day to day experience tells us that we interact differently with different people; with some we share an office throughout the working day; with some we share a house; others we see for only a few minutes at a time. There are some of our contacts whom we would expect to see daily and others that we might meet only once a week. Moreover, in some cases interaction is close, perhaps involving physical contact, whereas with others any interaction is distant.

To describe fully all the differences between links within a network is a Herculean task, but such differences can be summarised by the use of weighted networks, i.e. those where two individuals are not merely "linked" or "unlinked" but where links themselves are weighted according to the strength of the contact [9,42,43]. For example, the weight of a link may represent the number of hours spent together over the course of a week, or the number of days in a week on which an interaction takes place.

Fig 3a shows the distribution of link weights in the network shown in Fig 1a where here the weight is the number of times over the course of the survey individuals interacted. In a weighted network low-degree nodes may, through having links of high weight, have a large number of interactions, albeit with a small set of people. Fig 3b shows that, in this weighted network, where the rate of disease transmission is proportional to the weight of a link, degree becomes a much less effective predictor of infection risk whereas the total weight of all

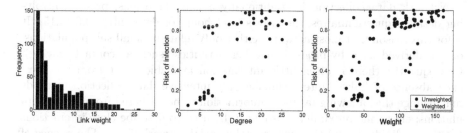

Fig. 3. a) The link weight distribution in the network in Fig 1a. b) Degree as predictor of risk on this weighted network. c) Weight as a predictor of risk on this weighted network (red) and in a model where transmission is unaffected by weight (blue).

links from a node is more useful (Fig 3c); in such a network an individual's risk of infection depends both on his degree and the strength of his interactions.

7   Other Considerations

Static networks provide a useful tool to describe interactions, but the social world is not static: new connections form and others dissolve [21,25,26,44]. Dynamic network modelling is a challenge to both modellers and data gatherers; without detailed and careful surveys it will prove impossible to parameterise a dynamic network model – even measuring a static network is extremely time consuming.

Even with the use of weighted or dynamic networks to give some representation of the variable nature of interactions, models must still make assumptions about the transmission process within such networks; if all links are equal it is reasonable to assume that all transmit infection at the same rate, but without studying a real epidemic on a measured network it is difficult to assess the relationship between time spent or proximity of contact and transmission probability; model validation requires detailed datasets that are seldom available.

8   Conclusions

Networks are useful tools for visualising patterns of human interactions. By employing these tools, and modelling approaches based upon them, more realistic representations of social mixing can be included in epidemic models. To maximise their use requires accurate and complex data, but even without detailed information network ideas can still be used to inform models.

The more complex a model becomes, and the more the assumptions of simpler models are contested, the more it becomes clear how many approximations any model must make; approximation is inevitable given the complexity of the world, and the art of model making is to find a way to include the necessary details whilst retaining tractability [27]. Issues such as network dynamics and weighted links are complexities that place more demands on data collection, but may well come to be seen as important considerations.

Network data collection is traditionally a long, laborious, process, involving detailed questionnaires, interviews, and hours of data-entry [3,6,14,15]. It is for this reason that most networks either only relate to small sub-populations, or are derived as a by-product of other activities (such as contact tracing). Consequently, there is often little information available apart from the presence/absence of links. More technologically intensive data collection methods may change this, allowing far more information to be gathered from larger populations; electronic proximity sensors can be used to describe close interactions between individuals and therefore reveal a social network [5,45]. The science of network-based epidemiology is relatively new and growing rapidly but its future success depends on obtaining a better understanding of human behaviour.

Acknowledgements

The authors thank: EPSRC (KTDE), Emmanuel College, Cambridge (KTDE) and the National Institute of Health (JMR) for financial support; John Edmunds for use of the Warwick social contacts survey; Matt Keeling and the Ecology and Epidemiology group (Warwick University), and the BIOWIRE delegates for helpful discussions.

References

1. Anderson, R.M., May, R.M.: Infectious diseases of humans. Oxford University Press, Oxford (1992)
2. Ferguson, N.M., Donnelly, C.A., Anderson, R.M.: The foot-and-mouth epidemic in Great Britain: pattern of spread and impact of interventions. Science 292, 1155–1160 (2001)
3. Gilbert, M., Mitchell, A., Bourn, D., Mawdsley, J., Clifton-Hadley, R., Wint, W.: Cattle movements and bovine tuberculosis in Great Britain. Nature 435, 491–496 (2005)
4. Cooper, B.: Poxy models and rash decisions. Proc. Natl. Acad. Sci. USA 103, 12221–12222 (2006)
5. King, D.A., Peckham, C., Waage, J.K., Brownlie, J., Woolhouse, M.E.J.: Infectious diseases: preparing for the future. Science 313, 1392–1393 (2006)
6. Edmunds, W.J., O'Callaghan, C.J., Nokes, D.J.: Who mixes with whom? A method to determine the contact patterns of adults that may lead to the spread of airborne infections. Proc. R. Soc. Lond. B 264, 949–957 (1997)
7. Doherty, I.A., Padian, N.S., Marlow, C., Aral, S.O.: Determinants and consequences of sexual networks as they affect the spread of sexually transmitted infections. J. Infect. Dis. 191, S42–S54 (2005)
8. Klovdahl, A.S.: Social networks and the spread of infectious diseases: the AIDS example. Soc. Sci. Med. 21, 1203–1216 (1985)
9. Wasserman, S., Faust, K.: Social network analysis. Cambridge University Press, Cambridge (1994)
10. Keeling, M.J., Eames, K.T.D.: Networks and epidemic models. J. R. Soc. Interface 2, 295–307 (2005)

11. Boots, M., Sasaki, A.: "Small worlds" and the evolution of virulence: infection occurs locally and at a distance. Proc. R. Soc. Lond. B 266, 1933–1938 (1999)
12. Keeling, M.J., Rand, D.A., Morris, A.J.: Correlation models for childhood epidemics. Proc. R. Soc. Lond. B 264, 1149–1156
13. Keeling, M.J.: The effects of local spatial structure on epidemiological invasions. Proc. R. Soc. Lond. B 266, 859–867
14. Bearman, P.S., Moody, J., Stovel, K.: Chains of affection: the structure of adolescent romantic and sexual networks. Am. J. Soc. 110, 44–91 (2004)
15. Johnson, A.M., Mercer, C.H., Erens, B., Copas, A.J., McManus, S., Wellings, K., Fenton, K.A., Korovessis, C., Macdowall, W., Nanchaha, K., Purdon, S., Field, J.: Sexual behaviour in Britain: partnerships, practices, and HIV risk behaviours. Lancet 358, 1835–1842 (2001)
16. De, P., Singh, A.E., Wong, T., Yacoub, W., Jolly, A.M.: Sexual network analysis of a gonorrhoea outbreak. Sex. Transm. Infect. 80, 280–285 (2004)
17. Jolly, A.M., Wylie, J.L.: Gonorrhoea and chlamydia core groups and sexual networks in Manitoba. Sex. Transm. Infect. 78, i145–i151 (2002)
18. Potterat, J.J., Philips-Plummer, L., Muth, S.Q., Rothenberg, R.B., Woodhouse, D.E., Maldonado-Long, T.S., Zimmerman, H.P., Muth, J.B.: Risk network structure in the early epidemic phase of HIV transmission in Colorado Springs. Sex. Transm. Infect. 78, i159–i163 (2002)
19. Barabási, A.-L., Albert, R.: Emergence of scaling in random networks. Science 286, 509–512 (1999)
20. Liljeros, F., Edling, C.R., Amaral, L.A.N., Stanley, H.E., Åberg, Y.: The web of human sexual contacts. Nature 411, 907–908 (2001)
21. Read, J.M., Keeling, M.J.: Disease evolution on networks: the role of contact structure. Proc. R. Soc. Lond. B 270, 699–708 (2003)
22. Watts, D.J., Strogatz, S.H.: Collective dynamics of "small-world" networks. Nature 393, 440–442 (1998)
23. Newman, M.E.J.: The structure of scientific collaboration networks. Proc. Natl. Acad. Sci. USA 98, 404–409 (2001)
24. Brockmann, D., Hufnagel, L., Geisel, T.: The scaling laws of human travel. Nature 439, 462–465 (2006)
25. Eubank, S., Guclu, H., Kumar, V.S.A., Marathe, M.V., Srinivasan, A., Toroczkai, Z., Wang, N.: Modelling disease outbreaks in realistic social networks. Nature 429, 180–183 (2004)
26. Ferguson, N.M., Cummings, D.T., Fraser, C., Cajka, J.C., Cooley, P.C., Burke, D.S.: Strategies for mitigating an influenza pandemic. Nature 442, 448–452 (2006)
27. May, R.M.: Uses and abuses of mathematics in biology. Science 303, 790–793 (2004)
28. Bauch, C., Rand, D.A.: A moment closure model for sexually transmitted disease transmission through a concurrent partnership network. Proc. R. Soc. Lond. B 267, 2019–2027
29. Eames, K.T.D., Keeling, M.J.: Modeling dynamic and network heterogeneities in the spread of sexually transmitted diseases. Proc. Natl. Acad. Sci. USA 99, 13330–13335 (2002)
30. Andre, M., Ijaz, K., Tillinghast, J.D., Krebs, V.E., Diem, L.A., Metchock, B., Crisp, T., McElroy, P.D.: Transmission network analysis to complement routine tubercolosis contact investigations. Am. J. Pub. Health 96, 1–11 (2006)
31. Friedman, S.R., Neagius, A., Jose, B., Curtis, R., Goldstein, M., Ildefonso, G., Rothenberg, R.B., Des Jarlais, D.C.: Sociometric risk networks and risk for HIV infection. Am. J. Pub. Health 87, 1289–1296 (1997)

32. Ghani, A.C., Garnett, G.P.: Risks of acquiring and transmitting sexually transmitted diseases in sexual partner networks. Sex. Transm. Dis. 27, 579–587 (2000)
33. Christley, R.M., Pinchbeck, G.L., Bowers, R.G., Clancy, D., French, N.P., Bennett, R., Turner, J.: Infection in social networks: using network analysis to identify high-risk individuals. American Journal of Epidemiology 162, 1024–1031 (2005)
34. Rothenberg, R.B., Potterat, J.J., Woodhouse, D.E., Muth, S.Q., Darrow, W.W., Klovdahl, A.S.: Social network dynamics and HIV transmission. AIDS 12, 1529–1536 (1998)
35. Eames, K.T.D., Keeling, M.J.: Contact tracing and disease control. Proc. R. Soc. Lond. B 270, 2565–2571 (2003)
36. Huerta, R., Tsimring, L.S.: Contact tracing and epidemics control in social networks. Phys. Rev. E 66, 056115-1–0561154 (2002)
37. Palla, G., Derényi, I., Farkas, I., Vicsek, T.: Uncovering the overlapping community structure of complex networks in nature and society. Nature 435, 814–818 (2005)
38. van Baalen, M.: Contact networks and the evolution of virulence. In: Dieckmann, U., Metz, J.A.J., Sabelis, M.W., Sigmund, K. (eds.) Adaptive Dynamics of Infectious Diseases: in Pursuit of Virulence Management. Cambridge University Press, Cambridge (2002)
39. Read, J.M., Keeling, M.J.: Disease evolution across a range of spatio-temporal scales. Theoretical Population Biology 70, 201–213 (2006)
40. Boots, M., Hudson, P.J., Sasaki, A.: Large Shifts in Pathogen Virulence Relate to Host Population Structure. Science 303, 842–844 (2004)
41. Frank, S.A.: Models of parasite virulence. Quarterly Review of Biology 71, 37–78 (1996)
42. Boccaletti, S., Latora, V., Moreno, Y., Chavez, M., Hwang, D.-U.: Complex networks: structure and dynamics. Phys. Rep. 424, 175–308 (2006)
43. Yan, G., Zhou, T., Wang, J., Fu, Z.-Q., Wang, B.H.: Epidemic spread in weighted scale-free networks. Chin. Phys. Lett. 22, 510–513 (2005)
44. Kossinets, G., Watts, D.J.: Empirical analysis of an evolving social network. Science 311, 88–90 (2006)
45. http://www.haggleproject.org

Epidemiology and Wireless Communication: Tight Analogy or Loose Metaphor?

Stephen Eubank, V.S. Anil Kumar, and Madhav Marathe

Network Dynamics and Simulation Science Laboratory,
VIrginia Bioinformatics Institute at Virginia Tech
{seubank,akumar,mmarathe}@vbi.vt.edu

Abstract. The analogy between viral dynamics in humans and in computers is a detailed and useful one. At first glance, the extension to infectious disease epidemiology on human social networks and communication in wireless networks is also a compelling analogy. Mathematical epidemiology has a long history and seems to offer a biological inspiration for communication network design. In this paper, however, we argue that while epidemiology as a metaphor may hold insights into communication networks, the relationship is not concrete enough to permit us to adapt solutions from one domain to another. Our conclusion is that it is certain new mathematics and methodologies, rather than the results themselves, that are most likely to generalize well to communication systems.

Keywords: epidemic, network, interaction-based, communication.

1 Introduction

1.1 Analogy and Metaphor

An analogy such as "computer virus is to computer as human virus is to human" postulates an isomorphism between the interactions and thus the dynamics of two different systems. For example, a human virus can only replicate by hijacking the resources of its host. In particular, a virus relies on a host cell's gene copying mechanisms, the host's energy and chemical resources, and the host's exchange of resources with the environment to copy and distribute its own genetic information. Similarly, a computer virus cannot replicate outside a host computer, and it uses the host's operating system, applications, and communications networks to replicate and distribute itself. The name "computer virus" (as opposed to, say, "computer bacterium") was chosen precisely because of these parallels.[1]

We expect that two systems that are isomorphic in some regard will share some behaviors. If we are lucky, solutions to problems arising in one system can be translated to the other, perhaps with an appropriate adaptation to the

---

[1] Wikipedia lists Fred Cohen as the author who first used the term in an academic publication, and David Gerrold as the person who coined the term.

P. Liò et al. (Eds.): BIOWIRE 2007, LNCS 5151, pp. 91–104, 2008.

specifics of the new context. In the context of a computer virus, we see vaccination based on "antigens" – some sort of properties specific to the virus. We also see diagnostic software that attempts to detect common strategies of hijacking computer resources, for example specific unusual sequences of system calls, and "antiviral treatments" aimed at blocking these strategies. There are also diagnoses that depend on "symptoms" – systemic consequences of infection.

A metaphor is less concrete, and certainly not as strong as an isomorphism. One would not expect to be able to map solutions directly from one side of the metaphor to the other. At most, one might hope that solution methodologies would be applicable in both domains. We argue below that the relationship between disease transmission and message transmission is better thought of as metaphor than analogy. We also indicate insights and mathematical approaches that should generalize from one to the other and give a few examples of particular analytical tools that offer promise in both domains.

1.2   Network Epidemiology

We begin by describing the specific kind of epidemiological model that provides the best metaphor for communication networks. The origins of mathematical epidemiology can be traced back at least as far as Bernoulli, who published sophisticated investigations into smallpox vaccination, as described by Dietz and Heesterbeek[1]. Modern epidemiological models have been dominated by sets of coupled rate equations, known as compartmental models, dating from the early 1900's. These are generally attributed to Ross and MacDonald[2,3] in the case of malaria or Kermack and McKendrick[4]. Compartmental models are typically deterministic models for the mean fraction of a population in a given state of health (the *compartments*) as a function of time. Probabilistic models, in particular the chain binomial approach of Reed and Frost (as described, for example, in [5]), have also contributed to a well developed theory for the final size of epidemics and the dynamics of endemic disease. For a careful, up to date description of these models, see Diekmann and Heesterbeek.[6]

In practice, both the compartmental and Reed-Frost models rely on the assumption of mass action – that the number of new infections is proportional to the product of the number currently infected and the number currently susceptible. That is, the interactions among people that lead to disease transmission are homogeneous.[2]

In contrast with these uniform-mixing models, researchers, including [7,8,9,10] have developed network models of infectious disease transmission. A network model of infectious disease includes entities that interact via a network to spread disease. We will assume for the purposes of this paper that the entities are individual people, although aggregations are certainly possible. In a network model, each potential host is represented by a vertex, and edges between two

---

[2] This assumption is relaxed somewhat in what are known as "structured population" models, but only to the extent that the single mixing term is replaced by a small mixing matrix.

vertices represent contact between the hosts that creates a non-zero probability of transmission. Associated with each vertex are two things:

1. a label that includes those features of the host relevant to the spread of disease;
2. a model that represents the host's state of health. This may be as simple as a finite state machine distinguishing the susceptible state from the infectious state.

Associated with each edge is a weight giving the probability of transmission across the edge conditioned on the source being infectious. The probability of transmission depends on the susceptibility and infectiousness of the people involved, which may be drawn from a distribution conditioned on any of their attributes, e.g. age or immunization status. The resulting network is in general a directed, weighted graph. It is also, in general, time dependent. Time dependence may be incorporated into edge labels or it may be represented by constructing an explicit unfolding of the sequence of graphs. Here, we will focus on the union of all the time dependent graphs on the assumption that the dynamics of disease progression in the host is slow relative to the changes in the graphs.

Network models are particularly valuable in areas such as HIV/AIDS, where the assumption of uniform mixing is clearly violated. As Lord Robert May has said, "Much relevant work remains to be done in teasing apart the social, genetic, age-related, and other complications that are smoothed out in the usual mass action assumption."[11] Network models can easily represent many kinds of heterogeneity (and correlations among them) that can be accommodated in compartmental models only with difficulty, if at all. For example, arbitrary heterogeneity in susceptibility that is correlated with age can be represented easily by assigning each person a susceptibility drawn from a distribution conditioned on age. More complicated correlation structures such as demographically determined compliance rates can be represented just as easily, as long as the data are available.

1.3 Information Diffusion across Communication Networks

The primary goal of algorithms/protocols for communication networks is rapid information diffusion based solely on local information. Their effectiveness can be measured by the total amount of information that is moved in the network to accomplish the task, the time it takes for updates to be complete, the kind of global information assumed (e.g. global clocks) and the nature of constraints (e.g. only certain types of operations allowed). Depending on the specific requirements; the goal of the information diffusion process might be:

- send specific information from a source s to a destination t, (routing [12,13,14,15]);
- send information from a source s to a subset of nodes $T \subseteq N$ (multicast, updating replicated databases[16,17,18]);
- send information from source s to all nodes in the network (broadcast, rumor spreading or worm propagation);

1.4   Questions and Answers in Epidemiology

The central question of epidemic (as opposed to endemic) disease is: given an initial distribution of infectious hosts, how will the disease spread through the population? This can be broken into several parts, in increasing order of difficulty:

1. How many hosts will have been infected when the outbreak dies out?
2. How many hosts are infected at any given time?
3. Which hosts are *vulnerable*, i.e. likely to be infected at any given time?
4. Which hosts are *critical*, i.e. contribute to the largest number of later infections, at any given time?

Compartmental models can answer the first and second of these questions, but by their nature cannot address the third and fourth. They are thus most useful for understanding the consequences of interventions applied broadly and early in the course of an outbreak. Consider, for example, the canonical result of these models: the existence of a phase transition at a critical value of the parameter R_0, known as the basic reproductive number. For $R_0 < 1$, outbreaks die out before spreading far; for $R_0 > 1$, outbreaks become epidemic with high probability. The parameter R_0 represents a combination of both the transmission and also the mixing rates. Formally, it is the expected number of people directly infected by a single infectious person introduced into a population of susceptibles. Because R_0 depends on the fraction of the population that is immune, the sharp phase transition leads immediately to the notion of *herd immunity*: disease cannot make significant inroads into a herd as long as the fraction immune is large enough that $R_0 < 1$, and the required fraction is less than unity.

Nowadays however, we demand not simply *a* control strategy, but the *optimal* control strategy under constrained resources obtained by targeting interventions at the most vulnerable and/or critical people. This requires network models, in which the *topology* of the interaction network, as well as the transmissibility of the disease, is crucial. For instance, the vulnerable and critical vertices sought in questions 3 and 4 above may be determined by network topology.

Unfortunately, network models resist analytical solution except in a few well-known examples[19,20,21]: Erdős - Rényi random graphs; lattices or grids; trees; scale-free networks[3]; and small world networks[4]. The problem is that solution methodologies often assume people's health states are independent, or that the probability of infection is additive.[21,22] A complete categorization must take into account global topology (as the small world property does) and not only local topology (as encoded in the degree distribution). The reason is simple. The probability that any vertex is infected depends on the joint probability that its neighbors are infected. Given a configuration of neighbors' states we can calculate a probability of infection, but that probability must be weighted

---

[3] Those whose vertex degrees are distributed according to a power-law with exponent less than -1.

[4] Those whose diameter grows only logarithmically with number of vertices.

by the probability of the neighbors being in those states simultaneously. If the joint distribution could be replaced by the product distribution – i.e. if the events of infecting the neighbors were independent – then the likelihood that any vertex is infected would be proportional to its weighted degree. However, if two neighbors were both potentially infected by the same vertex, there is an induced correlation between them. More generally, if there is a loop between a vertex and the initial condition, its neighbors will not be independent and degree will not necessarily be well-correlated with vulnerability. This can be made precise in the concept of d-separability[23]. It is not surprising that d-separability, an important concept in the analysis of Bayesian networks, is also important here: a network epidemiological model can be thought of as a Bayesian network representing the joint probability of infection of each vertex.

Enumeration of all the loops and their overlaps essentially gives a complete description of the global topology of a network. But this is not feasible analytically for any realistic – i.e. asymmetric, irregular – network. Instead, we rely on simulation to understand the behavior of network models in realistic networks. One advantage of simulation is that detailed results are available for intermediate times as well as for the long-time limit. Another advantage is that the simulation can be controlled as it runs, so that it is possible to represent dynamic, adaptive changes such as contact tracing and treatment. Indeed, it is hard to imagine how compartmental models could ever represent individually targeted, adaptive control strategies while maintaining analytical tractability.

2 Comparing and Contrasting Epidemiology and Communication

The dynamics of communication over wireless networks and epidemiology over social networks share many properties. In the following section we compare and contrast the problems of transmitting an infectious disease across a social network and transmitting messages across a wireless network. Specific aspects of the wireless network problem, such as its robustness against worms or its use in maintaining a distributed database, may exhibit more or less similarity with epidemiology, but we confine our comments to the general case.

2.1 Similarities

Dynamics of the Transmission Process. In both domains, a global configuration that lists the states of every vertex contains all the information needed to calculate the probability of any other configuration in the next instant. That is, the dynamics are Markovian in configuration space. In principle, then, the system's dynamics are linear and are fully characterized by the Markov matrix. However, configuration space (and hence the dimension of the Markov matrix) is truly enormous: if each of N vertices could be in any of m states, then the system can be in any of m^N configurations.

It is more natural to think of the dynamics as a local flow among states of individual vertices. But the dynamics in this representation are no longer linear. The flows that represent them are non-conservative. For example, random walkers modeling these flows must allow for both creation – one host infecting (or sending a packet to) more than one neighbor – and destruction – a host failing to transmit the disease (or packet) at all.

Dynamics of the Network Itself. The dynamics of the networks themselves also exhibit similarities in surprising detail. In both cases, of course, they are time dependent. There may also be feedback between transmission and the network structure. For example, a person whose neighbor becomes ill may break off contact with him, and the probability of this act may vary depending on the prevalence of disease in the network as a whole. Similarly, the content of a message transmitted by the communication network may lead one of the receivers to move, and this may depend on the density of messages in the network.

Existence of Solutions for Related, Simpler Problems. Shannon's information theory and ideas such as channel capacity provide an old, canonical solution to a simpler problem. Similarly, compartmental models provide old, canonical solutions to epidemiological problems. As noted above, there are traditional solutions in percolation theory (i.e. network models) for certain regular networks. These canonical solutions are often used to develop intuition about the problem, but just as often they can lead to poor intuitions because the more complex systems admit a greater variety of solutions.

Transmission Network vs Underlying Network. Each instance of an epidemic creates a trace or stain on the contact network – a subgraph representing the observed web of transmission. Similarly the trace of packets from one or more messages yields a subgraph of the communication network that participated in the actual transmission of the messages. One way to view the network or protocol design problem is to fix properties of the desired transmission network, then define a process that, given an underlying network with potentially very different properties, constructs subgraphs with the desired properties. This refocuses attention away from the properties of the underlying network and toward ways to select random subgraphs biased toward desired structures.

2.2   Differences

Single Pathogen vs Multiple Messages: Flooding. In a communication network, we are interested in supporting the ability to send multiple messages simultaneously from different sources to different targets. In contrast, for an epidemiological system we are chiefly interested in the ability of a single pathogen to diffuse across the network. This has important implications for the efficacy of different strategies. For example, because there is only a single genetic "message" carried by a pathogen, infectious diseases have evolved to make use of what amounts to flooding, i.e. transmitting themselves to every possible receiver. This is an inefficient use of resources for communication networks.

Single Pathogen vs Sequence of Packets: Timing. In order to transmit disease, only a single infectious dose of a single type of pathogen needs to be transmitted. Communication systems must not only deliver all the packets reliably, they must take into account any overhead required to sort them into the correct order as sequential parts of a message.

Time Scales and the Role of Mobility. The most obvious difference between biological and electronic transmission is the speed of the latter. Routers can forward packets orders of magnitude faster than humans forward pathogens. Mobility plays an important role in epidemiology, since it is responsible for long-range transmission, but it is likely to play an even greater role in communication over wireless networks. At any instant, the social network consists of a set of disjoint subgraphs, each of which is completely connected within itself. Mobility changes the number and membership of the complete subgraphs. But because incubation periods and the duration of infectivity are typically measured in days, the propagation of disease is much slower than changes in network topology. Thus the effects of mobility can be summarized in a static graph that is a union of the instantaneous graphs. In contrast, in networks of mobile nodes, the topology of the network typically changes on time scales that are slower than or comparable to the propagation of messages.

Passive Carriers. Some pathogens (e.g. staphylococcus) are harbored by large fractions of the population, but only result in illness under specific conditions. The network models of epidemiology we describe here are perhaps not the best representation of the spread of such pathogens. But communication networks routinely rely on nodes to forward packets to distant destinations. This leads to two types of interacting networks - the social contact network of a node, and the underlying infrastructure network.

Subadditive Synergy vs Interference. Without careful synchronization, radios interfere with each other, even if they are sending the same packet. Thus on one hand, simultaneous transmission of messages in a wireless network *reduces* the probability of successful transmission. On the other hand, multiple simultaneous exposures to a single infectious agent in different infectious people is likely to *increase* the probability of becoming infected, though subadditively. The differences between destructive and constructive interference often change the statement and difficulty of proving flow-based theorems.

Strong Coupling Limit. In the strong coupling limit, when the probability of transmission is near unity, epidemiology reduces to finding the shortest path between any vertex in the initially infected set and every vertex outside the set. The network can be classified as robust or fragile depending on the distribution of shortest path lengths between all pairs of vertices. In the same limit in a communication network, however, each radio floods its neighbors with packets. In this situation, no messages can reliably be communicated through the system.

Designed vs Evolved System: Role of Intention. The communication network is typically constructed of components designed and optimized for the particular purpose of communicating, whereas human social networks are designed to accomplish goals almost entirely unrelated to disease transmission. Furthermore, the components of contact networks, i.e. people, have intentions and their behavior is difficult or impossible to predict. Radios, by contrast, may perform as expected or fail, but it is unlikely that they will broadcast in an entirely unexpected way.[5]

Targets for Control: Protocol or Network. The analogy to the communications network's protocol layer in epidemiology is the mechanism of transmission, e.g. inhalation of aerosol or transfer of pathogen via direct contact. There is perhaps also a physical layer in epidemiology representing access to media, e.g. coating particles in a sneeze with virions. In any case, the primary target for control in public health is the contact network itself. Public health measures such as quarantine have the effect of removing edges or vertices from the contact network. In contrast, efforts to improve communication network performance[6] have often focused on the protocol stack and not network topology.

3   Useful Approaches from Epidemiology

3.1   Methodologies

As with any system of interacting components, the key to understanding global behavior is our ability to derive global semantics from local syntactic rules. There are, of course, many promising approaches to this problem, some of them also inspired by biology. In this section we discuss the following examples of such methodologies:

- A Mathematical Theory for Network Processes
- Simulation
- Characterizing Networks by Topology
- Designing Epidemiologically Inspired Algorithms

A Mathematical Theory for Network Processes. Formally, we model dynamics of social networks using discrete dynamical systems. We refer to our model as a *Stochastic Synchronous Dynamical System* (SSyDS). Each SSyDS S over a domain $\mathbb{D}$ is specified as a pair $S = (G, \mathcal{F})$. Here, $G(V, E)$ is an undirected graph with n nodes, with each node having a state value from the domain $\mathbb{D}$. This graph represents the topological structure of the social network. The set $\mathcal{F} = \{f_1, f_2, \ldots, f_n\}$ is a collection of stochastic interaction functions in the system. Here, f_i denotes the stochastic local transition function associated with node v_i, $1 \leq i \leq n$. A **configuration** of an SSyDS is an n-vector $(b_1, b_2, \ldots, b_n)$, where $b_i \in \mathbb{D}$ is the value of the state of node v_i $(1 \leq i \leq n)$.

---

[5] Although this may be an interesting parallel to adversarial control of parts of a network.

[6] Or to attack it, such as with worms that generate unnecessary traffic.

A single SSyDS transition from one configuration to another is obtained by updating the state of each node *synchronously* using the corresponding local transition function. For $1 \leq i \leq n$, the inputs to the function f_i are the state values of node v_i and those of the neighbors of v_i. For each combination of inputs to f_i and each element θ of $\mathbb{D}$, the function f_i specifies the probability that the next state value of v_i is θ. (For each combination of inputs, the sum of the probabilities assigned by f_i over the values $\theta \in \mathbb{D}$ must be 1.)

To further clarify the notion of stochastic local transition functions used here, consider a node v_i and let $v_{i_1}, v_{i_2}, \ldots, v_{i_r}$ represent the neighbors of v_i in G. For any j and t, let s_j^t denote the state of node v_j at time t. The local transition function f_i at node v_i satisfies the following equation:

$$\Pr\{s_i^t = \theta \mid s_i^{t-1} = \theta', s_{i_1}^{t-1} = \theta_{i_1}^1, \ldots, s_{i_r}^{t-1} = \theta_{i_r}^1\} = f_i(\theta', \theta_{i_1}^1, \ldots, \theta_{i_r}^1, \theta). \quad (1)$$

Simulation. As discussed above, it is not feasible to compute the exact dynamics, i.e. the ensemble of possible paths, in the full configuration space. Instead, we use simulation to estimate likely individual paths through the space. Simulation is required especially when parameter values for transmission lie near the epidemic phase transition, when the networks are highly irregular, or when there are large-scale topological structures in the network. Cases that do not fall into one or more of these groups are generally unrealistic.

Over the past few years there has been substantial interest in developing computational models for representing and understanding the spread of worms in IP networks [24,25,26,27]. However, this work still uses the basic SIR construct. High resolution models such as the network epidemic models incorporate a number of important features, including: (i) demographic attributes of hosts (e.g. whether the system is running MS windows or not), and (ii) the time that nodes spend in the infected stage. As in the case of human epidemiology, high resolution modeling allows us to study a wider variety of quarantining and mitigating policies and methods. For example, specific kinds of IP hosts within a subnet can be turned off if a worm attack is detected. In contrast, homogeneous mixing models cannot distinguish between individual IP hosts.

Worms that spread on Bluetooth enabled devices are relatively new. The primary difference between Bluetooth worms and older internet worms such as code Red is that Bluetooth worms are inherently spatial in nature: an infected bluetooth-enabled device can infect another bluetooth-enabled device if the two are in close proximity to one another. In a recent article, Kleinberg [28] convincingly argues that the advent of short range wireless nodes can potentially bring epidemiological models for transmission of diseases in biological systems and for communication networks closer.

Characterizing Networks by Topology. One approach to characterizing networks by the dynamics they support begins by determining the number and boundaries of *communities*: regions of a network that are more highly connected within themselves than they are to other communities. This is similar to the classical clustering problem, and similar methodologies for solving it have been proposed. Some are based on the max cut / min flow theorem. Others are based on the spectrum of the adjacency, Laplacian, or other closely related matrices. Still others are based on global, computationally hard to evaluate, statistics such as betweenness or expansion. It is crucial to develop good sampling techniques and algorithms with provable error bounds for approximating these statistics.

Designing Epidemiologically Inspired Algorithms. Epidemiologically inspired algorithms are specifically designed to work in environments when network connectivity is highly variable due to mobility or faults. Thus, these algorithms can often be viewed as extreme forms of distributed algorithms: (i) they assume no global knowledge and very little knowledge even about their neighbors; and (ii) the decision to (re)-transmit information is usually based on the local state at a node and information (packet) received from its neighbors (e.g. the time packet has been alive, the number of hops, etc.)

Alternative Perspectives on the Problem. On a more speculative note, evolutionary strategies that are beyond the scope of this paper represent methodologies for optimizing network diffusion that may generalize well from biology to communication networks. For instance, the problem of communication across a network, as is the case for infectious disease, is generally conceived in terms of parasitism: how can we trick nodes of the network into directing their resources towards conveying our information to our intended recipients? It may be that a more productive paradigm, also drawn from the biological world, is symbiosis: how can we arrange things so that delivering our information inherently benefits the nodes involved, so that there is an active competition to deliver messages.

3.2   Some Specific Tools

In addition to these general principles or methodologies that appear likely to be useful, we will mention here a few specific notions developed in the course of epidemiological studies. Answering questions 3 and 4 in Sec. 1.4 requires making precise the notions of *vulnerability* and *criticality*. We outline appropriate definitions of these quantities here. Next we give examples of epidemiologically inspired algorithms for updating replicated databases and for routing. Finally, we describe a local way to measure the importance of global network topology to the dynamics of a single vertex.

Stochastic Reachability. In deterministic discrete systems, we are familiar with the notion of *reachability*: whether a configuration is reachable (in finite time) from a given configuration. There is a natural generalization to stochastic systems: the probability that a certain configuration will be reached (at a particular time) from a given configuration. Given a dynamical system S, two configurations $\mathcal{I}$ and $\mathcal{B}$ and a probability value p; the question is whether S starting from $\mathcal{I}$ can reach $\mathcal{B}$ with a probability of at least p. An important variant of this problem is one in which the goal is to determine whether S starting from $\mathcal{I}$ can reach $\mathcal{B}$ in at most t steps with a probability of at least p. The computational complexity of these problems is known. Under reasonable complexity theoretic assumptions, reachability problems for stochastic systems are harder than the corresponding problems for deterministic systems.[29]

It is also useful to sum this probability over certain sets of configurations. For example, we might sum over all configurations B in which a particular vertex is in a particular state. In an epidemiological model, this gives us, for instance, the probability as a function of time that a person is infected given an initial condition. This probability is the person's vulnerability. The advantage of this perspective is that it makes clear that vulnerability is a function of the person, the time since the outbreak began, and the initial condition. The literature often confuses this general notion of vulnerability with the time $t = 1$ vulnerability, as in the common assertion that a vertex's vulnerability is equivalent to its weighted degree. At time $t = 2$, it is obvious that the correlation between neighboring vertices' degrees has an effect, and at longer times, the full global topology becomes important as discussed above. The vulnerability of a vertex serves as a useful measure (in the sense of weight) for calculating averages. That is, it can be used to calculate mean age of infected people, etc. Aggregate values of vulnerability – averaged over vertices, initial conditions, and time – can characterize the vulnerability of an entire network.

Criticality. Intuitively, vertex criticality compares what happens when a vertex is present and when it is absent. This can easily be made precise, but, like vulnerability, it will depend on the vertex, the initial condition, and the time that elapses before the vertex is removed. Moreover, it will depend on the time that has elapsed since removal when the comparison is made. As with vulnerability, if we remove the vertex at time 0 and make comparisons after only a single time step, then the criticality of a vertex is given by its weighted degree. The identity of criticality and vulnerability in this very restricted domain has led to an unfortunate confusion of the two distinct properties. It is clear from a more careful definition that the two are not necessarily related.

Updating Replicated Databases. Demers et al. [16], were the first to consider the use of epidemiologically inspired algorithms for updating databases. For additional discussion, see [30,31]. Managing a replicated database using synchronous methods is challenging when the underlying network is unstable or very large. As a result asynchronous methods, and in particular epidemiologically inspired algorithms, are often used. For example, every time the value of an entry changes at a node, it is locally propagated. Under reasonable assumptions about global information, simple rules can be used to ensure local and global consistencies. Maintaining serializability is more challenging. Agarwal et al. outline a sophisticated approach to ensure consistency and serializability.

Routing. In contrast to the database update problem, the primary goal of routing is to ensure that packets originating at a node s reach a given node t. Networks with mobile nodes and sensor networks in which nodes might be turned off (to conserve power) or fail, provide different kinds of challenges than database maintenance problems. Probabilistic local routing is a class of algorithms motivated by epidemiology; for recent papers and discussion, see `http://roland.grc.nasa.gov/~weddy/biblio/epidemic/`. Our recent work demonstrates that such algorithms are very competitive in highly mobile environments such as vehicular ad-hoc networks and in delay tolerant networks.

Generalized d(t)-separability.   As discussed above, there exist several solution techniques for network epidemiology under the assumption that vertex states are independent, or whether the vertices are *d-separable*. For networks with any interesting topology, especially when most edges are undirected, two vertices are unlikely to be d-separable. There are two important questions that are not addressed by d-separability, though:

1. How does the correlation of vertex states spread as a function of time?
2. How important is the correlation?

Answering the first requires looking beyond the static graph we have considered for most of this paper. Instead, to study the separability of two vertices at time t, consider a new graph made of t copies (or layers) of the vertices. In this new graph, two vertices are connected by an edge directed from person S in layer l_S to person T in layer l_T if and only if l_T is the layer immediately below l_S and there is an edge from person S to person T in the original graph. This new graph is a directed, acyclic graph, and it contains no directed paths longer than t. d-separability in this graph answers the first question above. To answer the second question, we pose the following hypothesis test: if there were no correlation, we could write down the dynamics (in terms of stochastic reachability) in a straightforward fashion. If we observe the simulated dynamics, we can determine whether the null hypothesis of no correlation can be rejected. Moreover, even without evaluating the statistical significance, we can determine how large an effect the correlation has on any vertex at any time. Together, these ideas give us a handle on an important question – does the topology of a given network have an important effect on the outcome, and if so, how sensitive is the effect to changes in the network structure?

4   Conclusion

We believe it is currently the intersection of simulation science, computer science, and discrete mathematics that is driving development and generalization of the most promising methodologies for studying network dynamics. While biology, and in particular infectious disease epidemiology, has indeed inspired some of these methodologies, the differences between problem domains is significant enough to render solutions difficult to translate from one to the other. Instead, it is the methodologies themselves and concepts such as reachability that can be readily translated from epidmiology to communication networks.

This work was supported by the National Institute of General Medical Sciences Models of Infectious Disease Agent Study (cooperative agreement 5U01GM070694).

References

1. Dietz, K., Heesterbeek, J.: Daniel bernoulli's epidemiological model revisited. J. Math. Biosci. 180, 1–21 (2002)
2. Ross, R.: The Prevention of Malaria. E. P. Dutton & Co., New York (1910)
3. Macdonald, G.: The analysis of equilibrium in malaria. Trop. Dis. Bull. 49(9), 813–829 (1952)
4. Kermack, W.O., McKendrick, A.G.: Contributions to the mathematical theory of epidemics–i,ii,iii. Bulletin of Mathematical Biology 53(1-2), 33–118 (1991)
5. Bailey, N.T.J.: The Mathematical Theory of Infectious Diseases and its Applications, London (1975)
6. Diekmann, O., Heesterbeek, J.A.P.: Mathematical Epidemiology of Infectious Diseases: Model Building, Analysis and Interpretation. John Wiley, Chichester (2000)
7. Elveback, L.R., Fox, J.P., Ackerman, E., Langworthy, A., Boyd, M., Gatewood, L.: An influenza simulation model for immunization studies. American Journal of Epidemiology 103(2), 152–165
8. Eubank, S., Guclu, H., Anil Kumar, V.S., Marathe, M.V., Srinivasan, A., Toroczkai, Z., Wang, N.: Modelling disease outbreaks in realistic urban social networks. Nature 429(6988), 180–184 (2004)
9. Longini, I.M.J., Nizam, A., Xu, S., Ungchusak, K., Hanshaoworakul, W., Cummings, D.A., Halloran, M.E.: Containing pandemic influenza at the source. Science 309(5737), 1083–1087 (2005)
10. Ferguson, N.M., Cummings, D.A., Fraser, C., Cajka, J.C., Cooley, P.C., Burke, D.S.: Strategies for mitigating an influenza pandemic. Nature 442(7101), 448–452 (2006)
11. May, S.R.: Enhanced: Simple rules with complex dynamics. Science 287(5453), 601–602 (2000)
12. Ganesan, D., Krishnamachari, B., Woo, A., Culler, D., Estrin, D., Wicker, S.: An Empirical Study of Epidemic Algorithms in Large Scale Multi-hop Wireless Networks, http://citeseer.ist.psu.edu/ganesan02empirical.html
13. Vahdat, A., Becker, D.: Epidemic Routing for Partially-Connected Ad Hoc Networks, Duke University Technical Report CS-200006 (April 2000)
14. Barrett, C.L., Eidenbenz, S.J., Kroc, L., Marathe, M., Smith, J.P.: Parametric probabilistic sensor network routing. In: Proceedings of the 2nd ACM international conference on Wireless sensor networks and applications, San Diego, CA, USA, September 19 (2003)
15. Barrett, C.L., Eidenbenz, S.J., Kroc, L., Marathe, M., Smith, J.P.: Parametric probabilistic routing in sensor networks. Mobile Networks and Applications 10(4), 529–544 (2005)
16. Demers, A., Green, D., Hauser, C., Irish, W., larson, J., Shenker, S., Sturgis, H., Swinehart, D., Terr, D.: Epidemic Algorithms for Replicated Database Maintainence. In: Proc. 6th Symposium on Principles of Distributed Computing, pp. 1–12 (1987)
17. Nath, S., Gibbons, P.B.: Synopsis Diffusion for Robust Aggregation in Sensor Networks. Technical Report ITR-03-08, Intel Research Pittsburgh (August 2003)
18. Agrawal, D., El Abbadi, A., Steinke, R.C.: Epidemic algorithms in replicated databases (extended abstract). In: Proceedings of the sixteenth ACM SIGACT-SIGMOD-SIGART symposium on Principles of database systems, Tucson, Arizona, United States, May 11-15, 1997, pp. 161–172 (1997)

19. Barabsi, A.L., Albert, R.: Emergence of scaling in random networks. Science 286(5439), 509 (1999)
20. Watts, D.J.: Small worlds. Princeton University Press, Princeton (1999)
21. Newman, M.: The structure and function of complex networks. SIAM Review 45, 167–256 (2003)
22. Ganesh, A., Massoulie, L., Towsley, D.: The effect of network topology on the spread of epidemics. IEEE INFOCOM 2, 1455 (2005)
23. Geiger, D., Verma, T., Pearl, J.: Identifying independence in bayesian networks. NETWORKS 20(5), 507–534 (1990)
24. Forrest, S., Hofmeyr, S., Somayaji, A.: Computer Immunology. Communications of the ACM 40(10), 88–96 (1997)
25. Staniford, S., Paxson, V., Weaver, N.: How to Own the Internet in Your Spare Time. In: USENIX Security Symposium 2002, pp. 149–167 (2002)
26. Weaver, N., Paxson, V., Staniford, S., Cunningham, R.: A Taxonomy of Computer Worms. In: Proc. ACM CCS Workshop on Rapid Malcode (October 2003)
27. Moore, D., Shannon, C., Claffy, K.: Code-Red: a case study on the spread and victims of an internet worm. In: Proc. 2nd ACM SIGCOMM Workshop on Internet measurment (2002)
28. Kleinberg, J.: The Wireless Epidemic. Nature (2007)
29. Barrett, C., Hunt III, H., Marathe, M., Ravi, S., Rosenkrantz, D., Stearns, R., Thakur, M.: Computational aspects of analyzing social network dynamics. In: International Joint Conference on Artificial Intelligence (IJCAI), pp. 2268–2277 (2006)
30. Bonnet, P., Gehrke, J.E., Seshadri, P.: Towards Sensor Database Systems. In: Tan, K.-L., Franklin, M.J., Lui, J.C.-S. (eds.) MDM 2001. LNCS, vol. 1987, pp. 3–14. Springer, Heidelberg (2000)
31. Madden, S., Franklin, M.J., Hellerstein, J.M., Hong, W.: TAG: a Tiny AGgregation Service for Ad-Hoc Sensor Networks. In: Proc. OSDI (2002)

Epidemic Spreading of Computer Worms in Fixed Wireless Networks

Maziar Nekovee[1,2]

[1] BT Research, Polaris 134, Adastral Park, Martlesham, Suffolk IP5 3RE
[2] Centre for Computational Science, University College London, 20 Gordon Street,
London WC1H 0AJ, UK
maziar.nekovee@bt.com

Abstract. Worms are stand-alone computer viruses which use networks
for their spreading among computing devices. The last few years have seen
the emergence of a new type of worms which specifically target portable
computing devices, such as smartphones and laptops. The novel feature of
these worms is that they do not necessarily require Internet connectivity
for their propagation but can spread directly from device to device using
a short-range communication technology, such as Bluetooth or WiFi. In
this paper we use a combination of large-scale simulations and mathematical modelling to explore epidemic spreading of wireless worms in fixed ad-
hoc networks. We show that the spreading of worms in these networks is
greatly affected by a combination of spatial correlations arising from network topology and temporal correlations resulting from the interference-
limited nature of communications in thees networks. Standard mean-field
and network mean-field models from mathematical biology, which are
widely used to model worm epidemics in computer networks, are inadequate for describing worm epidemics in wireless adhoc networks but spatial epidemic models provide a promising alternative.

Keywords: Wireless Computer Worms, Complex Networks, Epidemic
Spreading, Modelling and Simulations.

1   Introduction

Computer worms are self-replicating malicious software that can propagate in a
network without the need for any human intervention [1,2,3], and their spreading in computer networks shows important similarities to the spread of epidemic
diseases in populations. The last few years have seen the emergence of a new
type of worms which specifically targets portable computing devices, such as
smartphones and laptops. The novel feature of these worms is that they do not
necessarily require Internet connectivity for their propagation. They can spread
directly from device to device using a short-range wireless communication technology [5,6]. Although these types of worms have not yet achieved widespread
penetration, prototypes have successfully exploited vulnerabilities in wireless
protocols including Bluetooth [4] and WiFi. With wireless networks becoming

P. Liò et al. (Eds.): BIOWIRE 2007, LNCS 5151, pp. 105–115, 2008.

pervasive, many security experts predict that these networks will soon become a main target of attacks by worms and other type of malware [5].

From the perspective of network theory, the underlying contact network along which wireless worms spread is greatly different from the extensively studied random graph and scale free networks [28,29,30]. Unlike these networks, which are largely unconstrained by physical proximity, wireless networks are embedded in a metric space where links between the nodes is a function of their spatial distance. While epidemics in random graphs and scale-free networks have been the subject of much studies, investigations of epidemic processes in spatial networks is still at its infancy [16,17,18].

In a recent paper [16] we studied worm epidemics in a class of wireless networks which are created on the fly by devices equipped with, e.g., WiFi or Bluetooth, the so-called wireless adhoc networks. Our studies had as their main focus stationary properties of epidemics in these networks, such as the epidemic threshold and the epidemic prevalence. These properties were found to be greatly different from the much studied properties of worm epidemics on the Internet. In this paper we explore via a combination of modelling and stochastic simulations dynamic patterns of worm epidemics in fixed wireless adhoc networks. We perform extensive Monte Carlo simulations of worm spreading for a range of node densities, both in the absence and in the presence of an interactive immunisation process.

We find that the dynamic patterns of epidemics in our networks are greatly affected by a combination of spatial correlations arising from their spatial topology and temporal correlations resulting from the interference-limited nature of communications in these networks [8]. Due to these correlations the standard mean-field models from mathematical biology and their recent extension to include network effects [9,10,12], become inadequate in describing worm epidemics in wireless adhoc networks. On the other hand, we argue that spatial epidemic models, such as those applied to plant diseases [19], provide a promising alternative.

The rest of this paper is organised as follows. In section 2 we describe our models of network topology, medium access control and worm spreading in wireless ad hoc networks. In section 3 the relevant mean-field models of epidemics on networks are reviewed. In section 4 we use the models developed in section 2 to perform Monte Carlo simulation studies of worm propagation in adhoc networks for a range of device densities, and for different worm attack scenarios. We close this paper in section 5 with conclusions.

2   Modelling Preliminaries

2.1   Network Model

We consider a collection of nodes distributed in a two dimensional plane which communicate using short-range radio transmissions. The received radio signal strength at a device j resulting from a transmission by a device i decays with the distance between the sender and the receiver due to a combination of free-space attenuation and fading effects. Phenomenologically this effect is described using the so-called pathloss model [20], which ignores statistical fluctuations in the signal strength due to, e.g., shadowing and multiple reflections. Consequently, the mean value of the signal power at a receiving device j is related to the signal power of the transmitting node i via the following equation:

$$P^{ij} = \frac{P^i}{cr_{ij}^{\alpha}}. \tag{1}$$

In the above equation r_{ij} is the Euclidean distance between node i and node j, P^i and P^{ij} are the transmit power and the received power, respectively, and c is a constant whose precise value depends on a number of factors including the transmission frequency. For free space propagation $\alpha = 2$, but depending on the specific indoor/outdoor propagation scenario it is found empirically that α can vary typically between 2 and 5. A data transmission by node i is correctly received at node j, i.e. i can establish a communication link with j, provided that:

$$\frac{P^{ij}}{\nu} = \frac{P^i/cr_{ij}^{\alpha}}{\nu} \geq \beta_{th}. \tag{2}$$

In the above equation β_{th} is an attenuation threshold and ν is the noise level at node j.

Condition (2) translates into a maximum transmission range for node i:

$$r_t^i = \left(\frac{P^i}{c\beta_{th}\nu} \right)^{1/\alpha}, \tag{3}$$

such that each device can establish wireless links with only those devices within a circle of radius r_t^i. A communication graph is then constructed by creating an edge between node i and all other nodes in the plane that are within the transmission range of i, and repeating this procedure for all nodes in the network. Assuming that all devices use the same transmit power P, and a corresponding transmission range r_t, the topology of the resulting network can be described as a two dimensional random geometric graph (RGG) [21,22]. Like Erdős-Rényi random graphs (RG) [23], these graphs have a binomial degree distribution, $P(k)$, which peaks at an average value $\langle k \rangle$ and shows small fluctuations around $\langle k \rangle$. However, other properties of a RGG are radically different from a Erdős-Rényi random graph. Most notably, these networks are characterised by a large cluster coefficient, $C = 0.59$, which is a purely geometric quantity independent of both node density and $\langle k \rangle$ [22,27]. Furthermore, it has been shown numerically that the critical connectivity in these networks is at $\langle k \rangle = 4.52$ [22], which is much higher than the well-known $\langle k \rangle = 1$ value in RG.

2.2   Medium Access Control

In WiFi networks access to the available frequency channels is controlled by a coordination mechanism called the Medium Access Control (MAC) [24]. The function of the MAC is to ensure interference-free wireless transmissions of data packets in the network. This is achieved by scheduling in time the transmissions of nearby devices in such a way that devices whose radio transmissions may interfere with each other do not get access to the wireless channel at the same time. The presence of the MAC introduces novel spatio-temporal correlations in the dynamics of data communications in these networks which are absent in the Internet communications.

The MAC protocol used by WiFi-based wireless devices follows the IEEE 802.11 standard [24], which specifies a set of rules that enable nearby devices coordinate their transmissions in a distributed manner. The IEEE 802.11 MAC is a highly complex protocol and we do not attempt to fully model this protocol. Instead we focus on the most relevant aspect of this protocol, the so-called listen-before-talk (LBT) rule. This rule dictates that each device should check the occupancy of the wireless medium before starting a data transmission and refrain from transmitting if it senses that the medium is busy [16].

2.3 Worm Propagation Model

Following [16] we assume that wireless worms primarily utilise multihop forwarding for their propagation in adhoc networks, a mechanism which does not require any Internet connectivity. With respect to an attacking worm we use the so-called susceptible-infected (SI) and susceptible-infected-removed (SIR) models from mathematical biology, adapted to the context of wireless communications. We assume nodes in the network to be in one of the following three states: vulnerable, infected, or immune. Infected nodes try to broadcast the worm to their neighbours at every possible opportunity. Vulnerable nodes can become infected with probability λ when they receive a transmission containing a copy of the worm from an infected neighbour. Finally, in the SIR version of the model, infected nodes can get patched and become immune to the worm with probability δ. We denote by $S(t)$, $I(t)$ and $R(t)$ the population of vulnerable, infected and immune nodes, respectively.

We note that temporal characteristics of the underlying system such as processing delays are likely to have a significant effect on the propagation of worms. In the current study we model the processing time required by a worm to complete the infection of a node as a constant value of one clock tick, and assume that the wireless transmission time of worm packets can be considered instantaneous (or at least is much smaller than the processing time).

3 Mean-Field Theory of Epidemic Spreading on Networks

Ever since the first appearance of computer worms, attempts to understand and model their propagation patterns has drawn on parallels with biology. Initially, standard "homogeneous mixing" models from mathematical epidemiology were adapted and used by researchers in order to analyse the propagation of worms in computer networks [7].

The homogeneous mixing hypothesis is equivalent to the assumption that the underlying contact network along which a disease spread is a *complete graph*, i.e. each member of the population can establish a contact with every other member of the population, on a timescale which is much shorter than the timescale of the epidemic. In the last few years, however, significant progress has been made in understanding the impact of much more complex contact network topologies on the properties of epidemics [8,11]. Consequently, new network "mean-field" models have been put forward [9,10,12,13] which are capable of incorporating some of the structure of the underlying networks in the dynamics.

In the case of a network where no correlations are present, the mean-field equations for the SI model yield for the initial growth of the epidemic [14]

$$I(t) = I_0 \left[1 + \frac{\langle k \rangle - 1}{\kappa - 1}(e^{t/\tau} - 1) \right] \tag{4}$$

where

$$\tau = \frac{1}{\lambda(\kappa - 1)}, \tag{5}$$

and I_0 is the initial population of infected nodes. Furthermore, $\kappa = \langle k^2 \rangle / \langle k \rangle$ is a parameter defining the level of fluctuations in the degree distribution of the networks.

The above result implies an *exponential* initial growth rate of the epidemic. Furthermore, it shows that the growth timescale of an epidemics decreases with increased fluctuations in degree distribution. In networks with a Poisson degree distribution we have $\kappa = \langle k \rangle + 1$ and we obtain $\tau = (\lambda \langle k \rangle)^{-1}$. On the other hand, in the so-called power-law (or scale free networks), $\langle k^2 \rangle$ may diverge in the infinite systems size limit implying an *instantaneous* initial growth of the epidemic.

In the case of SIR epidemics the inclusion of the immunisation term δ does not change the initial exponential growth dynamics given by Eq. (4). However, the time-scale, τ, is found to be given by [14,10]

$$\tau \sim \frac{1}{\lambda \kappa - (\delta + \lambda)}. \tag{6}$$

The above discussion results holds only in the case of *uncorrelated* networks. Network correlations are usually encoded in the mean-field equations via the correlation function $P(k'|k)$ which gives the conditional probability that a node with degree k is connected to a node with degree k'. In the presence of correlations the mean-field equations become more complex. However, also in this case it can be shown that the dominant behaviour of the epidemic growth is of the exponential form [14,12]:

$$I(t) \sim e^{\Lambda_m t}, \tag{7}$$

where Λ_m is the largest eigenvalue of the matrix $\mathbf{C} = \{C_{k,k'}\}$ with elements:

$$C_{k,k'} = \lambda k \frac{k' - 1}{k'} P(k'|k) \tag{8}$$

4   Simulation Studies

We simulated the propagation of worms in wireless adhoc networks comprising N devices spread in a $L^2 = 1000 \times 1000 \ m^2$ area. The transmission range of all devices was set at 65 m. This value is somewhere between the typical minimum and maximum range of the WiFi systems, and also guarantees network connectivity for all device densities considered. For a given density, nodes were distributed randomly and uniformly in the simulation cell. The resulting RGG networks were constructed following the prescription of Sec. 2. 1, and periodic boundary conditions were used in order to reduce finite-size effects. We verified numerically that the the average degrees of the resulting networks were well-described by

$$\langle k \rangle = \pi r_t^2 \rho, \tag{9}$$

where $\rho = N/L^2$ is the device density.

The above procedure generates wireless adhoc networks which have *static* network topologies. In reality the topology of ad hoc network can be highly dynamic due to the movement of nodes. However, the timescale at which the topology of the network changes is usually much slower than the timescale of worm propagation. This justifies our approach in using "frozen " topologies to study the spreading of worms in these networks. In the case of intermittently connected mobile adhoc networks, however, node mobility need to be taken into account [31].

The spreading dynamics was simulated on top of the above frozen networks using Monte Carlo simulations. Each Monte Carlo run starts by infecting a single randomly chosen node and proceeds following the rules described in Sec. 2.3 until the stationary state in reached. The results were averaged over 500 Monte Carlo runs and were also averaged over simulations starting from at least 5 different initial infected seeds.

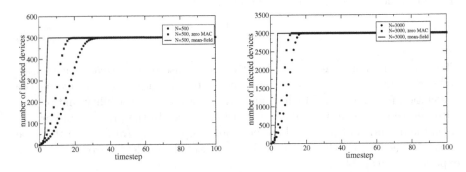

Fig. 1. Time evolution of the number of infected device is shown for the spreading of an unknown worm in networks consisting of $N = 500$ (left panel) and $N = 3000$ (right panel) devices. Simulation results are shown both in the presence and in the absence of MAC and are compared with the mean-field results.

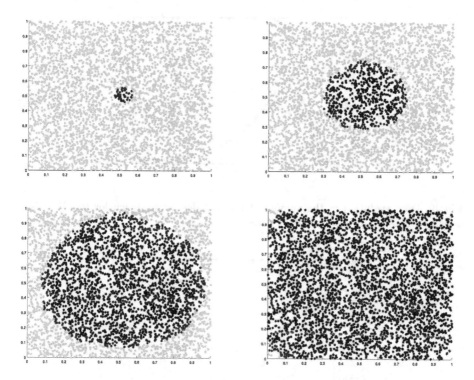

Fig. 2. The spreading of a worm epidemic in a wireless adhoc network is shown at (from top to bottom and left to right) $t = 1, 5, 10, 15$ simulation timesteps. The network size is $N = 3000$ and the parameters used are $\lambda = 1$ and $\delta = 1$. Susceptible nodes are shown in green, infected nodes in red and immunised nodes in blue.

4.1   The Spreading of Unknown Worms

First we consider the spread of an unknown worm in the network. In this case none of the nodes is immune against the worm and we also assume that there is no mechanism for interactive immunisation of nodes (i.e. $\delta = 0$). In the simulations reported here the infection rate was fixed at $\lambda = 1$. We refer the reader to [16] for results obtained for other values of λ. Figure 1 displays the propagation dynamics of the worm in our networks for device densities corresponding to $N = 500$ (left panel) and $N = 3000$ (right panel). The simulation results were obtained both in the presence of the MAC mechanism and in and idealised scenario where this mechanism is not required (we call this zero MAC). For comparison, we have also plotted the analytical results obtained from the mean-field SI model for these networks. It can be seen that for the $N = 500$ network the simulated propagation speed of the worm is significantly slower than the exponential growth predicted by the mean-field theory. Switching off the MAC protocol and increasing the density to $N = 3000$ somewhat increases the propagation speed. However, the resulting curves are still well below the mean-field prediction.

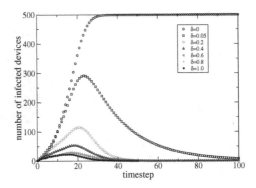

Fig. 3. Time evolution of the number of infected devices is shown for the $N = 500$ node networks for different patching rates

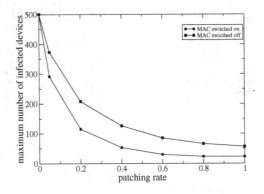

Fig. 4. The maximum number of infected devices is shown as a function of the patching rate the $N = 500$ metwork

The above results show that the slow growth of the epidemic on our networks is the result of two distinct effects. Firstly, it is caused by spatial correlations in our networks which are absent in, e.g., random graphs and scale-free networks. Such correlations effects are beyond the mean-field theory but can be accounted for in mathematical models of spatial epidemics [19]. Secondly, the presence of the MAC introduces a new *self-throttling* effect [16,18] of adjacent infective devices contending for accessing the shared wireless medium, which further slows down the progress of the epidemic.

4.2   Interactive Immunisation

Next we consider the scenario where nodes are patched against the worm in real-time while the worm is spreading in the network. These patches can be downloaded onto the network using, for example, WiFi access points or via a cellular link. Once a node downloads a patch it can send copies to other nodes in the adhoc network using multihop forwarding. In the following we shall assume that nodes can be patched at a given rate δ through one or a combination of the above mechanisms, without considering the patching mechanism itself in detail.

Fig. 2 shows, as an example, snapshots of the spatio-temporal patterns of the worm epidemic for the $N = 3000$ network as obtained from our simulations. These simulations were performed using $\lambda = 1$ and $\delta = 1$ hence, apart from the randomness introduced by network topology, they are deterministic. It can be seen that starting from a single infected device in the centre of the simulation cell, the epidemic spreads outwards along the network links, following a propagation pattern which closely resembles the spreading of a disease in, e.g., a plant population [25].

Figure 3 shows time evolution of the total number of infected devices, $I(t)$, in the network with $N = 500$ for different patching rates, δ, while the infection rate is kept fixed at $\lambda = 1$. It can be seen that while patching does not have a significant impact on the initial spreading rate, it mitigates very effectively a large scale spread of the worm in the network. Even when nodes are immunised at a relatively low rate of $\delta = 0.05$, the maximum fraction of infected nodes is reduced to $< 60\%$. Further details of the impact of patching can be seen in Fig. 4 where the maximum fraction of infected nodes is plotted versus the patching rate, both in the presence and absence of the MAC protocol. It can be seen that this quantity decreases monotonically and rapidly with increased patching rate in both cases. However, the impact of patching is much more pronounced when the MAC protocol is switched on.

5 Conclusions

In this paper we investigated the outbreak of worm epidemics in wireless ad hoc networks by means of modelling and simulations, and examined the effect of interactive patching on preventing such worm epidemics. Performing our simulation studies for a range of device densities we found that worms spread faster in networks with a higher device density. For all densities considered, however, the initial spreading rate is much slower than the exponential spreading rate that has been observed in worm attacks on the Internet. Mean-field epidemiological models on networks also predict an exponential growth rate and are therefore inadequate for describing worm propagation in our networks. The dynamic patterns of worm epidemics in our networks show interesting similarities to propagation patterns of spatial epidemics in, e.g., a plant population [19,6]. We are currently investigating how these analogies could be exploited in devising analytical models for the spreading of wireless epidemics [32]. In addition to fully incorporating the network topology, in our simulations we also took into account the limited bandwidth available to wireless nodes due the MAC contention mechanism. We found that introducing this factor further slows down the initial growth of the epidemic. This is due to a *self-throttling mechanism* introduced by the MAC protocol which, to our knowledge, has no analogy in biological systems.

An important feature of worm epidemics in wireless networks, which was not addressed in the current paper, is the role of device mobility [33]. However, we have recently investigated the impact of mobility in the context of epidemic-style information dissemination in highly dynamic vehicular adhoc networks [17]. These networks consist of isolated clusters that merge and disintegrate dynamically as vehicles move around. We have found that epidemic spreading in such *intermittently connected* wireless networks is greatly affected by mobility. It consists of fast worm propagation within isolated network clusters alternated by much slower inter-cluster worm propagation, which are mediated by vehicular (or human) mobility.

Acknowledgements

M.N. acknowledges support from the Royal Society through an Industry Fellowship and thanks the Centre for Computational Science at UCL for hospitality.

References

1. Szor, P.: The Art of Computer Virus Research and Defense. Symantec Press (2006)
2. Stantiford, S., Paxton, V., Weaver, N.: Proc. of the 11th USENIX Security Symposium (Security 2002) (2000)
3. Klephart, J., White, S.: Directed-graph epidemiological models of computer viruses. In: Proc. of the IEEE Computer Symposium on Research in Security and Privacy, May 1991, pp. 343–359 (1991)
4. Levitt, N.: IEEE Computer 38(4), 20–23 (2005); Dagon, D., Martin, T., Starner, T.: IEEE Pervasive Computing 3, 11–15 (2004)
5. Hypponen, M.: Scientific American, 70–77 (November 2006)
6. Kleinberg, J.: Nature 449, 287–288 (2007)
7. Zhou, C.C., Gong, W., Towsley, D.: Proc. 9th ACM Conference on Computer and Communication Security, Washington D.C. (2002)
8. Pastor-Satorras, R., Vespignani, A.: Phys. Rev. Lett. 86, 3200 (2001)
9. Pastor-Satorras, R., Vespignani, A.: Phys. Rev. E 63, 066117 (2001)
10. Moreno, Y., Pastor-Satorras, R., Vespignani, A.: Eur. Phys. J. B 63, 521 (2002)
11. Ganesh, A., Massoulie, L., Towsley, D.: Proc. IEEE Infocom (2005)
12. Boguna, M., Pastor-Satorras, R., Vespignani, A.: Lecture Notes in Physics 625, 127–147 (2003)
13. Nekovee, M., Moreno, Y., Bianconi, G., Marsili, M.: Physica A 374, 457–470 (2007)
14. Bathélemy, M., Barrat, A., Pastor-Satoras, R., Vespignani, A.: Phys. Rev. Lett. 92, 178701 (2004)
15. Newman, M.E.J., Forrest, S., Balthrop, J.: Phys. Rev. E 66, 035101(R) (2002)
16. Nekovee, M.: New J. Phys. 9, 189 (2007)
17. Nekovee, M.: Proc. IEEE VTC Spring 2006, Melbourne, Australia (2006)
18. Cole, R.G., Phamdo, N., Rajab, M.A., Terzis, A.: Proc. 19th IEEE Workshop Parallel Distr. Simul., New York, pp. 207–214 (2005)
19. Bolker, B.M.: Bull. Math. Biol. 61, 849–874 (1999)
20. Rappaport, T.: Wireless Communications, Principle and Practice. Prentice-Hall, Englewood Cliffs (2000)

21. Penrose, M.: Random Geometric Graphs. Oxford University Press, Oxford (2003)
22. Dall, J., Christensen, M.: Phys. Rev. E 66, 016121 (2002)
23. Bollobas, B.: Modern Graph Theory. Springer, New York (1998)
24. Gast, M.S.: 802.11 Wireless Networks, 2nd edn. O'Reily (2005); Stalling, W.: Wireless Communications Networks. Prentice Hall, Englewood Cliffs (2005)
25. Brown, D.H., Bolker, B.: Bull. Math. Biol. 66, 341–371 (2004)
26. Hekmat, R.: Adhoc networks: Fundamental properties and network topologies. Springer, Heidelberg (2006)
27. Glauche, I., Krause, W., Sollacher, R., Geiner, M.: Physica A 325, 577–600 (2002)
28. Albert, R., Barabási, A.-L.: Rev. Mod. Phys., 74, 47 (2002)
29. Newman, M.E.J.: SIAM Rev., 45, 167 (2003)
30. Boccaletti, S., Latora, V., Moreno, Y., Chavez, M., Hwang, D.-U.: Physics Reports 424 (2006)
31. Nekovee, M.: Proc. IEEE VTC Spring 2007, Dublin, Ireland (2007)
32. Nekovee, M.: BT Internal Technical Report
33. Frasca, M., Buscarino, A., Rizzo, A., Fortuna, L., Boccaletti, S., Latora, V.: Phys. Rev. E 74, 036110 (2006) arXiv0707.1673

Wireless Epidemic Spread in Dynamic Human Networks

Eiko Yoneki, Pan Hui, and Jon Crowcroft

University of Cambridge Computer Laboratory
Cambridge CB3 0FD, United Kingdom
{firstname.lastname}@cl.cam.ac.uk

Abstract. The emergence of Delay Tolerant Networks (DTNs) has culminated in a new generation of wireless networking. New communication paradigms, which use dynamic interconnectedness as people encounter each other opportunistically, lead towards a world where digital traffic flows more easily. We focus on human-to-human communication in environments that exhibit the characteristics of social networks. This paper describes our study of information flow during epidemic spread in such dynamic human networks, a topic which shares many issues with network-based epidemiology. We explore hub nodes extracted from real world connectivity traces and show their influence on the epidemic to demonstrate the characteristics of information propagation.

Keywords: Time Dependent Networks, Connectivity Modelling and Analysis, Network Measurement, Delay Tolerant Networks, Social Networks.

1 Introduction

Increasing numbers of mobile computing devices form dynamic networks in daily life. In such environments, the nodes (i.e. laptops, PDAs, smart phones) are sparsely distributed and form a network that is often partitioned due to geographical separation or node movement. We envision new communication paradigms, using dynamic interconnectedness between people and urban infrastructure, leading towards a world where digital traffic flows in small leaps as people pass each other [15]. Delay Tolerant Networks (DTNs) [9] are a new communication paradigm to support such network environments, and our focus is a type of DTN that provides intermittent communication for humans carrying mobile devices: the Pocket Switched Network (PSN) [2].

Efficient forwarding algorithms for such networks are emerging, mainly based on epidemic protocols where messages are simply flooded when a node encounters another node. Epidemic information diffusion is highly robust against disconnection, mobility and node failures, and it is simple, decentralised and fast. However, careful tuning to achieve reliability and minimise network load is essential. Traditional naïve multiple-copy-multiple-hop flooding schemes have been empirically shown to work well in dense environments, and they provide fair performance in sparse settings – such as city-wide communications – in terms of delivery ratio and delay [2]. However, in terms of delivery cost, the naïve approach is far from satisfactory, because it creates a large amount of unwanted traffic as a side-effect of the delivery scheme. To reduce the overhead of epidemic routing, various approaches have been reported, ranging from

P. Liò et al. (Eds.): BIOWIRE 2007, LNCS 5151, pp. 116–132, 2008.

count-, timer- or history-based controlled flooding to location-based strategies (see Section 7 for further details).

We have previously reported an approach that uses a logical connection topology, and that uncovers hidden stable network structures, such as social networks [14] [34], from the human connectivity traces. In PSNs, social networks could map to computer networks since people carry the computer devices. We have shown improved performance by applying these extracted social contexts to a controlled epidemic strategy [13]. During this work, we have realised that further understanding of network models is essential, because the properties of human contact networks – such as community and weight of interactions – are important aspects of epidemic spread. Recently, online-based social networks have been studied; however, understanding network structures and models hidden in pervasive dynamic human networks is a still-untouched research area.

Networks represent flows of information and make it possible to characterise the complex systems of our world. A network is a map of interactions, because communication is fundamental in our society. These networks are often neither regular lattices, nor are all units connected randomly, but the interaction patterns are complex. This paper shows a preliminary study of patterns of information flow during epidemic spread in complex dynamic human networks, which share many issues with network-based epidemiology. Many studies have been conducted, and these are based either on simulation or a small collection of data. Our study uses real world data, and we believe that it gives an interesting insight on real human interactions. We consider a model for time paths based on graph evolution, called *Time-Dependent Networks*, in which links between nodes depend on a time window. We explore epidemic change by exploiting device connectivity traces from the real world and demonstrate the characteristics of information propagation. We describe preliminary empirical results, but further mathematical modelling work is outside the scope of this paper.

The rest of this paper is structured as follows. We introduce the experimental data sets in Section 2, and then describe the complexity of real world connectivity data in Section 3. We discuss the result of the epidemic spread experiments in Section 4, and the influence of hub nodes for the epidemic spread in Section 5. We describe a summary of community detection in Section 6, which is followed by the related work. Finally, we conclude the paper with a brief discussion.

2   Real World Human Connectivity Traces

The quantitative understanding of human dynamics is difficult and has not yet been explored in depth. The emergence of human interaction traces from online and pervasive environments allows us to understand details of human activities. For example, the Reality Mining project [7] collected proximity, location and activity information, with nearby nodes being discovered through periodic Bluetooth scans and location information from cell tower IDs. Several other groups have performed similar studies. Most of these [7] [6] [22] use Bluetooth to measure device connectivity, while others [12] rely on WiFi. The duration of experiments varies from 2 days to over one year, and the numbers of participants vary. We have analysed various traces from the Crawdad database [3] listed below, and Table 1 summarises the configuration.

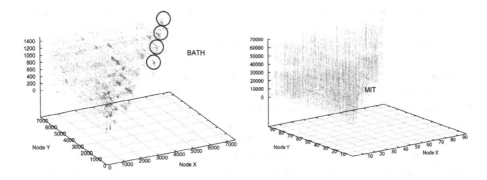

Fig. 1. Node Contact: BATH and MIT traces

MIT: in the MIT Reality Mining project [7], 100 smart phones were deployed to students and staff at MIT over a period of 9 months. These phones were running software that logged contacts.

UCSD: in the UCSD Wireless Topology Discovery [29], approximately 300 wireless PDAs running Windows CE were used to collect WiFi access point information periodically for 11 weeks.

CAM: in the Cambridge Haggle project [18], 40 iMotes were deployed to 1st year and 2nd year undergraduate students for 11 days. iMotes detect proximity using Bluetooth.

INFC06: 78 iMotes were deployed at the Infocom 2006 conference for 4 days [2].

BATH: in the Cityware project, 9 Bluetooth scanners across the city of Bath were deployed to monitor the presence of mobile devices within an approximate 10 metre radius [23]. The co-location of a device pair is identified from the log data. Also part of devices are equipped with a Bluetooth scanning program [22] and detected device information is collected via GPRS. This leads to the construction of a connectivity graph for each time unit.

Note that it is a complex task to collect accurate connectivity traces using Bluetooth communication, as the device discovery protocol may limit detection of all the devices nearby. Bluetooth inquiry can only happen in 1.28 second intervals. An interval of $4 \times 1.28 = 5.12$ seconds gives a more than 90% chance of finding a device. However, there is no data available when there are many devices and many human bodies around.

Table 1. Characteristics of the experiments

Experimental data set	MIT	UCSD	CAM	INFC06	BATH
Device	Phone	PDA	iMote	iMote	PC
Network type	Bluetooth	WiFi	Bluetooth	Bluetooth	Bluetooth
Duration (days)	246	77	11	3	5.5
Granularity (seconds)	300	600	120	120	Continuous
Number of Experimental Devices	97	274	36	78	7431

The power consumption of Bluetooth also limits the scanning interval, if devices have limited recharging capability. The iMote connectivity traces in Haggle use a scanning interval of approximately 2 minutes, while the Reality Mining project uses 5 minutes. The advantage of BATH data is that scanning is done continuously. The ratio of devices with Bluetooth enabled to the total number of devices is around 7%. Because of the uniqueness of urban-scale human connectivity data, we focus on analysis using the BATH trace in this paper. Fig. 1 depicts all contact points between two nodes along the timeline in 3D form. The z-axis represents time, with 300 seconds per unit. This depicts the same node pair encountering repeatedly, which is marked with circles. The Bath data dictates 5 days repeating contact patterns, while the MIT trace shows as a vertical line during 9 months.

3   Complexity of Real World Networks

In general, to understand the network structure one requires three key metrics: the average path length to show the distance between a pair of nodes, the cluster coefficient to indicate how well nodes are clustered, and the degree distribution. In DTNs, the topology changes every time unit and data paths, which may not exist at any one point in time, potentially arise over time. Thus, existing metrics for static networks are difficult to apply. Previously, the characteristics of a pair of nodes – such as inter-contact and contact distribution – have been explored in several studies [2] to which we refer the reader for further background information. We also described the extraction of information related to levels of clustering or network transitivity, and strong community structure in our previous work [34] [14].

As PSNs are formed by humans, it is assumed that social networks take a major role in epidemic spread. Most social networks are neither random nor regular but complex. The properties of nodes include fixed states, variable states, neighbour nodes and network positions (i.e. centralities). Understanding a complex system requires not only understanding of the elements in the system, but also of the patterns of interactions between the elements. Thus, observing communication over the network is expected to give some information about the network structure and, vice versa, the network structure affects the communication. In this paper, we focus on information flow during epidemic spread, including the impact of hub nodes. In the following subsections, we discuss various metrics that can be used in expressing dynamic time-dependent networks.

Table 2. Average Hops and Cluster Coefficient

Experimental traces	Average Hop Count	Cluster Coefficient
MIT	1.6	0.44
UCSD	2.2	0.41
CAM	1.2	0.66
INFC06	1.5	0.52
BATH	3.3	0.45

Fig. 2. Evolution of Connection Map and Edge Characteristics (UCSD Trace)

3.1 Node Distance and Clustering

The average shortest path length between any two randomly chosen people on the planet (i.e. 6.5 billion people) is 6. This is easy to explain if social ties are highly random. However, real social networks are not random, as they exhibit a great deal of clustering, and the average distance between two nodes is small. There are also shorcuts between clustered groups. A network with small average degrees, high clustering, and small average distances has been called a small world network by Watts [31]. Table 2 summarises the average hop counts and cluster coefficient values for each trace. The cluster coefficient value of the MIT trace – 0.44 – is the probability that, if node A knows nodes B and C, nodes B and C know each other. The BATH trace, where proximity data is collected in city scale, shows an average hop count of 3.3 and cluster coefficient value of 0.45.

3.2 Weighted Graph

The connectivity traces can be represented by weighted graphs – also called contact graphs – in which the weight of an edge represents the *contact duration* and *contact frequency* for the two end vertices. Understanding human interaction can then be tackled in the domain of weighted network analysis. Possible outcomes from studying of the weighted contact graphs include community detection and determining node centrality. Many real world networks are weighted, but due to complexity, little analysis has been done in this area. The seminal work is a weighted network analysis paper by Newman [20]. A weighted graph can be converted into a multi-graph with many unit edges. Here, we only consider symmetric edges. In reality, edges can be symmetric (undirected) or asymmetric (directed), possibly with a different strength in either direction. Fig. 2 depicts network evolution over a period 15 minutes in the UCSD trace (taken from our visualization work [35]). The network exhibits a small-world-like formation at first, which breaks down into two groups, each forming a star topology. See Section 6 for further community detection.

3.3 Node Centrality

Understanding a network and a node's participation in the network is important. Centrality measurements give insight into the roles and tasks of nodes in a network. The

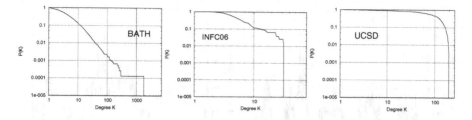

Fig. 3. Aggregated Degree Distribution

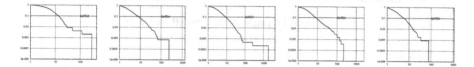

Fig. 4. Degree Distribution: Gates in Bath Trace

centrality of a node in a network is a measure of the structural importance of that node. Freeman defined several centrality metrics [10] and three of the best-known metrics are described below:

Degree. centrality C_D of a node a measures the number of direct connections d. It indicates how active a node is in the network.

$$C_D(a) = d_a \qquad (1)$$

Social networks in general exhibit small average degree compared to the number of nodes, where people have limited connections to the other people. There are over 200 million web sites, with an average degree of only 7.5, and most sites with less than 10 links, but some sites have thousands of links. In time-dependent networks, the degree centrality should ideally be calculated within an appropriate time-window (see further discussion in Section 5). Fig. 3 depicts the degree distribution of BATH, INFC06 and UCSD traces: the BATH and INFC06 traces exhibit a power-law distribution, whereas the UCSD trace shows that most nodes have a similar degree. The UCSD data is based WiFi and may not have as precise proximity information unlike the other traces. Fig. 4 shows the degree distribution at the scanner locations in the BATH trace that exhibit power-law distribution.

Betweenness. centrality C_B indicate that a node acts as a bridge between two nonadjacent nodes. Thus, a node with high betweenness potentially has control over these two nonadjacent nodes. A high-betweenness node in the network may impact on the data flow between two groups of nodes.

$$C_B(a) = \sum_{b<c} [g_{bc(a)}/g_{bc}] \qquad (2)$$

where g_{bc} is the number of geodesics between b and c, and $g_{bc(a)}$ is the number of geodesics between b and c that contain a. In other words, the betweenness centrality is

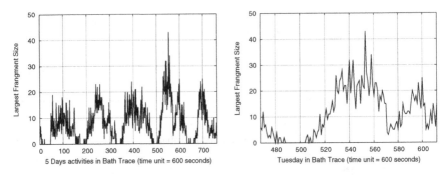

Fig. 5. Largest Fragment in Timeunit (Bath Trace)

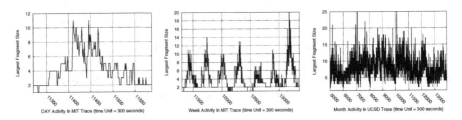

Fig. 6. Largest Fragment in Timeunit (MIT and UCSD Traces)

a sum over all pairs (b, c) of the proportion of geodesics linking the pair that contain node a. Betweenness centrality in time-dependent networks may be calculated using traffic simulation to establish the role of each node (see further discussion in Section 5).

Closeness. centrality C_C indicates the visibility of a node in the network and subnetwork. Maximising closeness centrality yields the node with the shortest path to all others and the best visibility. We have used closeness centrality to build an overlay over the communities [34]. It is a measurement of how long it will take data to spread the others in the community. The closeness centrality, $C_C(a)$, for a vertex a is inverse of the sum of distances to all other nodes:

$$C_C(a) = 1 / \sum_b d_{ab} \qquad (3)$$

3.4 Dynamic Human Behaviour

Analysing the structural properties of growing networks could be relevant for social networks. In each time unit t_i, several nodes appear or disappear, and each selects or deselects k possible counter parts from the existing networks. They join or leave the network with probability p. Identifying the values k and p from the empirical trace defines the form of network evolution. When p is large, over many time steps the network transition is significant.

Fig. 5 depicts the size of the largest connected subgraph in each time unit, based on the BATH trace, which shows the network dynamics over **4 days (Sunday through**

Wednesday). The snapshot on Tuesday depicts a single day's activity, and distinct day- and night-time dynamics can be observed. Fig. 6 shows the same dynamics, including monthly periodicity, based on the MIT and UCSD traces. Note that the size of the largest fragment in the BATH trace is slightly larger than in the MIT/UCSD traces, because the BATH trace covers all devices in the city of Bath, wheres the MIT/UCSD traces only consider a known group of 100–270 participants. The cause of larger fragment sizes in the BATH trace raises an interesting question: are these due to temporal/spatial connections or tighter social connections?

4   Epidemic Dynamics

Epidemiology can be used to deal with intermittent connectivity in DTN environments. The small-world topology of interpersonal connection and its hierarchical structure yields a two-level structure that has a strong impact on epidemic spread in a population. DTNs bring a further complex new network structure, because devices can either communicate through the communication mechanism like the Internet, or directly when they are in the communication range using short-range wireless communication.

Pastor-Satorras has conducted an analytical and numerical study on a large-scale dynamical model on epidemic spread in synthetic networks [26][25]. In this section, we show various epidemic characteristics from our experiments using the real world traces.

4.1   SIS Model

For epidemic spread, we use the *Susceptible-Infected-Susceptible (SIS)* model. Each node in the network represents an individual, and each link is a connection along which the virus infection can spread from between individuals. The SIS model is defined as follows:

1. Each node can be in one of two states:
   - Susceptible (not currently infected)
   - Infectious (infected)
2. The initial infectious nodes may be drawn from the following groups. These nodes do not participate in epidemic propagation until they appear in the trace.
   - Top percentile of the degree distribution
   - 50th percentile of the degree distribution (i.e. average)
   - Bottom percentile of the degree distribution
3. When a node is infectious, it can infect the other nodes with probability λ, where $\lambda = 1$. At each time unit, if that node has a link with a susceptible node, the susceptible node becomes infectious.
4. In each infectious node, the virus has a time-to-live (TTL). When the TTL expires, the node reverts to the susceptible state.

Fig. 7 depicts how the infectious nodes change with time, based on the BATH trace, and using the nodes in the top percentile of the degree distribution as the original infectious nodes. When the TTL is set to 6 hours, we do not observe epidemic spread at all; while, if the TTL is set to 1 day, the effect is similar to having an infinite TTL.

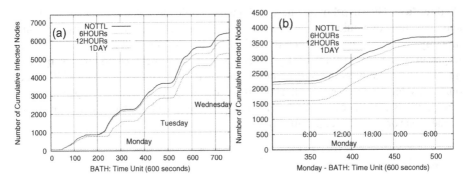

Fig. 7. BATH: Epidemic Spread

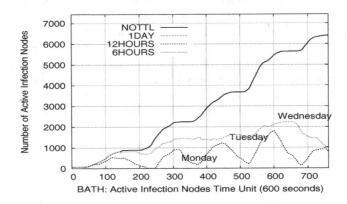

Fig. 8. BATH: Active Infected Nodes

Fig. 8 depicts the number of infectious nodes during epidemic spread. With a TTL of 12 hours, a circadian cycle can be observed, with an increase in the number of infectious nodes during day-time and the virus dying out at night. However, the resilient epidemic comes back during the next day. The trace is not long enough to see the trend towards the end of Wednesday and we plan to conduct extended experiments that will yield traces ranges from a month to a year in length.

We have conducted experiments to investigate the impact of selecting different source nodes. As stated above, we base our selection on the distribution of node degrees. When the bottom percentile of this distribution (75 nodes) is selected to give source nodes, epidemic spread only begins after 1 day, whereas starting with high degree nodes causes epidemic spread to begin immediately. Once the epidemic spread has begun, the spread proceeds at a similar rate in either case.

Fig. 7(b) shows three stages of epidemic spread during a 24-hour period. The stages are (1) a rapid increase at first where propagation may take place within clusters, (2) slow climbing when infectious nodes encounter external clusters, and (3) exhaustion of infection as the epidemic spread hits the upper limit of infection. During the first stage, linking between clusters may occur, and this accelerates the increase of infected nodes.

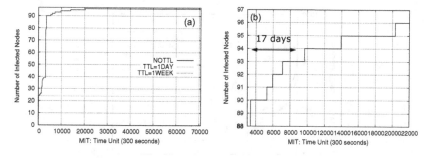

Fig. 9. MIT: Epidemic Spread

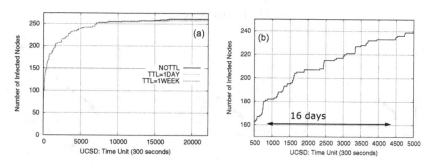

Fig. 10. UCSD: Epidemic Spread

This three stages can be observed in the MIT and UCSD traces (see Fig. 9 and Fig. 10). Fig. 9(b) and Fig. 10(b) depict the second stage with an enlarged time unit scale.

5   Influence of Hub Nodes

In this section, we investigate hub nodes and their influence on epidemic spread using the BATH trace. We have defined hubs based on the following centralities and extracted the top 100 such hub nodes from the trace. We then ran the epidemic spread simulation described in Section 4 but excluding the hub nodes, in order to observe how much influence they have on the spread.

DEGREE Hub: The total degree of each node over the entire duration of the trace indicates the popularity of the node (*Degree Centrality*). With this metric, it is not possible to distinguish two types of hubs: the node has high degree within a short time window (*party hub*) or a larger time window (*date hub*) [11]. Most nodes interact with only a few other nodes while a small number of hub nodes may have many interactions.

In [13], we examined the degree per unit time (e.g. the number of unique nodes seen per 6 hours). We chose a 6-hour time window based on our intuition that daily life is divided into 4 main periods: morning, afternoon, evening and night. This is similar to the approach described in [33]. However, it is sensitive to starting the time window at

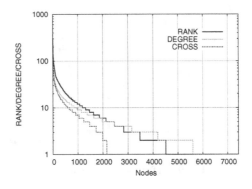

Fig. 11. Hub Nodes

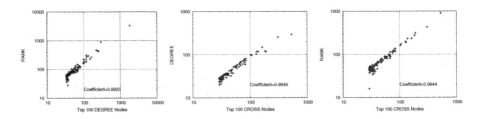

Fig. 12. Correlation of 100 Hub Nodes

different absolute times of the day. As Fig. 5 shows, the day cycle could be a more efficient time window in the urban space.

RANK Hub: The frequency that a node is used to relay data to other nodes indicates the centrality of the node. We simulated flooding over the temporal graph extracted from the trace and counted the number of times each node is used for relaying the data. We exploited different counting schemes, such as counting any time a node relays data or only when the node is on the shortest path from the source to the destination. Different schemes result in a similar ranking. This metric is equivalent to *Betweenness Centrality* in time-dependent networks.

CROSS Hub: The appearance of a node at different locations indicates that it has *Mobility Centrality*. With the BATH trace, 9 locations are extracted and the rate of appearance at each location is measured. Fig. 11 depicts the distribution of all nodes with extracted centrality metrics. The y-axis shows a centrality metric on a logarithmic scale.

5.1   Hub Nodes Similarity

Fig. 12 and Table 3 show the correlation between the sets of hub nodes using different metrics. The coefficient values are greater than 0.95 in every case. The correlation between *RANK HUB* and *CROSS HUB* has the highest value. Table 4 depicts the

Table 3. Hub Nodes Correlation

Category	All Nodes	Top 100 Nodes	Top 50 Nodes	Top 30 Nodes
Rank/Degree	0.99	0.99	0.99	0.99
Degree/Cross	0.97	0.96	0.96	0.96
Corss/Rank	0.99	0.99	0.99	0.99

Table 4. Hub Node Membership Similarity

Top n Nodes	Rank/Degree	Rank/Cross	Degree/Cross
100	0.79	0.43	0.44
70	0.92	0.41	0.41
50	1.00	0.43	0.49
30	1.00	0.46	0.46
10	1.00	0.33	0.33

membership similarity of hub nodes. The *RANK HUB* and *DEGREE HUB* sets share many nodes, while the *CROSS HUB* set has only around 50% of nodes in common with the *DEGREE HUB* or *RANK HUB* sets.

5.2 Inactivation of Hub Nodes

Fig. 13 depicts the impact of deactivating hub nodes during the epidemic spread. Fig. 13 shows inactivation of the top 50 nodes. Removing the top 50 *DEGREE HUB* or *RANK HUB* nodes significant reduces the epidemic spread. Both *DEGREE HUB* and *RANK HUB* nodes have a similar impact. On the other hand, removing the *CROSS HUB* nodes does not show as dramatic an impact as does removing the other two types of hub nodes. Randomly selected 1% of top 30% of high degree nodes are used as the source of the infection. The result indicates the strong influence of hub nodes. We are further investigating what differentiates static hubs from dynamic hubs in a pair interaction.

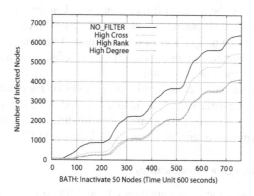

Fig. 13. Inactivation of Hub Nodes

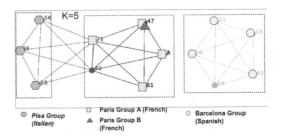

Fig. 14. k-CLIQUE Community Detection in INFC06 Trace

6   Inferring Human Communities

People inherently form groups, yielding social structures in which prominent patterns or information flow can be observed. We have worked on uncovering the structure and dynamics of social communities from human connectivity traces, in which social groups must be embedded [14] [34]. We have shown various community detection mechanisms which can be applied to human connectivity traces in both a centralised and a decentralised way.

Community detection in complex networks has attracted a lot of attention in recent years. In the Internet, community structures correspond to autonomous systems. It is crucial to construct efficient algorithms for identifying the community structure in a generic network. Many community detection methods have been proposed and examined in the literature (see the recent review papers by Newman [21] and Danon *et al.* [5]).

We have exploited different algorithms [13]. The k-CLIQUE method has been designed for binary graphs, and we therefore need to threshold the edges of the contact graphs in the traces [24], while *Weighted Networks Analysis* [20] can work on weighted graphs directly without any threshold.

6.1   *K*-CLIQUE Community Detection

Palla *et al.* define a community as a union of all k-cliques (complete sub-graphs of size k) that can be reached from each other through a series of adjacent k-cliques, where two k-cliques are said to be adjacent if they share $k - 1$ nodes. As k is increased, the k-clique communities shrink, but on the other hand become more cohesive since their member nodes have to be part of at least one k-clique. An advantage of this approach is that it allows overlapping communities, which is useful as, in human society, one person may belong to multiple communities.

Fig. 14 depicts the detected communities in the INFC06 trace, when $k = 5$. Three distinct communities are detected, which include two nodes that belong to two communities. Fig. 15(a) shows the detected communities with different k values and Fig. 15(b) depicts the community size distribution when $k = 5$ in the BATH trace.

6.2   Inter- and Intra-gate Communities

It is known that the location is an important attribute for social community structure. The BATH trace includes the location of scanners (i.e. Gate 1 to 9), and Fig. 16 depicts

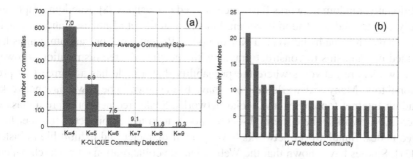

Fig. 15. Community Detection in Bath City Trace

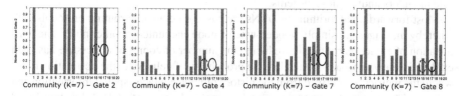

Fig. 16. Bath Trace: Communities in Gates

the appearance rate of community members at Gates 2, 4, 7 and 8. For example, the members of community 17 (with solid circle) are observed at Gate 2, but almost never at the other locations; whereas the members of communities 1 and 15 (with dashed circle) appear at every gate. We refer to the former type of community as an *Intra-Community*, and to the latter type as an *Inter-Community*. Intra-communities may have a strong tie with the location, while inter-communities may indicate a group of people moving together.

We ran a simulation to investigate the effect of deactivating 100 nodes of each community type. Communities 1 and 15 were selected at random to represent inter-communities, and 7 and 17 are selected as intra-communities. Removing intra-communities causes an up-to-10% reduction in infectious nodes, while removing inter-communities has no effect in this scenario. It is well known in social networks that inter-relationship within a group is stronger than external links. The experiment result indicates the characteristics of social networks.

The communities detected in the traces may be static social communities or transient communities, such as a group of people who happen to be in the same location. Our current approach does not distinguish between these two different community concepts and further refinement of community concepts, along with membership management is part of our ongoing work.

7   Related Work

The recent discovery of complex network properties in the structure of biological and social systems [28] has brought different perspectives on real world networks.

Traditionally, random networks have been studied extensively [8]. Random graphs are usually constructed by randomly adding links to a static set of nodes. Random graphs tend to have short paths between a pair of nodes. Recent work on random graphs has provided mechanisms to construct graphs with specified degree distributions. Power-law networks are networks where the probability that a node has a degree k is proportional to k. Many real-world networks have been shown to be power-law networks, including Internet topologies and social networks. Scale-free networks are a class of power-law networks where the high-degree nodes tend to be connected to other high degree nodes. Small world networks have a small diameter and exhibit high clustering [31]. Studies have shown that the Web, scientific collaboration on research papers, film actors, and general social networks have small world properties []. It has become clear that this pattern of interactions, which forms the network, plays a fundamental role in understanding these systems.

Most forwarding algorithms in DTNs are based on epidemic routing protocols [30], whereby messages are simply flooded when a node encounters another node. The optimisation of epidemic routing by reducing the number of copies of a message has been explored. Many approaches calculate the probability of delivery to the destination node, where the metrics are derived from the history of node contacts, spatial information and so forth. The pattern-based Mobyspace Routing by Leguay *et al.* [17], location-based routing by Lebrun *et al.* [16] and PROPHET Routing [19] fall into this category. The Message Ferry approach of Zhao *et al.* [32] takes a different approach by controlling the movement of each node. Recently, attempts to uncover hidden stable network structure in DTNs and social networks have emerged. For example, SimBet Routing [4] uses ego-centric centrality and its social similarity. Messages are forwarded towards a node with higher centrality to increase the possibility of finding the potential carrier to the final destination.

Emerging wireless technologies are creating physical network in the actual physical space along online communication (e.g. social network services, email). Understanding this new pervasive network as a time-dependent dynamic human network is still an open research area. Social relationships and interactions (i.e. social context) is gaining importance. New results in the area of complex network theory [1] give new insight on social networks.

8   Conclusions and Future Work

In this paper, we have shown our study of epidemic spread in dynamic human networks from human connectivity traces. The human networks exhibit periodic activity. Daily circulation is significant, and epidemic spread demonstrates that if the virus has over one day of life, the spread rate reaches almost the same level as when the virus has infinite life. Removing the top 100 hub nodes (using various definitions of "hub") out of over 7500 nodes from consideration yields a significant reduction in the rate of epidemic spread.

In the BATH trace, some communities exhibited strong ties to particular locations. Therefore the local network structure could possibly form a type of small-world network, and a small number of nodes could connect with external nodes, forming a scale free network. We are currently working to prove this assumption by constructing corresponding synthetic networks that can be compared with real world networks. Our future work includes investigating an asymmetric communication model (i.e. forming a directed graph) and defining new network measurement criteria such as time-dependent centralities and cluster coefficient values. We are taking an empirical approach and therefore obtaining accurate and fine grained trace data is essential. We are planning to deploy several urban scale experiments for data collection and information diffusion.

Pervasive DTNs are dynamic, and we are particularly interested in how network structure affects information flow, and vice versa: how the ongoing communication affects the network structure. Pairwise communication and social structure need to be integrated and modelled alongside dynamic interactions. The social network reflects access to information and change of social activities can be seen as seeking better information access. Our ultimate goal is a complete understanding of human-to-human network models in the urban space.

Acknowledgment

This research is funded in part by the Haggle project under the EU grant IST-4-027918. We would like to acknowledge the EPSRC Cityware project for providing the City of Bath trace, and the CRAWDAD project [3] for their hosting and sharing of the connectivity/mobility data.

References

1. Albert, R., Barabasi, A.-L.: Statistical mechanics of complex networks. Reviews of Modern Physics 74, 47 (2002)
2. Chaintreau, A., et al.: Impact of human mobility on the design of opportunistic forwarding algorithms. In: Proc. INFOCOM (April 2006)
3. Dartmouth College: A community resource for archiving wireless data at dartmouth (2007), http://crawdad.cs.dartmouth.edu/index.php
4. Daly, E., Haahr, M.: Social network analysis for routing in disconnected delay-tolerant manets. In: Proceedings of ACM MobiHoc (2007)
5. Danon, L., Duch, J., Diaz-Guilera, A., Arenas, A.: Comparing community structure identification (2005)
6. Diot, C., et al.: Haggle Project (2008), http://www.haggleproject.org
7. Eagle, N., Pentland, A.: Reality mining: sensing complex social systems. Personal and Ubiquitous Computing 10(4), 255–268 (2006)
8. Erdos, P., Renyi, A.: On random graphs i. Mathematicae 5 (1959)
9. Fall, K.: A delay-tolerant network architecture for challenged internets. In: Proc. SIGCOMM (2003)
10. Freeman, L.C.: A set of measuring centrality based on betweenness. Sociometry 40, 35–41 (1977)
11. Han, J.-D.J., Bertin, N., Hao, T., Goldberg, D.S., et al.: Evidence for dynamically organized modularity in the yeast protein-protein interaction network. Nature 430 (2004)
12. Henderson, T., et al.: The changing usage of a mature campus-wide wireless network. In: Proc. Mobicom (2004)

13. Hui, P., Crowcroft, J., Yoneki, E.: BUBBLE Rap: Social Based Forwarding in Delay Tolerant Networks. In: MobiHoc (2008)
14. Hui, P., Yoneki, E., Chan, S., Crowcroft, J.: Distributed community detection in delay tolerant networks. In: Proc. MobiArch (2007)
15. Kleinberg, J.: The wireless epidemic. Nature 449(20) (2007)
16. Lebrun, J., Chuah, C.-N., et al.: Knowledge-based opportunistic forwarding in vehicular wireless ad-hoc networks. In: VTC 2005, pp. 2289–2293 (2005)
17. Leguay, J., et al.: Evaluating mobility pattern space routing for DTNs. In: Proc. INFOCOM (2006)
18. Leguay, J., et al.: Opportunistic content distribution in an urban setting. In: ACM CHANTS (2006)
19. Lindgren, A., Doria, A., Schelen, O.: Probabilistic routing in intermittently connected networks. In: Dini, P., Lorenz, P., Souza, J.N.d. (eds.) SAPIR 2004. LNCS, vol. 3126, pp. 239–254. Springer, Heidelberg (2004)
20. Newman, M.: Analysis of weighted networks. Physical Review E 70, 056131 (2004)
21. Newman, M.: Detecting community structure in networks. Eur. Phys. J. B 38, 321–330 (2004)
22. Nicolai, T., Yoneki, E., Behrens, N., Kenn, H.: Exploring social context with the wireless rope. In: Meersman, R., Tari, Z., Herrero, P. (eds.) OTM 2006 Workshops. LNCS, vol. 4277, pp. 874–883. Springer, Heidelberg (2006)
23. O'Neill, E., et al.: Instrumenting the city: Developing methods for observing and understanding the digital cityscape. In: Dourish, P., Friday, A. (eds.) UbiComp 2006. LNCS, vol. 4206, pp. 315–332. Springer, Heidelberg (2006)
24. Palla, G., et al.: Uncovering the overlapping community structure of complex networks in nature and society. Nature 435(7043), 814–818 (2005)
25. Pastor-Satorras, R., Vespignani, A.: Epidemic dynamics and endemic states in complex networks. Phys. Rev. E. 64(066117) (2001)
26. Pastor-Satorras, R., Vespignani, A.: Epidemic spreading in scalefree networks. Phys. Rev. Lett. 86(14) (2001)
27. Rahul, S.J., Shah, C., Roy, S., Brunette, W.: Data mules: Modeling a three-tier architecture for sparse sensor network. In: IEEE Workshop on Sensor Network Protocols and Applications (SNPA) (May 2003)
28. Strogatz, S.H.: Exploring complex networks. Nature 410, 268–276 (2001)
29. UCSD. Wireless topology discovery project (2004), http://sysnet.ucsd.edu/wtd/wtd.html
30. Vahdat, A., Becker, D.: Epidemic routing for partially connected ad-hoc networks. Technical Report CS-200006, Duke University (April 2000)
31. Watts, D.J.: Small Worlds – The Dynamics of Networks between Order and Randomneess. Princeton University Press, Princeton (1999)
32. Wenrui Zhao, M.A., Zegura, E.: A message ferrying approach for data delivery in sparse mobile ad-hoc networks. In: ACM Mobihoc (May 2004)
33. Winters, P.: Forecasting sales by exponentially weighted moving averages. Management Science 6, 324–342 (1960)
34. Yoneki, E., Hui, P., Chan, S., Crowcroft, J.: A socio-aware overlay for multi-point asynchronous communication in delay tolerant networks. In: Proc. MSWiM (2007)
35. Yoneki, E., Hui, P., Crowcroft, J.: Visualizing Community Detection in Opportunistic Networks. In: ACM MobiCom - CHANTS (2007)

Stochastic Spreading Processes on a Network Model Based on Regular Graphs

Sebastian V. Fallert[1] and Sergei N. Taraskin[1,2,*]

[1] Department of Chemistry, University of Cambridge, Cambridge, UK
[2] St. Catharine's College, University of Cambridge, Cambridge, UK
snt1000@cam.ac.uk

Abstract. The dynamic behaviour of stochastic spreading processes on a network model based on k-regular graphs is investigated. The contact process and the susceptible-infected-susceptible model for the spread of epidemics are considered as prototype stochastic spreading processes. We study these on a network consisting of a mixture of 2- and 3-fold coordinated randomly-connected nodes of concentration p and $1 - p$, respectively, with p varying between 0 and 1. Varying the parameter p from $p = 0$ (3-regular graph of infinite dimension) to $p = 1$ (2-regular graph - 1D chain) allows us to investigate their behaviour under such structural changes. Both processes are expected to exhibit mean-field features for $p = 0$ and features typical of the directed percolation universality class for $p = 1$. The analysis is undertaken by means of Monte Carlo simulations and the application of mean-field theory. The quasi-stationary simulation method is used to obtain the phase diagram for the processes in this environment along with critical exponents. Predictions for critical exponents obtained from mean-field theory are found to agree with simulation results over a large range of values for p up to a value of $p = 0.95$, where the system is found to sharply cross over to the one-dimensional case. Estimates of critical thresholds given by mean-field theory are found to underestimate the corresponding critical rates obtained numerically for all values of p.

Keywords: Network Epidemics, SIS Model, Contact Process, Critical Exponents.

1 Introduction

The spread of epidemics poses a threat to biological populations as well as to computer networks and investigations into its dynamics and mechanisms are therefore of great current interest. One common class of epidemic models considers individuals to be in one of two possible states: susceptible (S) or infected (I). In this paper, we consider both the Contact Process (CP) [1] and the SIS model, two models of disease propagation via nearest neighbour contact, in which a disease is passed on to healthy nearest neighbours stochastically at a rate λ specific to the model while infected sites spontaneously recover at rate ϵ.

---

* Corresponding author.

P. Liò et al. (Eds.): BIOWIRE 2007, LNCS 5151, pp. 133–144, 2008.
© Springer-Verlag Berlin Heidelberg 2008

These Markovian spreading processes have attracted wide attention in the past due to their applicability to phenomena as diverse as autocatalytic chemical reactions, spreading of rumours and transport in disordered media [2]. As the rates λ and ϵ are varied, an epidemic will be in one of two distinct states: an invasive regime (active state) in which it is present with a non-zero probability of ultimate survival and one in which this probability is zero thus leading to a state which allows no further evolution because the disease has died out (absorbing state).

These two regimes are known to be connected by a continuous phase transition thereby rendering them of conceptual interest for investigations into this kind of critical phenomenon of non-equilibrium statistical mechanics (see [3] for a review). The critical behaviour for these models in one-, two- and three-dimensional lattices has been investigated very accurately [4] and is found to be characteristic of the Directed Percolation (DP) universality class. From a range of studies, critical thresholds for the phase transition as well as critical exponents of predicted power-law scaling relations are known to high precision.

With the growing interest in complex networks among the statistical physics community in recent years [5,6], the question of the behaviour of dynamic processes on such topologically disordered structures has arisen [7,8]. Particularly motivated by the fact that networked structures are ubiquitous in nature, the effects of these environments on, for example, the spread of a disease are of immediate interest. In a series of papers [9,10,11,12,13], the behaviour of the CP and the SIS model on a range of networks has been considered and even comparisons with data of computer virus outbreaks have been attempted [11]. As networks in general are infinite-dimensional objects, the dynamical mean-field (MF) approximation is expected to become exact in these cases in principle rendering many models tractable by analytical means. Both Monte Carlo (MC) simulations and the MF approximation have been used in previous investigations and produced such astonishing results as the absence of an epidemic threshold infection rate for infinite scale-free networks [11].

In this paper, we propose to investigate the behaviour of the CP and the SIS model as two paradigmatic stochastic spreading processes on networks of k-regular graph topology. The model network considered in this investigation consists of a mixture of 2- and 3-fold coordinated randomly-connected nodes of concentration p and $1 - p$, respectively. Varying the parameter p from $p = 0$ to $p = 1$ transforms the system from a 3-regular graph of infinite dimension to a 2-regular graph, i.e. a 1D chain. While both the CP and the SIS model are expected to exhibit mean-field features for $p = 0$, the processes effectively take place in a one-dimensional environment for $p = 1$ which is a very well-studied regime of the DP universality class. It is our aim to investigate the behaviour of both the critical rates and accessible critical exponents for this crossover from an infinite- to a one-dimensional case thereby probing the validity of the MF approximation in this setting. The analysis is undertaken by means of Monte Carlo simulations using the quasi-stationary (QS) simulation method [14] and the application of mean-field theory [10].

This paper is structured as follows. Section 2 outlines the definitions and some properties of the processes considered. The MF approximation and the QS simulation method are described in section 3. We present and discuss our results in section section 4 and summarise our findings in section 5.

2   Background

As outlined in the previous section, both the CP and the SIS model are simple toy models for the spread of an infectious disease by nearest-neighbour contact. In these models defined on a network, nodes represent susceptible or infected individuals surrounded by their neighbours connected via links along which the epidemic may spread. Proliferation of the disease to nearest neighbour sites happens at a transmission rate λ while recovery is spontaneous at rate ε making the sequence of events an individual can cycle through *Susceptible* $\rightarrow$ *Infected* $\rightarrow$ *Susceptible*.

The CP and the SIS model are very similar, the difference being the exact mechanism of the spreading of infection. In the case of the CP, a site attempts to transmit its disease at rate λ/k to a randomly selected neighbour where k denotes the number of nearest neighbours. If the selected neighbour is already infected, proliferation fails. For the SIS model, transmission to any non-infected neighbour happens, in contrast, at rate λ independent of the connectivity of the nodes. Thus, the spreading mechanism in the CP effectively compensates for the local connectivity present in the network through a suitable reduction of the spreading rate through a particular link.

Once suitable initial states for all sites have been chosen, the above rules dynamically evolve the spread of a disease in the network. A typical initial condition is the state of a fully-infected system from which the system relaxes very quickly. For very long times, and formally as time $t \rightarrow \infty$ and for an infinite number of sites in the network $N \rightarrow \infty$, the system is expected to be in one of two states: An *active state* in which there remains a finite density of infected sites or an *absorbing state* in which the disease has died out and that therefore admits no further time evolution. Depending on the value of the transmission and recovery rates λ and ϵ, ultimately the system will be in one of the two possible states. More precisely, there exists a continuous phase transition between these regimes as one fixes one of the rates and varies the other. This transition takes the system from a phase where the density of infected sites (order parameter) ρ is zero to one where it continuously grows from zero as the transmission rate (control parameter, assuming ε fixed) is increased.

Without loss of generality, one can perform a rescaling of time and set one of the two rates to unity for convenience. In the following, the recovery rate is assumed to be $\epsilon = 1$ and the critical point is therefore characterised by a critical transmission rate λ_c alone.

There exist a range of well-established scaling relations for various observables in these models of which we present those relevant for this investigation. The density of infected sites in the thermodynamic limit as $t \rightarrow \infty$, the order parameter, is expected to scale as

$$\lim_{t\to\infty} \langle\rho(t)\rangle = \overline{\rho} \sim |\lambda - \lambda_c|^{\beta} \tag{1}$$

thereby defining the order parameter critical exponent β where $\langle\dots\rangle$ denotes averaging over realisations of the process. Order parameter fluctuations are known to follow

$$V = N \left(\overline{\rho^2} - \overline{\rho}^2\right) \sim |\lambda - \lambda_c|^{-\gamma} \tag{2}$$

Both the models under consideration are known to belong to the directed percolation (DP) universality class [2]. Accordingly, the critical exponents defined above are those characteristic of this universality class. Above the upper critical dimension, $d_u = 4$, of these models, fluctuations are expected to be Gaussian and MF theory should be exact. Therefore, these processes taking place in infinite-dimensional networks are expected to exhibit exponents predicted by mean-field theory.

3   Methods

3.1   Mean-Field Approximation

Both the CP and the SIS model can be described by the master equation which reflects the conservation of probability flow [3]. In the following, we will first outline the case of the SIS model and then consider the extension to the simpler case of the CP.

In the dynamical MF approximation, which neglects density fluctuations and statistical correlations between the densities at different sites, and, for the moment, disregarding the structure of the network completely, the master equation for the SIS model takes the form

$$\partial_t \rho(t) = -\rho(t) + \lambda\, k\, (1 - \rho(t))\, \rho(t) \tag{3}$$

where $\rho(t)$ denotes the density of infected sites at time t averaged over realisations of the process which is identical to the probability of a site of the system to be infected at time t. This equation describes the rate of change of the average density in the network which is equal to the flow of density into and out of any site with time and makes for the destruction and the creation terms above. The destruction term due to the vanishing of infection at unit rate is proportional to the density $\rho(t)$. The creation term is due to the possible infection by infected neighbouring sites in the case that the vertex under consideration is not infected. Accordingly, it is proportional to the probability that a site is not infected, $(1 - \rho(t))$, the probability that a neighbouring site is infected $\rho(t)$, the local connectivity k and the spreading rate λ.

The master equation (3) can be extended in order to take into account the structure of the underlying network at the level of the node degree (connectivity) distribution (as developed by Pastor-Satorras and Vespignani [9]). It is clear that, unless one assumes a homogeneous network with $\langle k \rangle \approx k$ for all k, the expression will decouple into a set of equations for the densities of infected vertices characterised by a certain connectivity k, We can write for each k,

$$\partial_t \rho_k = -\rho_k + \lambda \, k \, (1 - \rho_k(t)) \, \Theta_k(t) \,, \tag{4}$$

where $\Theta_k(t)$ is the probability that an edge emanating from a vertex of degree k is connected to an infected site. The infection term as described above now incorporates the probability that a site of degree k is connected to an infected vertex $\Theta_k(t)$. One can interpret $\Theta_k(t)$ as the mean density of neighbouring infected nodes and consequently $k\Theta_k(t)$ as the mean number of infected nearest neighbours [9].

Networks which are Markovian are statistically described by their degree distribution $P(k)$ and the conditional probability $P(k'|k)$ that an edge of a node of degree k is connected to a vertex of degree k'. For such systems, we can write

$$\Theta_k(t) = \sum_{k'} P(k'|k) \, \rho_{k'}(t) \tag{5}$$

where the sum runs over all degrees k'. For uncorrelated networks, which we will exclusively consider in this investigation, this becomes [7]

$$\Theta(t) = \sum_{k'} \frac{k' P(k') \, \rho_{k'}(t)}{\langle k \rangle} \tag{6}$$

which does not depend on k any longer. Substituting this expression into the rate equation (4) and imposing stationarity ($\partial_t \rho_k = 0$) one obtains

$$\overline{\rho}_k = \frac{k\lambda\overline{\Theta}}{1 + k\lambda\overline{\Theta}} \tag{7}$$

where $\overline{\rho}_k$ and $\overline{\Theta}(\lambda)$ are the time-independent values for the mean density at sites of degree k and the mean density of infected neighbouring sites. Multiplying by $\frac{P(k)k}{\langle k \rangle}$ and summing over k yields

$$\frac{1}{\lambda} = \frac{1}{\langle k \rangle} \sum_k \frac{P(k) \, k^2}{1 + k\lambda\overline{\Theta}} \equiv f(\lambda\overline{\Theta}) \tag{8}$$

where $f(x)$ is a monotonically decreasing function of x. This equation only has a (unique) solution for $\overline{\Theta}(\lambda)$ different from zero for $\lambda > \lambda_c$ where λ_c is the threshold value for the transmission rate that makes $\overline{\Theta}$ smallest. As by definition the mean density of infected nearest neighbours $\overline{\Theta} \geq 0$ in the active regime and $f(x)$ is a monotonically decreasing function we have (effectively setting $\overline{\Theta} = 0$

$$\lambda_c^{\text{SIS}} = \frac{1}{f(0)} = \frac{\langle k \rangle}{\langle k^2 \rangle} \tag{9}$$

for the critical threshold [15].

In order to obtain an expression for the order paramter we start by combining equations (6) and (7) and arrive at a self-consistency equation for $\overline{\Theta}$ (equivalent to equation (8))

$$\overline{\Theta} = \sum_k \frac{kP(k)}{\langle k \rangle} \frac{k\lambda\overline{\Theta}}{1 + k\lambda\overline{\Theta}} \qquad (10)$$

that can in principle be solved for $\overline{\Theta}$ which in turn allows one to obtain the order parameter in the MF approximation from

$$\overline{\rho} = \sum_k \overline{\rho_k} . \qquad (11)$$

Critical exponents can be extracted from MF theory by considering the leading behaviour of the relevant expressions. For example, combining the last expression Eq. (11) and Eq. (7) and expanding in $\lambda - \frac{\langle k \rangle}{\langle k^2 \rangle}$ in analogy to the scaling form $\overline{\rho} \sim (\lambda - \lambda_c)^\beta$, one obtains the MF value for the order parameter exponent $\beta = 1$. Similarly, one obtains $\gamma = 0$ for the corresponding fluctuations.

In the case of the CP where the effective spreading rate is inversely proportional to the number of links connected to an infected site, Eq. (5) has to be modified and reads

$$\Theta_k(t) = \sum_{k'} \frac{P(k'|k) \, \rho_{k'}(t)}{k'} \qquad (12)$$

which for uncorrelated networks leads to $\Theta^{nc}(t) = \rho(t)/\langle k \rangle$ where ρ is the average density of infected sites averaged over degrees. Following the procedure as for the SIS model, the critical threshold rate is found to be degree independent and given by [12]

$$\lambda_c^{CP} = 1 \qquad (13)$$

while the critical exponents are identical to those for the SIS model.

3.2 Monte Carlo Simulation: Quasi-Stationary Simulation

Both processes under consideration can be simulated effectively via time-dependent Monte Carlo simulations. Once initial conditions have been chosen, the system is evolved according to the appropriate rules with a simulation that selects possible events according to their prescribed rates and ensures that they happen in exponentially-distributed time intervals.

In contrast to lattices, networks are characterised by the existence of long range links. This implies that a dynamical process will very strongly feel the size of the system in a finite representation of the network leading to much stronger finite-size effects than experienced in lattices. While the most precise method of determining the critical point in lattices is spreading from a single seed while ensuring that the infection never reaches the boundary of the system, this is virtually impossible in networks. Thus, finite-size effects have to be systematically exploited in order to make a prediction about the infinite system using networks of different sizes. This is possible using the finite-size scaling hypothesis which predicts that values of observables in systems of size L are controlled by the ratio $L/\xi_\perp$, where $\xi_\perp$ is the spatial correlation length. In order to use this scaling behaviour, one requires a stationary average value of these observables and analyses their values for different system sizes. Problematically, due to the existence of the absorbing state no such true stationary state exists in a finite system. Fortunately, the processes under consideration evolve such that some observables attain quasi-stationary (QS) values.

Most notably, the density of infected sites averaged over surviving realisations of the process $\langle \rho(t) \rangle$ exhibits this behaviour after an initial transient starting from the initial state of a fully-infected system [2]. This quasi-stationary density $\bar{\rho}$ is expected to scale systematically in accordance with the finite-size scaling hypothesis. According to recent investigations into the finite-size scaling behaviour above the upper critical dimension [16,13] the prediction is

$$\bar{\rho} \sim L^{d\beta/2} \, G\left(L^{d/2}(\lambda - \lambda_c)\right) = N^{\beta/2} \, G\left(N^{1/2}(\lambda - \lambda_c)\right) \qquad (14)$$

where $L^d = N$, the number of sites in the system and $G(x)$ an appropriate scaling function. A similar expression with β replaced by γ is valid for the associated fluctuations.

In principle the QS state can be investigated via a conventional simulation in which the system is stochastically evolved in time from a fully-infected initial condition. The density of infected sites conditioned on survival can then be analysed and a temporal average over the duration of the QS state is an estimator for $\bar{\rho}$. This method is however plagued by a range of problems [17] which led de Oliveira and Dickman to propose a simulation method which samples the QS state directly [18].

In this QS simulation method, the absorbing state is eliminated and its probability weight is redistributed over the active states according to the history of the process. It can then be shown that the true stationary state of the resulting modified process corresponds to the quasi-stationary state of the original one. This method is ideally suited for our study as a single realisation of the network is investigated in one QS simulation run enabling us to analyse sample-to-sample fluctuations between realisations and find a critical point by use of the scaling relation Eq. (14).

4   Results and Discussion

4.1   MF Solution

The mean-field theory for the CP and the SIS model on networks can be applied to our network with two types of nodes of connectivity k_1 and k_2 present with probabilities $P(k_1)$ and $P(k_2)$ respectively. For the SIS mode, the self-consistency

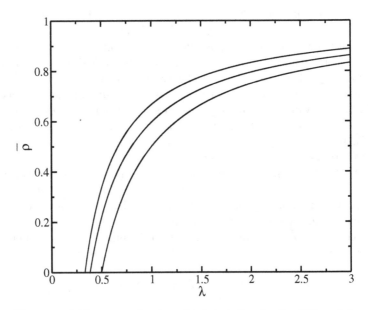

Fig. 1. The mean-field average density of infected sites for the SIS model on a binary random network with $k_1 = 2$, $k_2 = 3$ for (from top to bottom) p=0, 0.5, 1

equation, Eq. 10, for the stationary value of Θ, the average density of infected nearest neighbours, takes the form

$$\frac{\langle k \rangle}{\lambda} = \frac{k_1^2 P(k_1)}{1 + k_1 \lambda \overline{\Theta}} + \frac{k_2^2 P(k_2)}{1 + k_2 \lambda \overline{\Theta}} \tag{15}$$

which can be solved for $\overline{\Theta}$ giving

$$\overline{\Theta} = \frac{1}{2\lambda}\left(\lambda - \frac{k_1 + k_2}{k_1 k_2}\right) + \frac{1}{\lambda}\sqrt{\frac{1}{4}\left(\lambda - \frac{k_1 + k_2}{k_1 k_2}\right)^2 + \frac{\langle k^2 \rangle \lambda - \langle k \rangle}{k_1 k_2 \langle k \rangle}} \tag{16}$$

where $\overline{\Theta}$ is only defined for transmission rates $\lambda > \lambda_c = \frac{\langle k \rangle}{\langle k^2 \rangle}$ as explained above. Using the definition of the average stationary density $\rho = \sum_k P(k)\rho_k$ and finally substituting $P(k_1) = p$ and $P(k_2) = 1 - p$, the order parameter in the MF approximation is given by

$$\overline{\rho} = \lambda \overline{\Theta}\left(\frac{pk_1}{1 + k_1 \lambda \overline{\Theta}} + \frac{(1-p)k_2}{1 + k_2 \lambda \overline{\Theta}}\right) \tag{17}$$

The critical exponents given by the leading order contributions of the relevant expressions are found to be the standard MF exponents [12] as outlined above.

This solution is plotted for the special case $k_1 = 2$ and $k_2 = 3$ with p varying from 0 to 1 in Fig. 1. As expected, MF theory qualitatively reproduces the features of the phase transition: There exists a critical rate λ_c below which the

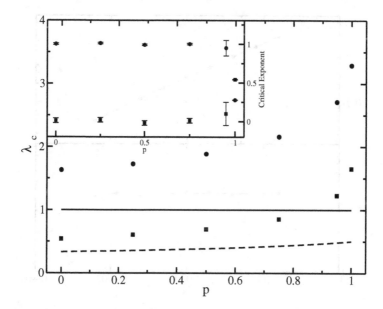

Fig. 2. The critical threshold λ_c obtained from MF theory (lines) and MC simulation (circles and squares) for the CP (circles and solid line) and the SIS model (squares and dashed line) for various values of p. Inset shows the critical exponents β (circles) and γ (squares) as a function of p.

stationary density is zero while it grows continuously for values above. For the CP, an analogous analysis yields a similar solution.

No direct comparison with numerical predictions for the order parameter is shown in Fig. 1 because of severe finite-size effects for networks which shift the simulation curve well above its true asymptotic position. We will however compare the critical threshold rate as well as some critical exponents in the next section.

4.2   Simulation Results

The critical thresholds λ_c for a range of values of p for both the CP and the SIS model were obtained via the QS simulation method outlined above. We used networks of sizes ranging from $N = 256 - 32768$ in QS simulation runs up to 10^8 time steps averaging over no less than 100 and up to a maximum of 1000 network realisations (the latter were required to minimise errors in light of strong sample-to-sample fluctuations for large values of p). The resulting critical thresholds are shown in Fig. 2 along with the MF predictions obtained from Eqs. (9) and (13). As expected from the definition of the two models, the CP threshold exceeds the one of the SIS model for a particular value of p which can be attributed to the reduction of the effective transmission rate by the local coordination number as in Eq. (12). Also, in the two cases of homogeneous connectivity, $p = 0$ and $p = 1$, the thresholds for the two models are expected to be simply related by a factor

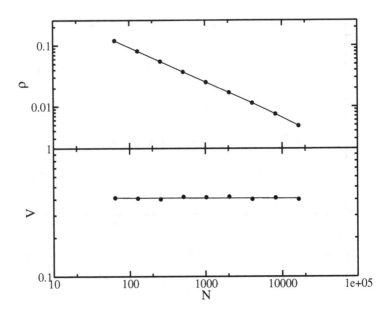

Fig. 3. The density of infection ρ (upper panel) and the corresponding fluctuations $V = N \left(\overline{\rho^2} - \overline{\rho}^2 \right)$ (lower panel) for various network sizes for the case $p = 0.5$ in the SIS model. Solid lines are best-fit regression lines to the scaling forms defined in Eq. (14).

of 3 and 2 (equal to the connectivity), respectively, as can be seen from Fig. 2. Note that the critical threshold for the SIS model in the quasi one-dimensional case ($p = 1$), $\lambda_c = 1.65$, is almost identical to the threshold for the CP on the 3-regular random network ($p = 0$) $\lambda_c = 1.63$.

Turning to the MF predictions in comparison to the MC results, one notes that they underestimate the true critical thresholds for all values of p and for both models. The difference between the MF approximation and the simulation results is more pronounced for the CP as compared to the SIS model.

The exponents β and γ were obtained by fitting data for the density of infected sites ρ and the corresponding fluctuations $V = N \left(\overline{\rho^2} - \overline{\rho}^2 \right)$ to the finite-size scaling form of Eq. (14). A typical set of data points is shown in Fig. 3 for the case $p = 0.5$ for the SIS model along with best-fit regression lines. Both quantities as a function of network size N show power-law behaviour with the expected MF exponents $\beta/2 = 0.5$ and $\gamma = 0$ indicating the validity of the MF approximation for this case. These values of critical exponents are plotted in the inset of Fig. 2. As p is further increased, strong sample-to-sample fluctuations for values beyond $p = 0.95$ render a precise analysis very complicated. For $p = 1$, the well-established 1D finite-size scaling exponents are recovered ($\beta/\nu_\perp = 0.253$ for $\rho(N)$ and $\gamma/\nu_\perp = 0.498$ for $V(N)$ [2]) as can be seen from the figure. In the transition region, error bars for exponents are large and our results give slight preference to the scenario of a discontinuous change in exponent values.

However, we feel that a very rapid yet continuous change of exponents towards the 1D values cannot be excluded.

5   Conclusion

We have investigated the CP and the SIS model, two paradigmatic stochastic spreading processes, in a network model which interpolates smoothly between an infinite-dimensional 3-regular random network and a linear chain through the variation of a single parameter p. The MF approximation yields a prediction for the critical threshold rate of an epidemic outbreak and critical exponents associated with the corresponding absorbing state phase transition. For no value of p does MF theory predict the true critical threshold as calculated from MC simulations. The predictions for critical exponents agree perfectly with simulations for a very wide range of p up to $p = 0.95$. Beyond this point, the analysis is complicated by strong sample-to-sample fluctuations. For $p = 1$ one recovers the established exponents for the one-dimensional case indicating a sudden crossover. While not being able to investigate the nature of this transition precisely, our simulations favour the scenario of a discontinuous change in the scaling exponents which reflects the abrupt change of the dimensionality of the network.

Acknowledgements. MC simulations were performed on the Cambridge University Condor Grid. SVF acknowledges financial support from the EPSRC and the Cambridge European Trust.

References

1. Harris, T.E.: Contact Interactions on a Lattice. Ann. Prob. 2, 969 (1974)
2. Marro, J., Dickman, R.: Nonequilibrium Phase Transitions in Lattice Models. Cambridge University Press, Cambridge (1999)
3. Hinrichsen, H.: Non-equilibrium critical phenomena and phase transitions into absorbing states. Adv. Phys. 49, 815 (2000)
4. Grassberger, P.: Directed percolation in 2+1 dimensions. J. Phys. A 22(17), 3673–3679 (1989)
5. Dorogovtsev, S.N., Mendes, J.F.F.: Evolution of networks. Adv. Phys. 51(4), 1075 (2002)
6. Albert, R., Barabáshi, A.L.: Statistical mechanics of complex networks. Rev. Mod. Phys. 74, 47 (2002)
7. Dorogovtsev, S.N., Goltsev, A.V., Mendes, J.F.F., Samukhin, A.N.: Spectra of complex networks. Phys. Rev. E 68, 046109 (2003)
8. Dorogovtsev, S.N., Goltsev, A.V., Mendes, J.F.F.: Critical phenomena in complex networks. eprint: arXiv:cond-mat, 0705.0010 (2007)
9. Pastor-Satorras, R., Vespignani, A.: Epidemic dynamics and endemic states in complex networks. Phys. Rev. E 6306(6), 066117 (2001)
10. Pastor-Satorras, R., Vespignani, A.: Epidemic spreading in scale-free networks. Phys. Rev. Lett. 86(14), 3200–3203 (2001)
11. Boguna, M., Pastor-Satorras, R., Vespignani, A.: Absence of epidemic threshold in scale-free networks with degree correlations. Phys. Rev. Lett. 90(2), 028701 (2003)

12. Castellano, C., Pastor-Satorras, R.: Non-mean-field behavior of the contact process on scale-free networks. Phys. Rev. Lett. 96(3), 038701 (2006)
13. Hong, H., Ha, M., Park, H.: Finite -size scaling in complex networks. Phys. Rev. Lett. 98, 258701 (2007)
14. de Oliveira, M.M., Dickman, R.: How to simulate the quasistationary state. Phys. Rev. E 71(1), 016129 (2005)
15. Joo, J., Lebowitz, J.L.: Behavior of susceptible-infected-susceptible epidemics on heterogeneous networks with saturation. Phys. Rev. E 69(6), 066105 (2004)
16. Lübeck, S., Janssen, H.K.: Finite-size scaling of directed percolation above the upper critical dimension. Phys. Rev. E 72(1), 016119 (2005)
17. Lübeck, S., Willmann, R.: Universal finite-size scaling behavior and universal dynamical scaling behavior of absorbing phase transitions with a conserved field. Phys. Rev. E 68, 056102 (2003)
18. de Oliveira, M., Dickman, R.: How to simulate the quasistationary state. Phys. Rev. E 71, 016129 (2005)

Weighted and Directed Network on Traveling Patterns

J.I.L. Miguéns[1] and J.F.F. Mendes[2]

[1] Economics, Management and Industrial Engineering Department, Aveiro
University, 3810 - 193 Aveiro, Portugal
joana.miguens@ua.pt
[2] Physics Department, Aveiro University, 3810 - 193 Aveiro, Portugal
jfmendes@fis.ua.pt

Abstract. The importance of weighted and directed networks is
brought into discussion. On this study we analyze the arrivals of international tourism (edges) over 206 countries and territories (nodes)
around the world, on the year 2004. Using tools from network theory
we characterize the topology and weighted properties of the resulting
network. International tourist arrivals are analyzed over *in* strength and
out strength flows, resulting on a highly directed and heterogenetic network. Remarkably the random network of connectivity is converted into
a power-law network of intensities. It is also shown how strategic positioning particularly benefit from market diversity and that interactions
among countries prevail on a technological and economic pattern, ques-
tioning the backbones of traveling driving forces. The network structure
may influence how tourism hubs, distribution of flows, and centralization
can be explored on strategic positioning.

Keywords: social networks, complex networks, traveling patterns, directed and weighted networks.

1 Introduction

The movement of tourists on a worldwide scale is responsible for a traveling
mobility of hundred millions tourist arrivals every year, representing the largest
movement of humans ever out of their usual environment, strongly influencing
local, regional, national and international economies, being one of the fastest
growing economic sector. Tourism is a consequence and a dynamic force on the
integration of world trade and markets, forming the global economy. But how is
this integration evolving? The nature of the connecting flows among countries
add some understanding about the dynamics of this network. Regardless the
crucial role of tourism, there is a lack of quantitative considerations of its flows,
although it is essential for understanding the self-organization of human traveling
patterns, and global wealth net flows.

Research on social networks has around 50 years, empirical and theoretically,
partly because social life is relational [1,2]. These studies contributed much for

P. Liò et al. (Eds.): BIOWIRE 2007, LNCS 5151, pp. 145–154, 2008.
© Springer-Verlag Berlin Heidelberg 2008

the clarity of the importance of relational systems. Such networks are represented as a set of *nodes* denoting people, companies, or other social actors, which are joined by *edges* the patterns of the relational structure, representing friendships, partnerships, collaborations, etc. A large variety of real world systems are structured in the form of networks, from social, biological, economic, infrastructure and information networks [3,4,5,6,7,8,9], also airline connections [10], financial relations [11,12,13], companies partnerships, ecological networks, movies actors, world trade, WWW [14], scientific collaboration network [15], human acquaintance patterns [16], among others [17,18,19]. Network theory have been build up largely from observation of the properties of many real world networks, and by comparatione of their structures.

Different theoretical perspectives on tourism recognize clusters and networks as one of the main competitive factors in tourism. It is increasing the amount of research on whether network perspective can be used to conceptually understand tourism networks. The conceptual and analytical framework of international tourism networks has been studied and tourism researchers have been introducing network analysis on measuring relationships and networks on tourism [20,21].

The international arrival of tourist is yearly measured by the World Tourism Organization (WTO, the major intergovernmental body concerned with tourism) over 208 countries and territories around the world [22], reaching a record of 763 million in 2004 (see Fig. 1). In this research we use techniques and indicators of network approach to study international tourism on the year of 2004. International tourist arrivals are analyzed to study *inbound* tourism and *outbound* tourism. *Inbound* tourism, involving the non-residents received by a destination country from the point of view of that destination. *Outbound* tourism, involving residents traveling to another country from the point of view of the country of origin.

Most complex networks share common properties that have common underlying structural principles [4,7,14]. Firstly we analyze the centrality of nodes on

(a)   (b)

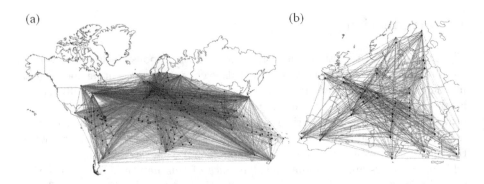

Fig. 1. Worldwide tourism departures and arrivals (a) on a country-to-country plot is displayed with an exponential grey scale according to the intensity of connections, and (b) tourist arrivals and departures in the European Union

a network, as more competitive nodes are recognize for having better strategic positions [9,23]. Centralization refers to the extent to which a network revolves around a single node, and also to the propensity of the node to diffuse information, knowledge or infections.

Degree centrality is one of the most used measures of node prominence [24], and equals the number of edges connected to it. The statistical characterization of real networks displays a large number of node degrees, k, and the appearance of hubs, nodes with large degree [25]. Additionally these networks show a power-law degree distribution, characterized by $P(k) \sim k^{-\theta}$ [7,14].

The techniques firstly applied to undirected and unweighted networks [19,18] are lately adapted to weighted and/or directed networks [5,26,27]. On social weighted networks is often relevant to assign a weight (strength) to each edge, measuring how good or strong is a relationship [2,16].

On this chapter the worldwide tourists arrivals network is analyzed. This chapter is organized as follows. The empirical analysis focuses on network topology (section 2), weighted analyze (section 3) and degree-degree correlation (section 3.1). Conclusion are drawn on section 4.

2   Network Topology

We used the data gathered by WTO over these 208 where countries and territories are considered *nodes*, N, and an *edge* exists from node i to node j when there are tourists from country i to country j. Notice that the network is directed, the edge from i to j is different from the edge from j to i, respectively $i \rightarrow j$ and $j \rightarrow i$. On our case we have 5775 edges, L, – representing arrivals of tourists from one country to another, on the year of 2004. On a directed network the nodes have *in* and *out* degree, where the *in* degree of a node i, $k_{in}(i)$, is the number of nodes directed to node i, and the *out* degree of i, $k_{out}(i)$, is the number of nodes that i is directed to. The *in* degree of a country is an indicator of its attractiveness has a destination country, *destination attractiveness indicator*, and the *out* degree of a country is an indicator of its emanation has a tourism origin country, *destination emanation indicator*.

On the average the shortest path length between countries is $l = 1.84$, and the diameter is 4, which are small values in accordance with a small–world effect $l \sim logN$. This means that any two countries have a high probability of being themselves connected, or that have very few intermediate country through each a connection is present. The small-world property has strong influence on the dynamics of the network, like spread of information, innovation, knowledge, promotion, or any other propagation process. The tourism international network is a giant component, so that all countries have a path or paths to any of the other countries. The fact of being a giant network and having a small shortest path length can imply fast transferring of knowledge and information.

An important statistical property to directed networks is reciprocity [9], meaning on the tourism network the appetency to exchange tourists. The links in the network are composed by 10% bidirectional links and 30% of asymmetric links. If

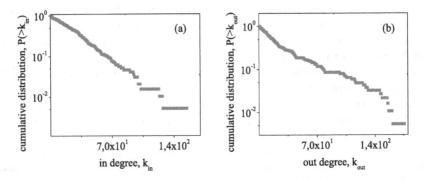

Fig. 2. Log-normal plot of degree distribution, for (a) *in* degree $P(> k_{in})$ and (b) *out* degree $P(> k_{out})$, with an exponential decay

country j has tourist arrivals from country i, then the probability that country i has tourist arrivals from j is only $\frac{1}{4}$, so the network is significantly directed. Notice also that 60% of all the pairs of countries are not connected to one another.

A fundamental aspect of real-world networks is the degree distribution [28], representing the distribution of the number of links of nodes. In binomial random graphs [7,8], nodes have similar degree, display an exponential network, with $P(k) \sim exp(k)$, decreasing exponentially fast, although many real-world networks have some nodes that are significantly more connected than others, many of those are scale free, having connectivity distributions that decay as a power law. A probable mechanism for this occurrence is preferential attachment [28], meaning that nodes with high degree are preferential. Network's topology displays the degree distribution which applied to tourist arrivals - directed network [29]- are studied two degree distribution functions, $P_{in}(k)$ representing the probability that a node has k nodes directed to itself, and $P_{out}(k)$ representing the probability that a node has a total of k edges to other nodes. Most networks have a power-law degree distributions [28], with $P(k) \sim k^{-\theta}$.

In our case, the *in* and *out* degree distributions decrease exponentially fast. Their cumulative distribution functions are represented on Fig. 2 (*a*) and (*b*), respectively $P_{in}(k)$ and $P_{out}(k)$. The topological network does not displaying power-law behavior, similar result on [18]. Contrarily, in other examples, in social [16], technological [14], economic [30], and biological networks [7], it was found a power-law degree distribution.

3   Weighted Analysis

The weighed analysis is essential because of weights heterogeneity. The network can be expressed by its adjacency matrix $A = \{a_{ij}\}$, dimension N × N , where $a_{ij} = 1$ if and only if there is an edge from i to j, and $a_{ij} = 0$ otherwise. The weighted adjacency matrix is $W = \{w_{ij}\}$, where w_{ij} equals the flow from i to j. Notice that w_{ij} represents the weight of the edge $i \to j$ and w_{ji} represents

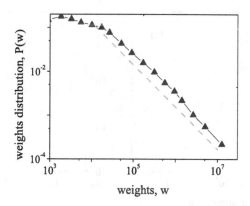

Fig. 3. On the plot is showed the flow of tourists with a power-law behaviour with $P(w) \sim w^{-\gamma}$ where $\gamma = 1.55$, with dominance of hubs

the weight of the edge $j \rightarrow i$, so w_{ij} and w_{ji} are different. The range of the weights goes from 0 to 19.369.677 with an average value of 81.813, revealing a high heterogeneity of weights. See Fig. 1.

The probability distribution function of the weights, $P(w) \sim w^{-\gamma}$ has a power-law behavior, with exponent $\gamma = 1.55$, see Fig. 3.

It is also relevant to study the strength of the nodes, which on a directed network each node has *in* strength, $s_{in}(i)$ (eq. 1), and *out* strength, $s_{out}(i)$ (eq. 2). It measures the strength of the nodes on relation to the total weight of their connections. On the tourist arrivals network *in* strength represents the *inbound* tourism, and *out* strength represents the *outbound* tourism. Strength is a measure of centrality for weighted networks:

$$s_{in}(i) = \sum_{j \in v(i)} w_{ij}, \qquad (1)$$

$$s_{out}(i) = \sum_{j \in v(i)} w_{ji}. \qquad (2)$$

The *in* strength distribution and *out* strength distribution functions are also fitted by a power-law, respectively $P(s_{in}) \sim s_{in}^{\gamma_{in}}$ and $P(s_{out}) \sim s_{out}^{\gamma_{out}}$, where $\gamma_{out} = 1.95$ and $\gamma_{in} = 1.9$, represented on Fig. 4.

Scale free networks, that follow a power-law distribution, have the ability to change scale in order to meet any level of demand. Tourism, among economic sectors has one of the fastest grow rates, and WTO forecasts that international arrivals are expected to reach nearly 1.6 billion [22]. So, two consequences are expected, the network is growing due to a scaling up, with an increase of flows intensity and/or due to a scaling out by new connections between countries.

A power-law behavior of $P(w)$, $P(s_{in})$ and $P(s_{out})$ have a strong structural meaning of the network, describing the way weights, and strength centrality, *inbound* and *outbound* tourism, are distributed. The weights and strengths range

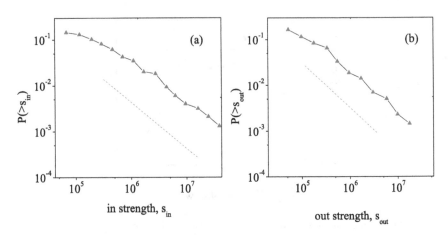

Fig. 4. Inbound distribution (a) $P(s_{in}) \sim s_{in}^{\gamma_{in}}$ with $\gamma_{in} = 1.9$, and (b) outbound distribution, $P(s_{out}) \sim s_{out}^{\gamma_{out}}$ with $\gamma_{out} = 1.95$

on a large spectrum of values, and the heavy-tailed distribution implies that nodes have a certain probability of having large strength values, where the average of all intermediate values has no meaning.

The observations on topological and weighted network reveal different structural results, therefore the relation of topological and weighted flows is studied in more detail, $s(k_{in})$ and $s(k_{out})$. The result is depicted on Fig. 5. On the *in* function:

$$s(k_{in}) = (k_{in})^{\beta_{in}}, \tag{3}$$

where $\beta_{in} = 1.1$. For $\beta = 1$ degree and weight are independent [19]. So $S(k_{in})$ and k_{in} are close to independent, revealing a very small relation between them. On the other side, $s(k_{out})$:

$$s(k_{out}) = (k_{out})^{\beta_{out}}, \tag{4}$$

$\beta_{out} = 1.75$, revealing a strong relation between *out* strength and *out* degree. This means that *outbound* tourism increases with *out* degree.

Interestingly, when analyzing the diversity of the market and its strength, comes out that inbound and outbound tourism have distinguished outcomes on Fig. 5. Even so, both have a power-law behaviour, $s(k) = k^{\beta}$, and unavoidable fluctuations. The diversification of outbound markets ($> k_{out}$) has a strong and positive increase on total outbound tourism $s(k_{out}) = k_{out}^{\beta_{out}}$, with a power of $\beta_{out} = 1.75$, meaning that the flow grows 1.75 faster than the degree. On the relation between the inbound tourism and its market diversification, $s(k_{in}) = k_{in}^{\beta_{in}}$ with $\beta_{in} = 1.1$, the relation is close to linear and it comes out that both quantities carry almost the same information [19]. It is concluded that the outbound tourism particularly benefits from market diversity.

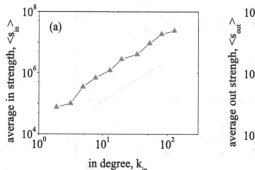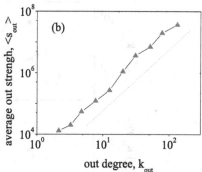

Fig. 5. Intensity plays an important role on network behaviour. The relation between degree and strength is closely independent on (a) inbound tourism, $s(k_{in}) = k_{in}^{\beta_{in}}$ with $\beta_{in} = 1.1$, but has a (b) strong relation on outbound tourism, $s(k_{out}) = k_{out}^{\beta_{out}}$ with $\beta_{out} = 1.75$.

3.1 Degree-Degree Correlations

We turn now to question in which sense do countries couple with one another. Is it in some sort of random choice, or is there a preference on the way they link with each others, meaning a choice that makes some connections more probable than others. In a social context is usually observed an assortative mixing [31], observed when the nearest neighbours of nodes with high degree have also high degree. On economic, technological and biological context is generally observed disassortative mixing, observed when the nearest neighbours of nodes with high degree have low degree.

In evolving network, degree-degree correlations are almost always strong. To measure the correlation on the network over degree, one may also study the average nearest-neighbors degree. This measures the tendency of node i to be connected to nodes with the same degree,

$$k'_{nn}(i) = \frac{1}{k_i} \sum_{j \in v(i)} k_j, \qquad (5)$$

where $v(i)$ denotes the set of neighbors of i. Considering that our network is directed, we correlate the *in* degree of node i with the *out* degree of its neighbors,

$$k_{nn}(i) = \frac{1}{k_i^{in}} \sum_{j \in v(i)} k_j^{out}. \qquad (6)$$

We can also average the over nodes of the same degree:

$$k_{nn}(k) = \frac{1}{NP(k)} \sum_{k_i=k} k_{nn}(i). \qquad (7)$$

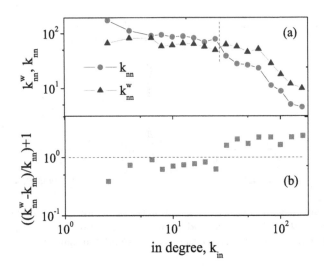

Fig. 6. Log-log plot of *in* degree – *out* degree correlations, over **(a)** *in* degree, both unweighted $k_{nn}(k)$ and weighted $k_{nn}^w(k)$ correlations, over k_{in}. **(b)** Comparing weighted and topological degree correlation. For low degrees $k_{nn}^w(k) < k_{nn}(k)$ and for high degrees $k_{nn}^w(k) > k_{nn}(k)$. Low (high) degree nodes have their edges with large weight directed from nodes with low (high) degree.

The assortativity mixing is represented by a growth of $k_{nn}(k)$ with k and disassortative mixing if represented by a decreasing of $k_{nn}(k)$ with k. This happens when nodes with high degrees have mainly neighbors with low degree. The international tourism network displays disassortative mixed. This behavior is mostly detected on transportation networks, providing a pattern where the hubs connect to the small degree nodes at the periphery of the network [31].

Degree-degree correlation for a weighted network is given by [18],

$$k_{nn}^w(i) = \frac{1}{s_i^{in}} \sum_{j \in v(i)} w_{ji} k_j^{out}. \tag{8}$$

$k_{nn}^w(k)$ measures the local weighted average of neighbors degree. The spectrum of the worldwide tourist flows on topological (equation 7) and weighted degree-degree correlations (equation 8) if represented on Fig. 6 (a).

For $k_{nn}^w(k) > k_{nn}(k)$ the edges with the larger weight are directed to the neighbors with larger degrees, and $k_{nn}^w(k) < k_{nn}(k)$ the edges with the larger weight are directed to the neighbors with lower degrees [18]. The weighted degree-degree correlation is slightly decreasing (Fig. 6 (b)), following the same behavior as the topological correlation, but with a slower slop. For low degrees $k_{nn}^w(k) < k_{nn}(k)$ and for high degrees $k_{nn}^w(k) > k_{nn}(k)$, meaning that low degree nodes have their edges with large weight directed from nodes with low degree, and high degree nodes have their edges with large weight directed from nodes with high degree.

4 Conclusion

In this study was addressed the importance of weighted and directed measurements, applied to the worldwide tourist arrivals network. The research shows a power-law behaviour on the weights covering 4 orders of magnitude. It describes short travelling range to long travels, on a global scale, surprisingly having affinity correlations typical from technological and economic networks which question the cultural backbone of tourism and travel. The scaling behavior of tourism flows, on a power-law refers to the self-organization of world trends, where disassortative correlations particularly reveal the influence of economic flows and spread of technologic and knowledge across international borders. The power-law nature of the weighted analyses contrary to the random topology opens a new class of networks. This brings us to a more general question; on how highly heterogenic and directed real-world networks hide some sort of preferential growing and hub-like structure on a random topological structure.

References

1. Travers, J., Milgram, S.: An Experimental Study of the Small World Problem. Sociometry 32, 425–443 (1969)
2. Granovetter, M.S.: The Strength of Weak Ties. The American Journal of Sociology 78, 1360–1380 (1973)
3. Albert, R., Barabasi, A.-L.: Statistical mechanics of complex networks. Rev. Mod. Phys. 74, 47–97 (2002)
4. Dorogovtsev, S.N., Mendes, J.F.F.: Evolution of networks. Adv. Phys. 51, 1079–1187 (2002)
5. Newman, M.E.J.: The Structure and Function of Complex Networks. SIAM Review 45, 167–256 (2003)
6. Jeong, H., Mason, S.P., Barabasi, A.-L., Oltvai, Z.N.: Lethality and centrality in protein networks. Nature 411, 41–42 (2001)
7. Dorogovtsev, S.N., Mendes, J.F.F.: Evolution of networks: From biological nets to the internet and WWW. Oxford Univ. Press, Oxford (2003)
8. Strogatz, S.H.: Exploring complex networks. Nature 410, 268 (2001)
9. Wasserman, S., Faust, K.: Social Network Analysis. Cambridge University Press, Cambridge (1994)
10. Guimerà, R., Mossa, S., Turtschi, A., Amaral, L.A.N.: The worldwide air transportation network: Anomalous centrality, community structure, and cities global roles. Proc. Natl. Acad. Sci. USA 102, 7794–7799 (2005)
11. Garlaschelli, G., Battiston, S.: The scale-free topology of market investments. Physica A 350, 491 (2005)
12. Caldarelli, G., Battiston, S., Garlaschelli, D., Catanzaro, M.: Emergence of Complexity in Financial Networks. Lecture Notes in Physics 650, 399 (2004)
13. Tibely, G., Onnela, J.-P., Saramaki, J., Kaski, K., Kertesz, J.: Spectrum, Intensity and Coherence in Weighted Networks of a Financial Market. Physica A 370, 145–150 (2006)
14. Albert, R., Jeong, H., Barabasi, A.-L.: Internet: Diameter of the World-Wide Web. Nature 401, 130–131 (1999)

15. Barabasi, A.L., Jeong, H., Neda, Z., Ravasz, E., Schubert, A., Vicsek, T.: Evolution of the social network of scientific collaborations. Physica A 311, 590–614 (2002)
16. Newman, M.E.J.: The structure of scientific collaboration networks. Proc. Natl. Acad. Sci. USA 98, 404–409 (2001b)
17. Brockmann, D., Hufnagel, L., Geisel, T.: The scaling laws of human travel. Nature 439, 462–465 (2006)
18. de Montis, A., Barthelemy, M., Chessa, A., Vespignani, A.: The structure of Inter-Urban traffic: A weighted network analysis. Environment and Planning B: Planning and Design 34(5), 905–924 (2007)
19. Barrat, A., Barthelemy, M., Vespignani, A.: The architecture of complex weighted networks. Proc. Natl. Acad. Sci. USA 101, 3747–3752 (2004)
20. Gibson, L., Hall, M., Lynch, P., Mitchell, R., Morrison, A., Schreiber, C.: Micro-Clusters and Networks: The Growth of Tourism (Advances in Tourism Research Series) (2006)
21. Shih, H.-Y.: Network characteristics of drive tourism destinations: an application of network analysis in tourism. Tourism Management 27, 1029–1039 (2006)
22. WTO Pbcn: New Yearbook of Tourism Statistics (World Tourism Organization Pbcn) (2006)
23. Burt, R.S.: Structural Holes: The Social Structure of Competition. Harvard University Press (1995)
24. Freeman, L.C.: Centrality in Social Networks Conceptual Clarification. Social Networks 1, 215–239 (1979)
25. Goltsev, A.V., Dorogovtsev, S.N., Mendes, J.F.F.: Critical phenomena in networks. Phys. Rev. E 67, 026123, 1–5 (2003)
26. Park, S.M., Kim, B.J.: Dynamic behaviors in directed networks. Phys. Rev. E 74, 026114 (2006)
27. Barrat, A., Barthelemy, M., Vespignani, A.: Weighted Evolving Networks: Coupling Topology and Weight Dynamics. Phys. Rev. Lett. 92, 228701 (2004)
28. Albert, R., Barabási, A.-L.: Emergence of scaling in random networks. Science 286, 509–512 (1999)
29. Krapivsky, P.L., Rodgers, G.J., Redner, S.: Degree Distributions of Growing Networks. Phys. Rev. Lett. 86, 5401–5404 (2001)
30. Chowell, G., Hyman, J.M., Eubank, S., Castillo-Chavez, C.: Scaling laws for the movement of people between locations in a large city. Phys. Rev. E 68, 066102 (2003)
31. Newman, M.E.J.: Mixing patterns in networks. Phys. Rev. E 67, 026126 (2003)

Communication Networks in Insect Societies

Stamatios C. Nicolis*

Department of Zoology, Universit of Oxford
South Parks road, OX1 3PS Oxford
snicolis@math.uu.se

Abstract. We show in this paper how communication networks can form spontaneously in social insects through self-organisation. Different models associated to food recruitment and clustering behaviour are analysed giving rise to temporal and spatio-temporal patterns. The conditions under which the response is optimised are also identified.

Keywords: Social insects, food recruitment, clusters, pattern formation.

1 Introduction

In this work we show how communication networks can form spontaneously in social insects through a mechanism of self-organisation. We shall deal with two representative situations in which "wiring" of network nodes is associated, succesively, to a process of decision making when a population is confronted to several options; and a process giving rise to pattern formation. Actually, these two processes share some common features such as:

- Competition between different sources of information.
- Amplifying interactions between constituting units reflected by the presence of positive feedback loops.

As a result, their study can be carried out using similar methodologies. In each case a key objective will be to establish the link between the characteristics of single individuals, the collective response at the scale of the network and the environmemental constraints. Furthermore, the conditions under which the collective response can be optimized will be identified.

The philosophy our approach is as follows: We first identify from experimental data the principal actors likely to play a role, and the nature of the interactions present. We translate this information in the form of a mathematical model describing the evolution of the relevant variables on the basis of hypotheses made on the underlying mechanisms. The model is analysed or simulated numerically using, typically, the tools of nonlinear science [1,2] and stochastic processes [3], and the results are confronted to the observations. Once validated, the model is extended to new types of situations or used to design new experiments. Eventually, the iterative process leads to a qualitative understanding and to a quantitative characterisation of the phenomenon at hand.

---

* *Present address*: Uppsala University, Mathematics department, P.O. Box 480, SE-751 06 Uppsala, Sweden.

P. Liò et al. (Eds.): BIOWIRE 2007, LNCS 5151, pp. 155–164, 2008.

2 Collective Decision Making Associated with Food Recruitment

Our first case study is collective decision-making associated with food recruitment in ant colonies. In nature there exist two types of communication in social insects. Direct interactions between individuals, like in bees [4,5,6] and interactions by chemical means, like in ants [7,8,9]. We will be interested in this latter case.

The mechanisms of recruitment can be described in the following manner: An ant discovers one food source, eats and returns to the nest laying down a chemical substance known as pheromone. The resulting "pheromone trail" has two functions : alert the other individuals to get out the nest and lead them to the food source.At each trip ants reinforce the trail and the source ends thus being exploited in a collective manner.

As the case of the presence of only one source is not common in nature ant colonies are usually confronted to the choice and the competition between multiple food sources. We first neglect individual and environmental variability and focus on the nature of the "traffic" established along the trails leading to the food sources. The key point allowing us to model this situation is to realise that the direct contacts between individuals can be neglected compared to their response to the pheromone concentration present in a given trail. The principal variables are thus the pheromone concentration C_i rather than the number of individuals present on the various trails i at a given time. A generic model capturing the main features of competition between the sources can then be written as [10,11].

$$\frac{dC_i}{dt} = \phi q_i \frac{(k + C_i)^\ell}{\sum_{j=1}^s (k + C_j)^\ell} - \nu_i C_i \qquad i = 1, ...s \qquad (1)$$

The first, positive, term corresponds to the attractiveness of trail i over the others. Its mathematical function has been applied and quantified for various ant species, in particular *Lasius niger* [12,13,14], *Linepitema humile* [10,15], army ants [16,17] and *Messor pergandei* [18,15]. Here ϕ is the flux of individuals getting out the nest (related to the size of the colony), q_i the quantity of pheromone laid down by an ant on the trail i, k a concentration threshold beyond which the pheromone is effective and ℓ the sensitivity of the choice of a particular trail. The latter parameter is also viewed as the strength of cooperativity between individuals. The second, negative, term corresponds to the disapperance of the pheromone on the trail i through, for instance, evaporation (parameter ν_i),

Resolving eq. (1) in the simplest case of two sources in competition and using the parameters associated with the species *Lasius niger* leads to the bifurcation diagrams depicted in Figs 1a,b, according to whether the two sources have the same or have different richness. For the case of equivalent sources, we see that there is an equal exploitation of the two sources for small values of pheromone deposition. After a threshold value the system switches to a preferred exploitation of one or other source. For different sources there is a preferred exploitation of the richest source for small values of q_1. After a threshold value, the system switches to the possibility to exploit the richest source or the poorest one.

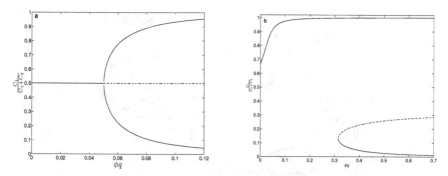

Fig. 1. Bifurcation diagrams of the steady-state solutions of equations (1) as a function of q_1 in the case $q_2/q_1 = 1$ and $q_2/q_1 = 0.5$. Parameter values $k = 6, \emptyset = 0.01s^{-1}, v_1 = v_2 = 1/2400s^{-1}$ and $s = 2$.

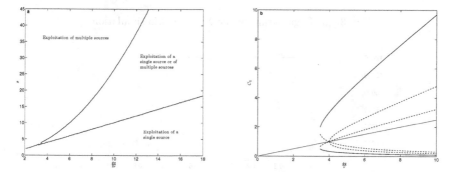

Fig. 2. (a) State diagram representing the parameter regions of different modes of exploitation of ressources in the case of $\emptyset/v = 10$. Parameter values as in Fig. 1a. (b) Bifurcation diagram of the steady-state solutions of eq. (1) in the case of four equal sources in competition. Parameter values as Fig. 1a.

In a more realistic situation where more than two sources are present, we have been able to build the state diagram of the different strategies of exploitation. Fig. 2a shows the number sources against a set of parameters. We see that for small values of ϕ and high number of sources there is an equal exploitation of all the sources. In intermediate values of the parameters and rather high number of sources the colony selects with some probability one source or all sources. For still higher values of the parameter but a low number of sources, the colony finally selects one source preferentially. As an example, the bifurcation diagram in the presence of four sources is shown in Fig. 2b.

3   Optimising the Exploitation of the Resources

We now extend the above scheme to account for variability. The question is, whether by incorporating fluctuations in the framework of a situation where we

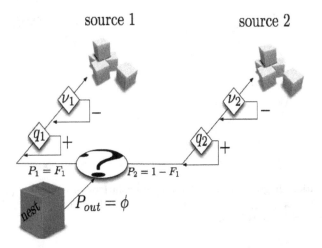

Fig. 3. (a) Organigram of the Monte Carlo simulation

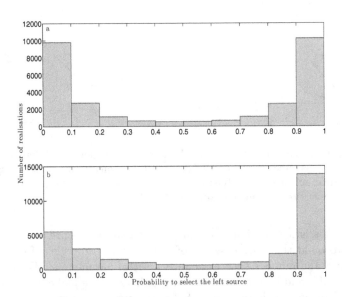

Fig. 4. Probability histograms corresponding to the case of sources of equal richness (a) and unequal sources (b). In (b), the left source is the richest one. Parameter values as in Fig. 1b.

have two food sources of different richness in competition, we obtain access to behaviours not amenable to a mean-field description.To this end we adopt a Monte Carlo approach, in which the process of interest is simulated directly on the basis of a certain set of rules [19,20].

Let us take for simplicity the case of two food sources. Fig. 3 summarises the organigram of the simulation. The individuals get out of the nest with a

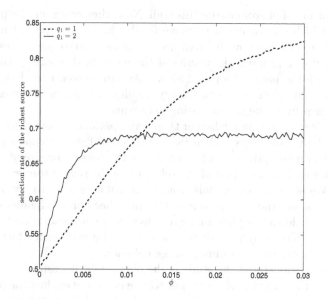

Fig. 5. Selection rate of the richest source as a function of the parameter ø for q_2/q_1 = 0.75. Parameter values as in Fig. 1.

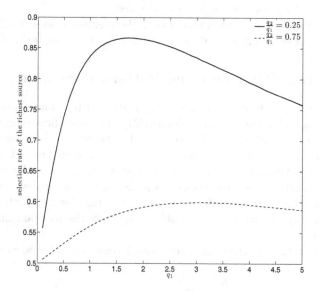

Fig. 6. Selection rate of the richest source as a function of the parameter q_1 for ø = $0.1s^{-1}$. Parameter values as in Fig. 1.

probability equal to *phi*. At the bifurcation point, the first ant has an equal probability to go to the source 1 or 2. Suppose it goes to source one, then drops a certain quantity of pheromone on the trail 1, which has a positive feedback on the concentration of pheromone on this trail. Now, time going on, the pheromone evaporates, which constitutes a negative feedback. The second ant arriving at the bifurcation point has now different probabilities to go to one of these sources.

We first try to reproduce the results of the mean field model by running the simulation with the parameters used above. As can be seen from Fig. 4 there is an agreement. The plots represent the probability histograms of the selection of one particular source, the sources being here equal.

We now raise the question, whether there exist parameter ranges for which the selection of the richest source when two sources of different richness are offered to the colony can be optimised. First let us see the role of the size of the colony, paremeter ϕ. Fig. 5 shows a plot of the selection of the richest source against this parameter. We see that individuals from small colonies have to lay down more pheromone to select the richest source. On the other hand individuals from big colonies may lay down less pheromone to select the richest source and, moreover, in a better way. This may provide a rationale for the well known fact that trail recruitment in ants mainly occurs in large colonies.

But the key parameter of the study is the pheromone deposition. In the graph of Fig. 6 where the selection of the richest source is plotted against the pheromone deposition, we see the existence of an optimal value of the parameter for which the exploitation is maximal. We also see that the maximal exploitation is higher when the difference between the source quality is larger. This reflects the fact that in this limit competition is less pronounced.

The existence of an optimal q shows that there exists a noise level that maximises the response in terms of efficiency-an at first sight counterintuitive result.

4   Clustering Behaviour and Pattern Formation

We now proceed to a second case study, the spatial pattern formation associated with clustering behaviour in ant colonies [21]. The specific context chosen is that of clustering of dead ants in preferred locations, to which one may refer as "cemeteries". Ant corpses are dropped in the circumference of an arena of a certain size. Living ants are entering in the arena through the center of it. They pick the corpses and drop them in another area. The process ends up with different clusters on the circumference, thereby giving rise to a pattern. Thanks to a mean field model we have been able to identify the mechanisms presiding in its formation.

The model pertains to the class of reaction-diffusion models and can be written as

$$\frac{\partial c}{\partial t} = \Omega(c, a)$$

$$\frac{\partial a}{\partial t} = -\Omega(c, a) + D\frac{\partial^2 a}{\partial x^2} \tag{2}$$

As can be seen, it involves two variables, $c(x, t)$, the density of corpses and $a(x, t)$, the density of carrying ants. D is a mobility coefficient and $\Omega(c, a)$ is the sum of three different terms

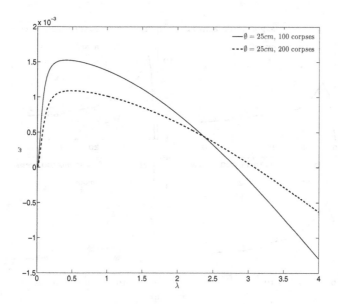

Fig. 7. Stability analysis of the homogeneous steady sate of eqs. (2). Solution of the characteristic equation, providing the rate of growth of small perturbations, as a function of the wave number λ for two different experimental conditions. For parameter values see [21].

$$\Omega\left(c,a\right) = v\left(k_d a + \frac{\alpha_1 a \phi_c}{\alpha_2 + \phi_c} - \frac{\alpha_3 \rho c}{\alpha_4 + \phi_c}\right) \qquad (3)$$

corresponding, respectively, to the spontaneous dropping, density-dependent dropping and density dependent picking. Here, k_d represents the spontaneous dropping rate per laden ants, ρ, the density of non carrying ants , v, the mean velocity of ants and α_1, α_2, α_3, α_4, empirical constants. ϕ_c is a nonlocal term introducing a short range interaction between workers and corpses:

$$\phi_c = \frac{1}{2\Delta} \int_{x-\Delta}^{x+\Delta} c\left(z\right) dz \qquad (4)$$

where Δ is a small radius of perception within which workers can detect corpses. All The parameters have been measured from experiments. Resolving eq.(2) leads to a unique, spatially uniform steady state solution(random deposition of corpses). Testing the stability of this solution leads to the graph of Fig. 7 where we see the existence of a finite range of unstable modes corresponding to positive values of the solution of the characteristic equation(indicative of the growth of small perturbations). In other words, for the two experimental conditions shown here there is an instability leading to a spatially inhomogeneous solution possessing a characteristic wavelength.

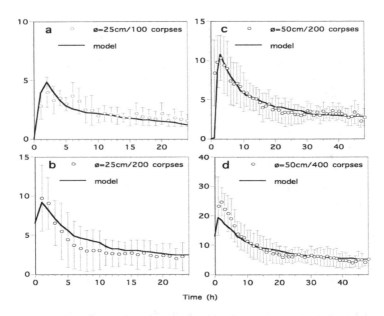

Fig. 8. Mean number of clusters (a cluster contains at least five corpses) as a function of time obtained from 20 integrations of the model equations (full lines) and from experiments (average and SD are given for six experiments per condition in four experimental conditions). The initial conditions(spatial distribution of corpses) are randomly set around the value steady state c_s. For parameter values see [21].

Finally we integrate the model and plot the number of clusters containing at least five corpses against time. As can be seen from Fig. 8, for the two conditions shown there is a very good agreement between experment and the model.

5   Conclusions

Amplifying communications play an important role in the organisation of animal societies, particularly in social insects. Amplification implies the presence of a nonlinear element in the dynamics. As seen in this work, one of the principal manifestations of such nonlinearities is the emergence of complex self-organising behaviours at a collective scale resulting from the interactions between individuals each of which has access only to local information [22]. As a rule the population has the choice between several options, a fact reflected by the multiplicity of the solutions of the underlying evolution laws. This confers to the response a marked plasticity against the environmental constraints. It is instructive to view the collective organisation achieved in this way as a network, in which the nodes are the values of the state variables at different spatial locations and the edges (accounting for the way these nodes are wired) represent the interactions. In this representation the wiring in the network associated to the problem considered in Sec. 4 is a nearest neighbour and hence a short-range one, whereas in the problem of Secs 2 and 3 connections are mediated by the pheromone and are thus long-ranged. In both cases the nonlinearities are manifested in the form of feedback loops, connecting a node to itself either directly or through a circuit involving other nodes.

Although obtained in the specific context of social insect biology, our results are in many respects paradigmatic. As such, they are expected to apply to a variety of other biological [22] or artificial processes [23]. Of special interest is the possibility of building and controlling mixed societies composed of animals and of artificial agents [24,25] with potentially far reaching applications in, among others, agriculture and farming.

References

1. Nicolis, G., Prigogine, I.: Self-organization in Non-Equilibrium Systems. Wiley, New York (1977)
2. Nicolis, G.: Introduction to non-linear science. Cambridge University Press, Cambridge (1995)
3. Gillespie, D.T.: Markov processes. Academic Press, London (1992)
4. Camazine, S., Sneyd, J.: A model of collective nectar source selection by honey bees. Self-organization through simple rules. J. Theor. Biol. 149, 547–571 (1991)
5. Seeley, T.D., Camazine, S., Sneyd, J.: Collective decision making in honey bees: how colonies choose among nectar sources. J. Theor. Biol. 28, 547–571 (1991)
6. Seeley, T.D.: The wisdom of the hive. Harvard University Press (1995)
7. Sudd, J.H.: Communication and recruitment in *Monomorium pharaonis*. An. Behav. 5, 104–109 (1957)
8. Holldobler, B., Wilson, E.O.: The ants. Springer, Heidelberg (1991)
9. Robson, J.F.A., Traniello, S.K.: Trail and territorial communication in social insects. In: The Chemical Ecology of Insects. Chapman & Hall, New York (1995)
10. Deneubourg, J.L., Goss, S.: Collective patterns and decision making. Behav. Ecol. Sociol. 1, 277–290 (1989)
11. Nicolis, S.C., Deneubourg, J.-L.: Emerging Patterns and Food Recruitment in Ants: an Analytical Study. J. Theor. Biol. 198, 575–592 (1999)
12. Beckers, R., Deneubourg, J.L., Goss, S.: Trail laying behavior during food recruitment in the ant *Lasius niger*. Insectes Sociaux 39, 59–72 (1992)
13. Beckers, R., Deneubourg, J.L., Goss, S.: Trails and U-turns in the selection of a path by the ant *Lasius niger*. J. Theor. Biol. 159, 397–415 (1992)
14. Beckers, R., Deneubourg, J.L., Goss, S.: Modulation of trail laying in the ant *Lasius niger* (Hymenoptera: Formicidae) and its role in the collective selection of a food source. J. Ins. Behav. 6, 751–759 (1993)
15. Goss, S., Deneubourg, J.L.: The self-organising clock pattern of *Messor pergandei*. Insectes Sociaux 36, 339–346 (1989)
16. Franks, N.R.: Teams in social insects: group retrieval of prey by army ants (*Eciton Burchelli* Hymenoptera: Formicidae). Behav. Ecol. Sociol. 6, 425–429 (1986)
17. Franks, N.R., Gomez, N., Goss, S., Deneubourg, J.L.: The blind leading the blind in army ant raid patterns: testing a model of self-organization (Hymenoptera: Formicidae). J. Ins. Behav. 4, 583–607 (1991)
18. Rissing, S.W., Wheeler, J.: Foraging responses of *Veromessor pergandei* to changes in seed production. Pan-Pacific Entomology 52, 63–72 (1976)

19. Nicolis, S.C.: Dynamique du recrutement alimentaire et de l'agregation chez les insectes sociaux. PhD thesis (2003)
20. Nicolis, S.C., Detrain, C., Demolin, D., Deneubourg, J.-L.: Optimality in Collective Choices: a Stochastic Approach. Bull. Math. Biol. 65, 795–808 (1991)
21. Theraulaz, G., Bonabeau, E., Nicolis, S.C., Sole, R.V., Fourcassie, V., Blanco, S., Fournier, R., Joly, J.L., Fernandez, P., Grimal, A., Dalle, P., Deneubourg, J.L.: Spatial patterns in ant colonies. PNAS 99, 9645–9649 (2002)
22. Camazine, S., Deneubourg, J.L., Franks, N.R., Sneyd, J., Bonabeau, E., Theraulaz, G.: Self-organized Biological Superstructures. Princeton University Press, Princeton (2001)
23. Caprari, G., Colot, A., Siegwart, R., Halloy, J., Deneubourg, J.-L.: InsBot: Design of an Autonomous Mini Mobile Robot Able to Interact with Cockroaches. In: Proceedings of IEEE International Conference on Robotics and Automation. ICRA 2004, New Orleans, pp. 2418–2423 (2004)
24. De Shutter, G., Theraulaz, G., Deneubourg, J.-L.: Animal-robots collective intelligence Annals of Mathematics and Artificial Intelligence 31, 223–238 (2001)
25. (2006), http://leurre.ulb.ac.be/index2.html

The Topological Fortress of Termites

Andrea Perna[1,2], Christian Jost[1], Sergi Valverde[1,3], Jacques Gautrais[1,2],
Guy Theraulaz[1], and Pascale Kuntz[2]

[1] Centre de Recherches sur la Cognition Animale, CNRS UMR 5169, Université Paul
Sabatier, 118 route de Narbonne, 31062 Toulouse Cedex 4, France
andrea.perna@cict.fr
[2] Laboratoire d'Informatique de Nantes Atlantique, Site Ecole Polytechnique de
l'Université de Nantes, La Chantrerie, BP50609, 44306 Nantes cedex 3
[3] ICREA-Complex Systems Lab, Universitat Pompeu Fabra, Dr. Aiguader 80, 08003
Barcelona, Spain

Abstract. Termites are known for building some of the most elaborate architectures observed in the animal world. We here analyse some topological properties of three dimensional networks of galleries built by termites of the genus *Cubitermes*. These networks are extremely sparse, in spite of the fact that there is no building cost associated with higher connectivity. In addition, more "central" vertices (in term of betweenness or degree) are preferentially localised at spatial positions far from the external nest walls (more than in a null network model calibrated to exactly the same spatial arrangement of vertices). We argue that both sparseness and the particular spatial location of "central" vertices may be adaptive, because they provide an ecological advantage for nest defence against the attacks from other insects.

Keywords: spatial networks, social insects, morphogenesys, complex systems, patterns.

1 Introduction

Social insect societies have attracted much attention because of the complex level of coordination and organization of their collective activities. These allow them to perform complex tasks, such as finding the shortest path to a food source [9] and building elaborate nests [12]. These abilities do not result from planning or supervision, but emerge from the direct or indirect interactions between insects. Social insect colonies are cooperative distributed systems [18] that have known an extraordinary ecological success during the last 100 million years [13]. Understanding how social insect societies work can help us to design efficient artificial distributed systems that at some level of description share similar needs and constraints to those ruling social insect colonies.

Termites in particular are known for building some of the most amazing architectures observed in the animal world. The mounds built by some species can reach up to 6 meters of height against a size of the individual insect of the order of the millimeter[10]. Even when the nest is comparatively small in size,

P. Liò et al. (Eds.): BIOWIRE 2007, LNCS 5151, pp. 165–173, 2008.

it can present an extremely complex form and internal organization [6]. The architectural refinement of these structures reflects their ecological importance. In fact, the nests protect insects from dessiccation, and contribute to maintain a stable internal environment [20,15,14]. The nest is also important for protecting insects from attacks by a variety of natural enemies. Because of their biomass, their chemical composition, the absence of a hard exoskeleton and their concentration in a single place, termites represent an interesting food resource for several predators [11].

A nest is also a network of interconnected chambers and galleries inside which all the displacements and the activities of insects take place. These networks are completely self-organized and emerge from the work of thousands to millions of individuals [19]. Because of their ecological importance, these networks may present particular topological properties.

Nests built by the termite genus *Cubitermes* are constructions made out of clay whose shape resembles that of a mushroom of 20-30 cm of height (figure 1-A). Inside, the nests are filled with chambers of similar size, interconnected by openings and short corridors, (visible in figure 1-B, where the nest is represented in a virtually cut reconstruction). The diameter of corridors is constant everywhere and just a little bit larger than the size of a "soldier" (soldiers are large termites of the same species specialized for defence).

Chambers and corridors in these nests can be mapped respectively into vertices and edges of a network. These are spatial networks, where each vertex occupies a precise position in the three-dimensional space, and edges are real physical connections. Each vertex in the network can be characterised by its topological properties in relation to the network (degree, centrality, belonging to a network motif), but also by the characteristics associated to its spatial position or arrangement. For instance figure 1-D colors in red the vertices identifying chambers adjacent to the external nest wall and in white the others. Henceforth

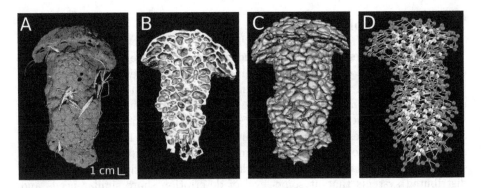

Fig. 1. A. Picture of one nest. B. Virtual cut of the nest. Chambers and corridors are clearly visible. C. Virtual cast of the same nest. Only the chambers and corridors are represented. D The connectivity network for the same nest. Each vertex corresponds to a chamber and each edge to a corridor. Nodes coloured in red indicate that the chamber is adjacent to the external nest surface.

we will call accordingly "internal" those vertices that are not adjacent to the external nest wall and "peripheral" the vertices that are adjacent to the external nest wall.

These attributes of a vertex only depend on the spatial position of the chamber, not on its topological localization inside the network. In the present paper we explore whether some network statistics correlate nevertheless with the internal or peripheral location of the vertices.

The rest of the paper is organized as follows. Section 2 presents our nest database and the segmentation process we developed to extract graphs from the tomographies of the nests. Section 3 introduces some indicators which characterize the properties of the nest transportation networks. In section 4 we define a null comparison model for networks with similar spatial constraints to those observed in our nests. Some strong effects of spatial embedding on network topology are described in section 5, while section 6 compares more directly some network properties of the real networks against the same properties in the null model. The results are discussed from an ecological viewpoint in section 7.

2   Network Extraction

Six *Cubitermes* nests were used. The nests, labeled M9, M10, M11, M12, M18 and M19 belonged to private or public collections (Natural History Museums of Paris and Toulouse) and originated from different locations in Central African Republic and Cameroon. One of the nests, M19, was still under construction when it was collected. This can be inferred from the fact that this nest still lacks the cap that is visible on top of all "complete" nests (old nests can also have more than one single cap, as is the case for nest M11 which has three caps).

Nests were imaged and reconstructed into 3-D virtual volumes using X-ray tomography with a medical scanner. For every nest, we extract its transportation network $G = (V, E)$ (fig 1.D). In this network, a vertex $v \in V$ represents a physical chamber and an edge $\{v_i, v_j\} \in E$ depicts a physical corridor between chambers v_i and v_j. We can reconstruct the network G as follows. "Cores" were defined as small empty regions located farther than about 1.5 mm from nest walls (internal or external). Given the narrow diameter of the corridors (less than ~ 0.5 mm in radius) these cores never belong to a corridor, but either to the space outside the nest or to a chamber. The chamber cores were identified as the network vertices. They were then concurrently dilated to progressively fill their surrounding empty space (stopping at walls). At some point, a dilated core also crams into its outgoing corridors and gets in touch with the others dilated cores coming from the other end of the corridor. In this case, an edge between the vertices was created, corresponding to the physical corridor.

To obtain the virtual network corresponding to the case when all physically adjacent chambers would be connected, the dilation of the same chamber cores was repeated in the pure 3D-space up to the complete space filling (neglecting now the nest walls). When two dilated cores got in touch, they were marked as adjacent. The results of this automatic segmentation were verified and manually corrected.

3   Network Measures

The topological properties of the graph G=(V,E) associated with a gallery network can be characterized by a variety of indicators [2]. We here focus on three features: the network sparseness, the communication efficiency and the betweenness centrality. Let us denote by k_i the degree of the vertex $v_i \in V$ defined by the number of edges incident to v_i. The average degree $\langle k \rangle = \sum k_i/N$ indicates the level of network sparseness (N is the total number of vertices).

We can also measure the patterns of connections involving more than one vertex. In particular we compute path length and betweenness centrality. Let d_{ij} be the number of edges on a shortest path between the vertices v_i and v_j i.e. a path between v_i and v_j with a minimum edge number. The average path length $< L >$ on G is defined as follows:

$$< L >= \frac{1}{N} \sum_{v_i,v_j \in V} d_{ij} \tag{1}$$

Average path length is a measure of network spread or compactness. For instance, networks with low $< L >$ can be efficiently navigated.

Normalized betweenness centrality $C_B(v)$ of vertex v is defined as follows:

$$C_B(v) = \frac{2}{(N-1)(N-2)} \sum_{v_i \neq v \neq v_j \in V} \frac{\sigma_{i,j}(v)}{\sigma_{i,j}} \tag{2}$$

where $\sigma_{i,j}$ is the number of shortest paths from v_i to v_j, and $\sigma_{i,j}(v)$ is the number of shortest paths from v_i to v_j that pass through v [1,3]. Vertices that have high betweenness centrality scores lie on important communication paths, and for this reason are important to guarantee fast displacements in the network.

Given the normalized betweenness centrality, one can compute the central point dominance [8], which is a measure of the maximum betweenness of any point in the graph: it will be 0 for complete graphs and 1 for "wheel" graphs (in which there is a central vertex that includes all shortest paths). Let v^* be the vertex with the largest betweenness centrality; then, the central point dominance is defined as

$$C_B' = \sum_{v_i \in V} \frac{C_B(v^*) - C_B(v_i)}{N-1} \tag{3}$$

4   Definition and Generation of Random Spatial Networks

The chambers of Cubitermes nests completely fill the space without leaving significant gaps for the passage of long distance corridors. As a consequence, in the corresponding transportation network, connections exist only between vertices representing physically adjacent chambers. The classical models of random graphs [7] are not well-suited here for a null comparison test because they do not take into account spatial constraints. We here propose a model which fits the physical specifities of the termite nests.

Let us define the Maximal Embedded Graph (MEG) as the network $G_M = (V, E_M)$ with the same set of vertices as in the gallery network G and where there is an edge $(v_i, v_j) \in E_M$ if chambers v_i and v_j are adjacent (separated by a single wall), independently whether they were also physically connected by a corridor or not. When there are no long distance connections between non-adjacent chambers (i.e. long chains in the associated graph), a MEG contains the whole edge set allowed by the constraints of the spatial embedding. Hence, all the possible networks compatible with these constraints can be generated as subgraphs of a MEG. We here restrict ourselves to graph spanners of G_M [16] i.e. connected graphs $G = (V, E_S)$ with the same vertex set V, and an edge set $E_S \subset E_M$ subset of E_M.

The topological assumptions required by our model have been checked in the real *Cubitermes* nests. There are no long-range connections in these nests; all the connections take place between physically adjacent chambers. In addition, the edges of the real nests are a subset of the MEG edges in the vast majority of the cases. These properties were verified for nests M9, M11, M12, M18 and M19; in each of these nests around 99% of the edges were also edges of the MEG. The remaining about 1% of the edges connected chambers adjacent at a corner, and for this reason these edges were not marked as adjacent by the automatic segmentation procedure. We added these edges to the MEG. Nest M10 displays a different behaviour because termites have built some long-range corridors on the external surface of the nest that link to distant chambers. For this reason, nest M10 was not used in some of the analyses.

For each nest of our corpus, we have compared its internal topological properties with those of 10000 random spanners of the MEG (for the same nest). These were obtained by first generating random spanning trees [17] of G_M, and then inserting additional edges (chosen with uniform probability among the edges in G_M) until we reach the same number of edges as in the observed gallery network G.

5 Effect of the Spatial Embedding on the Graph Average Degree

In *Cubitermes* networks, the maximum degree of the vertices is limited by the physical constraints: no vertex can have a higher degree than its associated degree in the MEG. Consequently, the average network degree is smaller than the average degree of G_M. Figure 2-A reports for the six nests of the corpus, the degree in the MEG and in the real network, showing that the average degree of real networks is significantly smaller than the average degree of G_M. This indicates that the strong maximum limit to vertex connectivity imposed by spatial embedding has hardly any effect on these networks. Indeed, real termite networks have connectivity near the percolation threshold, suggesting that these networks tend to minimize connectivity.

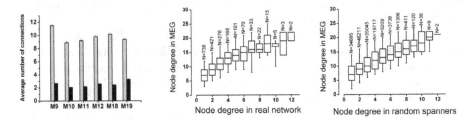

Fig. 2. Left graph: Average degree ($\pm SE$) in G_M (gray) and in real networks (black). Middle graph: For all vertices in the six nests, the degree of the vertex in G_M is binned according to the degree of the same vertex in the real network, in the abscissae. Right graph: The same as in the middle graph, but for the vertices of random spanners of each nest. The reported data for the random spanners are computed on a subset of 100 spanners for each nest.

However, spatial embedding can affect connectivity indirectly. For instance the degree of each vertex in the real network could correlate with its degree in the MEG. This is actually observed in our networks: figure 2-B bins together all the vertices with the same degree in the real networks (here, vertices from different nests are not differentiated) and reports a box-plot of the corresponding degree in G_M. The two are clearly correlated. If the vertices in G were simply a random sample from the vertices in G_M, a similar result should be found also for random spanners. The same statistics is reported for random spanners of G_M of each of the five nests in fig 2-C. The correlation of vertex degree with the degree in G_M is similar. However, real networks have a higher proportion of nodes with high degree (10 or more) than random spanners (see numbers N in the label associated to each box).

6   Effect of the Spatial Embedding on the Network Centrality

The correlation of the vertex degrees in real networks and in the corresponding G_{Ss} suggests that external vertices (representing chambers adjacent to the external nest surface, see page 167) should have lower degree than internal vertices, for the sole effect of spatial constraints. This tendency should be shared both by real networks and random spanners. However, we wonder if the same tendency is stronger in real networks than in random spanners. In order to investigate this issue, for each nest, we have computed two indicators for each vertex : (1) the ratio k^* between the degree the vertex has in the real nest and its average degree for 10000 random spanners of G_M defined on the same nest, (2) the ratio C_B^* between its betweenness centrality in the real nest and its average betweenness centrality for the random spanners. Internal and peripheral vertices are binned separately. If real networks do not particularly tend to segregate central vertices (in the network) across space, this ratio should be similar for both internal and

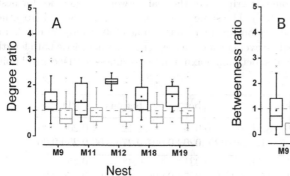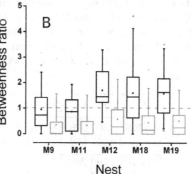

Fig. 3. Left. box plots representing the ratio, for each vertex, of the degree that the vertex has in the real network of galleries and its average degree in the spanners. In black: internal vertices; in gray, peripheral vertices. Right. the same as in the left layer, but with ratios of betweenness. The dashed line represents the expected ratio of 1 if there were no differences between internal and peripheral vertices.

peripheral vertices. Figure 3-A reports the distribution of the ratio k^* associated with the degree for internal and peripheral chambers in the five nests. In all the nests this ratio is higher in internal than in peripheral vertices. Higher degree does not necessarily correlate with a higher betweenness. Figure 3-B reports the distribution of the ratio C_B^* associated with the betweenness centrality. Again, the ratio is always higher in internal than in peripheral vertices.

Table 1 reports some additional statistics of real gallery networks, and of the random spanners for the same nest. Average path lengths in real networks are shorter than in random spanners with the same average degree, indicating that some optimization process is at work in termite gallery networks, which makes these networks more efficient than random networks.

7 Discussion: An Ecological Perspective

Networks of chambers and galleries inside termite nests are one of the few described examples of 3D self-organized spatial networks.

These networks are extremely sparse. In general, network sparseness is not an advantageous feature because it decreases the efficiency of displacement, increases the likelihood of traffic jams and decreases robustness to random failures or occlusions.

As a candidate rationale, the sparseness may result from a cost associated with adding new edges. In termite networks of galleries however, it is unlikely that there is such a building cost. First it is sufficient to dig a hole in a thin wall to obtain a new connection. Second, one of the nests, M19 which appeared to be still under construction has more connections (higher average degree in table 1) than the other nests. This indicates that the final topology is probably

Table 1. The table reports some descriptors of the real networks of galleries for nests M9, M11, M12, M18 and M19 (RN). The same descriptors are also reported (together with standard errors) for 10000 random spanners of the same nests (rand). V: number of vertices; E: number of edges; <k>: average node degre; <L>: characteristic path length; <bet>: average node betweenness; betmax: maximum betweenness; CPD: central point dominance.

nest	V	E	<k>	<L>	<bet>	betmax	CPD
M9-RN	507	676	2.67	8.51	0.015	0.22	0.20
M9-rand	507	676	2.67	11.18±0.42	0.020±0.001	0.27±0.06	0.25±0.06
M11-RN	260	280	2.15	9.11	0.031	0.48	0.45
M11-rand	260	280	2.15	15.66±1.56	0.057±0.006	0.52±0.04	0.47±0.04
M12-RN	183	233	2.55	8.19	0.040	0.36	0.33
M12-rand	183	233	2.55	9.92±0.65	0.049±0.004	0.40±0.07	0.35±0.06
M18-RN	287	342	2.38	8.40	0.026	0.32	0.30
M18-rand	287	342	2.38	11.00±0.69	0.035±0.002	0.34±0.07	0.31±0.07
M19-RN	268	437	3.26	7.89	0.026	0.35	0.33
M19-rand	268	437	3.26	9.47±0.38	0.032±0.001	0.31±0.05	0.28±0.05

reached by removing already existing connections, implying that the cost would be associated rather with edge removal than edge addition.

A better explanation of the sparseness of these networks could be their importance for defence. *Cubitermes* termites (like several other termite genera) are often attacked by ants that prey on the nests [5,11]. The reaction of *Cubitermes subarquatus* in response to attacks by the ant *Centromyrmex bequaerti* is accurately described [4]. In a first phase, if the ants find access to one chamber, each corridor leading from that chamber to other chambers of the nest is defended by a "soldier" termite. It is widely believed that the particular diameter of the galleries, exactly the size of a soldier, is an adaptive feature evolved to maximize the success of defence. Soldiers have specialized jaws and phragmotic heads (large heads that in some termite and ant species are be used to plug the nest entrance; Gr. $\phi\rho\alpha\gamma\mu\text{o}\varsigma \simeq fence, barrier$) that can effectively block narrow termite tunnels against ant entry. An additional defence strategy followed by termites consists in setting back to intact parts of the nest and plugging with earth the galleries that are still open to the invaded part.

The very low connectivity of the gallery network could hence present an high adaptive defence value, since it is often sufficient to close a single corridor to isolate an individual chamber, or a great part of the nest from outside. If this holds true, then the connectivity should be even sparser at the periphery of the nest (where defence ought to be the more efficient) than in the central part of the nest (where transportation ougth to be the prime concern). This hypothesis is clearly supported in the present case.

References

1. Anthonisse, J.M.: The rush in a directed graph. Technical report, Stichting Matematisch Centrum, Amsterdam (1971)
2. Boccaletti, S., Latora, V., Moreno, Y., Chavez, M., Hwang, D.: Complex Networks: Structure and Dynamics. Physics Reports-review section of Physics Letters 424(4-5), 175–308 (2006)
3. Brandes, U.: A Faster Algorithm for Betweenness Centrality. Journal of Mathematical Sociology 25(2), 163–177 (2001)
4. Dejean, A., Fénéron, R.: Predatory Behaviour in the Ponerine Ant, *Centromyrmex bequaerti*: a Case of Termitolesty. Behavioural Processes 47, 125–133 (1999)
5. Dejean, A., Durand, J.L., Bolton, B.: Ants Inhabiting *Cubitermes* Termitaries in African Rain Forests. Biotropica 28(4), 701–713 (1996)
6. Desneux, J.: Les Constructions Hypogées des *Apicotermes* Termites de l'Afrique Tropicale. Annales du Musée Royal du Congo Belge Tervuren 17, 7–98 (1952)
7. Erdös, P., Rényi, A.: On Random Graphs. Publicationes Mathematicae 6, 290–297 (1959)
8. Freeman, L.C.: A Set of Measures of Centrality Based on Betweenness. Sociometry 40, 35–41 (1977)
9. Goss, S., Aron, S., Deneubourg, J.L., Pasteels, J.M.: Self-organized Shortcuts in the Argentine Ant. Naturwissenschaften 76, 579–581 (1989)
10. Grassé, P.P.: Termitologia, Tome 2: Fondation des Sociétés, Construction. Masson, Paris (1984)
11. Grassé, P.P.: Termitologia, Tome 3: Comportement - Socialité - Écologie - Evolution - Systematique Masson, Paris (1986)
12. Hansell, M.: Animal Architecture. Oxford University Press, USA (2005)
13. Hölldobler, B., Wilson, E.O.: The ants. Belknap Press of Harvard University Press, Cambridge (1990)
14. Korb, J., Linsenmair, K.E.: Ventilation of termite mounds: new results require a new model. Behavioral Ecology 11, 486–494 (2000)
15. Lüscher, M.: Der Sauerstoffverbrauch bei Termiten und die Ventilation des Nestes bei *Macrotermes natalensis* (Haviland). Acta Trop. 12, 289–307 (1955)
16. Peleg, D., Schäffer, A.: Graph Spanners. Journal of Graph Theory 13, 99–116 (1989)
17. Propp, J., Wilson, D.: How to Get a Perfectly Random Sample from a Generic Markov Chain and Generate a Random Spanning Tree of a Directed Graph. Journal of Algorithms 27, 170–210 (1998)
18. Theraulaz, G., Bonabeau, E.: Coordination in Distributed Building. Science 269(5224), 686–688 (1995)
19. Theraulaz, G., Bonabeau, E., Deneubourg, J.L.: The Origin of Nest Complexity in Social Insects. Complexity 3(6), 15–25 (1998)
20. Turner, J.S.: The Extended Organism: the Physiology of Animal-built Structures. Harvard University Press, Cambridge (2000)

Evolutionary and Temporal Dynamics of Transcriptional Regulatory Networks

M. Madan Babu

MRC Laboratory of Molecular Biology, Hills Road, Cambridge CB2 0QH, UK
madanm@mrc-lmb.cam.ac.uk

Abstract. Transcriptional regulation is a key mechanism that allows cells to make the appropriate amount of proteins at the right time. This is mediated by transcription factors that respond to specific signals and regulate expression of the relevant genes. The set of all regulatory interactions in a cell can now be investigated and this is best represented as the transcriptional network where nodes represent transcription factors or target genes and edges represent regulatory interactions. In this manuscript, I will first discuss the current understanding of the organization of such networks from the model organisms *E. coli* and yeast. I will then demonstrate that such networks are extremely dynamic and adapt rapidly to changing environments. This will be illustrated by discussing the changes in the network structure in two different time scales: (i) those that occur during different cellular conditions and (ii) those that occur across different organisms living in diverse environments.

Keywords: network, evolution, dynamics, transcription and gene regulation.

1 Introduction

Over the last century, research in the area of biology has revealed that proteins in a cell rarely function in isolation. Instead, they interact with other macromolecules to form complex networks to co-ordinate various processes both in space and time. Though taking a reductive approach to investigate biological systems has undoubtedly provided us with a wealth of information, to understand how a complex system such as a cell functions, one needs to go beyond individual proteins and start to investigate the set of all interactions mediated by the components in a cell. There are some systems in biology where it is now possible to investigate such questions and one such system, which will be the subject of this chapter, is the transcriptional regulatory network.

Representing interactions between biological molecules as a network provides us with a conceptual framework that allows us to identify general principles that govern these complex systems [1]. A network is best represented as a graph that is made up of nodes, which denote the components, and links, which denote the interaction between the components. On the level of a whole cell, one could describe many different types of biological networks [1]. For instance, (i) the protein interaction network where nodes represent proteins and links represent physical interaction between the proteins, (ii) metabolic networks where nodes represent small molecules and links represent direct enzymatic conversion between the small molecules and (iii) the

P. Liò et al. (Eds.): BIOWIRE 2007, LNCS 5151, pp. 174–183, 2008.

transcriptional regulatory network, where nodes represent transcription factors (TFs) or target genes (TGs) and directed edges represent regulatory interaction where the transcription factor regulates the expression of the target gene [2]. In this chapter, I will discuss transcriptional regulatory networks from the model prokaryote *Escherichia coli* and the model eukaryote *Saccharomyces cerevisiae*. I will first describe the current understanding of the organization of such regulatory networks. I will then describe the general principles that emerge from the investigation of the changes in the structure of such transcriptional regulatory networks in two very different time scales: (i) the time scale of an individual generation, which is in the order of a few hours for most unicellular organisms, *i.e.*, the changes in network structure within an organism during different cellular conditions and (ii) the time scale of evolution of new organisms, which is in the order of millions of years. *i.e.*, the changes in network structure across different organisms.

2 Organization of the Transcriptional Regulatory Network

The transcriptional regulatory network is a representation of the blue print of gene expression program in an organism [2]. In the same way as in most complex systems, the transcriptional network is made up of a basic unit. The components of the basic unit consist of a transcription factor (TF) and a target gene (TG), with an edge between them denoting that the expression of the target gene is regulated by the factor (Fig 1a). The basic units do not occur in isolation but are interlinked to form small patterns of inter-connections. These small patterns of regulatory interactions referred to as network motifs form the building blocks of such networks (Fig 1b) [3, 4]. These motifs themselves are highly inter-connected and give rise to the transcriptional regulatory network, which is the set of all transcriptional interactions in a cell (Fig 1c).

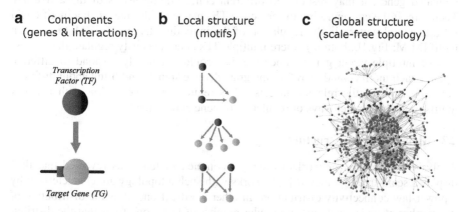

a Components
(genes & interactions)

b Local structure
(motifs)

c Global structure
(scale-free topology)

Transcription Factor (TF)

Target Gene (TG)

Fig. 1. Structure of the transcriptional regulatory network. (a) The basic unit consists of a regulatory interaction (gray arrow) between a transcription factor (black circle) and a target gene (gray circle). (b) The basic unit form small patterns of regulatory interactions called network motifs. The three most commonly occurring motifs are shown here: Feed-forward motif (FFM, top), Single input motif (SIM, middle) and Multiple input motif (MIM, bottom). (c) The set of all transcriptional regulatory interactions is referred to as the transcriptional regulatory network.

Analysis of the topological properties of such networks has revealed that they can be investigated at least at two levels: (i) the local level, where such networks have been shown to form ordered patterns of interactions referred to as network motifs, which have been proposed to perform specific information processing tasks [5], and (ii) global level, where it has been shown that such networks have a scale-free topology. Such a topology is characterized by the presence of a few transcription factors that regulate an unusually large number of target genes and a large number of transcription factors which regulate only a small number of target genes and is believed to confer robustness to such networks [1].

2.1 Local Network Structure

A network motif is defined as a small pattern of inter-connections that recur at many different parts of the network at frequencies much higher than what is expected by chance when compared to random networks of similar size [5]. Analysis of the transcriptional networks of *E. coli* and yeast has revealed the presence of three distinct motifs, each of which has distinct regulatory properties in the control of gene expression [5]. The three commonly occurring motifs in the transcriptional network are (i) Feed-forward motif (FFM; Fig 1b, top) where a top-level transcription factor regulates both the intermediate-level TF and the target genes, and the intermediate-level TF regulates the target gene. If both TFs are activators, such a connectivity pattern might ensure that the target gene is expressed only when persistent signal is received by the top-level transcription factor. Since the concentration of the intermediate TF should be built up for the regulation of the final target gene, random fluctuations and noise in activation of the top-level TF is filtered and does not get propagated. (ii) Single input motif (SIM; Fig 1b, middle) where a single TF regulates the expression of several target genes simultaneously. Depending on the promoter strength of the regulated genes, it may respond to different concentration levels of the active TF. Therefore, if the concentration of the active TF changes with time, such a motif could set a temporal pattern in the regulation of the individual targets. (iii) Multiple input motif (MIM; Fig 1b, bottom) where multiple TFs simultaneously regulate the expression of multiple target genes. Since the TFs could potentially respond to different signals, such motifs could therefore integrate diverse signals and bring about differential expression of the relevant targets. Thus regulation of genes via such network motifs provides distinct ways of regulation of gene expression [5].

2.2 Global Network Structure

Analysis of unrelated networks of several complex systems has revealed that they display a scale-free topology [6]. Networks with such a topology are characterized by a power-law connectivity distribution. In other words, if one plots the distribution of the number of nodes making a particular number of links, one finds that the distribution can be best fitted using a power-law equation of the type $y = ax^b$. Such a distribution is indicative of the presence of few highly influential TFs that regulate expression of several genes and a large number of TFs which regulate a few genes. The highly influential TFs are referred to as global regulators, or regulatory hubs and their presence contributes to the inherent robustness of such a topology [7, 8].

Robustness is the ability of complex systems to function even when the structure of the system is perturbed significantly [9]. A scale-free topology is robust because random inactivation of genes will most likely affect the TFs which regulate a few genes because these occur in very high numbers. This would still leave a central, highly connected sub-network that may still be functional. However, the downside of such a network structure is that they are vulnerable to targeted attacks of hubs. *i.e.,* targeted removal of the very highly connected nodes will result in the collapse of the system in to small sets of isolated fragments that no longer interact with each other [7]. Therefore, the highly connected proteins are believed to be crucial for the robustness and functioning of the regulatory network [7].

3   Temporal Dynamics of Transcriptional Regulatory Networks

Though investigations of regulatory networks in *E. coli* and yeast have uncovered key features of the local and global network structure [10-12], it should be realized that such features have been investigated largely on a static regulatory network. To investigate how the topology of the regulatory network changes within an organism across different conditions, we integrated gene expression data with the static regulatory network for yeast to obtain condition specific sub-networks [13]. This allowed the elucidation of active regulatory networks for five different conditions: cell cycle, and sporulation – both of which are developmental regulatory programs in a cell and DNA damage, stress response and diauxic shift – all three of which are regulatory programs that are important for survival (Fig 2a). Having identified the active sub-networks across the different conditions, we systematically investigated the changes in the local and global network topology to identify general principles behind the change in the topology of the active network structure.

3.1   Temporal Dynamics of Local Network Structure

Apart from the described functions of the network motifs, they also display distinct kinetic properties. The feed-forward motif is a slow-acting and indirect regulatory motif because this involves an intermediate-level TF for the regulation of the target genes. While the SIM and the MIM are fast acting and direct regulatory motifs because there are no intermediate TFs and hence the signal can be transmitted into a regulatory change relatively quickly [5]. Investigation of the active sub-networks across the different cellular conditions describing the two main types of regulatory program (development and survival) revealed a striking trend. The SIM motif was preferentially used in regulatory programs that enable survival and the FFM is preferentially used in the networks that govern developmental changes such as sporulation and cell-cycle (Fig 2b). This make intuitive sense as during stress conditions, a fast transfer of signal would allow efficient response and hence contribute to survival of the organism. Whereas in developmental regulatory programs, such as cell-cycle or sporulation, the regulation of genes under a feed-forward motif ensures that the next stage in the process is not initiated until a persistent signal from the previous stage is received [13]. Taken together, this suggests that the network motifs that allow for execution of regulatory events with distinct kinetic profiles are preferentially used in the different transcriptional programs.

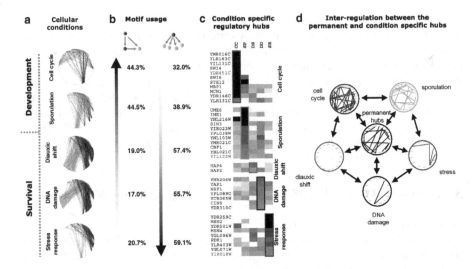

Fig. 2. Temporal dynamics of transcriptional regulatory network in yeast. (a) The active transcriptional regulatory network in a given cellular condition for the five major cellular processes is shown. These cellular conditions can be grouped into those that are involved in development and survival response. For each regulatory network that is active in a particular cellular condition, the transcription factors are shown in the top arc, the target genes are shown in the bottom arc and regulatory interactions are shown as a line connecting the two. (b) Preferential usage of network motifs in the regulatory programs governing development and survival. The numbers represent the fraction of active regulatory interactions forming a particular motif in that cellular condition. Feed-forward motif is preferentially used in developmental regulatory programs whereas single input motifs are used more frequently in transcriptional programs involved in survival. (c) Condition specific hubs under the five different conditions are shown. The gene name of the hubs is shown on the left. For each hub, the row to the right indicates the number of target genes regulated in each condition (column; CC: cell cycle; SP: sporulation; DS: diauxic shift; DD: dna damage; SR: stress response). Darker cells represent high number of regulated target genes. For instance, YMR016C is a transcription factor that regulates a large number of genes during cell cycle (black box under CC) but regulates almost no genes in the other conditions (white boxes under SP, SR, DS, DD) (d) Network of regulation between TFs in yeast reveals the extensive inter-regulation between the permanent and condition-specific hubs. More inter-regulation among TFs in CC and SP suggest a hierarchy in gene regulation whereas much less inter-regulation between TFs during survival response suggests a much flatter hierarchy.

3.2 Temporal Dynamics of Global Network Structure

Since the structure of the active transcriptional network could vary when cellular conditions change, it is important to understand the differences in the global network structure. In particular, do the active networks still display a scale-free topology? Do different proteins emerge as global regulatory hubs or does the same protein remains as a hub across different conditions? Investigation of the global structure of the active sub-networks across the different conditions revealed that they all still display a scale-free topology even when conditions change.

Investigation of the regulatory hubs across the different conditions revealed an important finding which is that there are two major classes of regulatory hubs: (i) permanent hubs and (ii) condition-specific hubs. While the permanent hubs are those regulators that affect expression of several genes independent of the cellular condition, the condition specific hubs only regulate a large number of genes under specific cellular conditions. The latter are generally TFs that initiate a developmental program or trigger a cellular response (Fig 2c).

An investigation of how often global regulatory hubs regulate each other revealed that there is an extensive inter-regulation between hubs that govern developmental regulatory programs (Fig 2d). This suggests a hierarchy in gene regulation which might be important for execution of the distinct phases in a developmental program. However, very little inter-regulation between the hubs were noticed in the active networks that govern survival, clearly suggesting a much flatter hierarchical structure which transfers signals rapidly into changes in gene expression. Though such a striking difference existed between the two major cellular programs, extensive inter-regulation between the condition-specific and permanent hubs were observed (Fig 2d). Taken together, this suggests that the transcriptional network of an organism is an extremely dynamic structure which has the potential to initiate cellular processes by triggering key regulatory hubs to respond to distinct cellular cues governing processes such as development and survival.

4 Evolutionary Dynamics of Transcriptional Regulatory Networks

While the analysis of the dynamic nature of the network structure provides information about the flexibility of network across different conditions, an investigation of which components are conserved between organisms provides us with a fundamental insight into how such complex regulatory systems evolve across different organisms. To understand the dynamics of changes in the network structure at the evolutionary time-scale, we used the *E. coli* transcriptional network and re-constructed sub-networks which are evolutionarily conserved in over 170 different prokaryotic genomes [14]. These organisms live in very different environmental conditions and are related to *E. coli* at varying levels of evolutionary distance. This therefore provided us with the information to investigate how the local and global network structure changes over time and adapts to changing environmental conditions.

4.1 Evolutionary Dynamics of Local Network Structure

In theory, network motifs may evolve in two major ways. They may (i) be retained or lost as a single unit or (ii) individual components which make up the network motif may be retained or lost, resulting in a partial motif being conserved. Previous analysis of the evolution of the protein interaction network have shown that network motifs tend to be retained or lost as a unit as retaining partial motifs is unlikely to be of any functional advantage [15]. Our analysis of the evolution of network motifs in transcriptional networks revealed that motifs are not retained or lost as whole units but individual components are retained or lost, thereby resulting in a partial motif being conserved [14]. A closer analysis of the patterns of loss and gain however suggested a very striking pattern, which is that by losing or gaining specific TFs, orthologous genes in different organisms can be embedded in different motifs and this change is dictated by the environment in which they live (Fig 3a).

a   Inter-conversion between motif via loss or gain of TFs

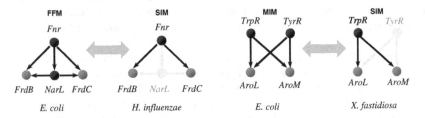

b   Evolutionary dynamics of local network structure

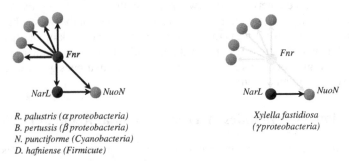

c   Evolutionary dynamics of global network structure

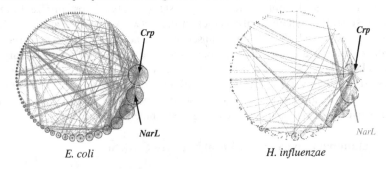

Fig. 3. Evolutionary dynamics of transcriptional regulatory networks in prokaryotes. (a) Loss or gain of TFs could result in the same gene being regulated under very different motif context. Light gray arrows represent absence of interaction. Light gray circles represent absence of a TF. (b) Organisms that are distantly related to *E. coli*, which is a γ proteobacteria, but share similar environments conserve network motifs. Whereas organisms that are closely related to *E. coli* but live in different environments rapidly lose components of network motifs. Light gray arrows represent absence of interaction. Light gray circles represent absence of a TF. (c) Condition specific hubs can be lost or replaced during evolution. In this representation of the network, TFs are shown in the centre of the concentric circles and TGs are shown around them. Lines represent regulatory interactions and the TFs in the larger concentric circles represent the hubs. The *E. coli* regulatory network is shown on the left and the conserved regulatory network in *H. influenzae* is shown on the right. Note the absence of NarL, a condition specific hub, in *H. influenzae*.

An analysis of the conserved networks across the different organisms living under different environmental conditions revealed two interesting principles (i) evolutionarily closely related organisms that have dissimilar environmental life-style do not conserve network motifs and (ii) evolutionarily very distant organisms that live in similar environments conserve network motifs (Fig 3b). In other words, the analysis revealed that organisms with similar lifestyle tend to conserve network motifs and may hence regulate orthologous target genes in a similar manner.

4.2 Evolutionary Dynamics of Global Network Structure

Since the local network structure displayed dynamic changes during the course of evolution, we investigated if the global network structure remains the same or differs in organisms living under different conditions. In principle, the global regulatory hubs, which are key elements in maintaining the overall network structure, could evolve in two possible ways: (i) the hubs could be conserved across the organisms simply because losing this may affect fitness and survival or (ii) the hubs could be lost or replaced by other regulators that may respond to a different signal in the environment. Previous studies on the protein interaction network have shown that hubs are essential proteins and removal can be deleterious to the organism [16]. Interestingly, for the transcriptional regulatory network, we identified that condition specific hubs can be lost or replaced during evolution (Fig 3c). Though such hubs can be lost or replaced in evolution, we observed that the conserved sub-networks still displayed a scale-free topology.

This suggests an important principle which is that orthologous TFs in organisms living in different environmental conditions may confer very different fitness effect on the same organisms, *i.e.*, very different adaptive values. Hence an orthologous protein which confers a higher fitness advantage due to efficient regulation under a particular environment is likely to emerge as a hub in one organism but not in the other which never experiences the same environment. Therefore, this would favor evolution of very different proteins as hubs in organisms living under different environmental conditions. Consistent with this general prediction, an analysis of the experimentally derived transcriptional network of *B. subtilis*, a gram positive bacterium, and *E. coli*, a gram negative bacterium, revealed that very different proteins emerge as global regulatory proteins. Taken together, these findings suggest that the scale-free structure emerged independently in evolution with regulatory hubs evolving according to requirements dictated by the environment of the organism.

5 Conclusions

In conclusion, by invoking the transcriptional regulatory network of living systems, I have discussed how such a representation allows us to investigate transcriptional regulation at a local level (in terms of the motifs) and the global level (in terms of the scale-free topology and regulatory hubs). By integrating gene expression data of five major regulatory programs which are involved in development or survival and by identifying the conserved parts of the networks across 170 prokaryotes living in different environmental conditions, I discussed the temporal and evolutionary dynamics of the local and global network structure.

At the local level of network structure, the temporal changes reveal that network motifs which display distinct kinetic properties are preferentially used in the different regulatory programs, which may allow for efficient response to changes in condition. On the evolutionary time scale, it was found that the environment and the life-style of an organism shape the regulatory motif content of an organism. At the global level of network structure, the analysis of temporal changes allowed the identification of permanent and condition specific hubs. Such hubs are not isolated, but regulate each other to allow for transition between the different regulatory programs by triggering the condition specific hubs. At the time-scale of the evolution of species, an analysis of the conservation of the global network structure revealed that condition-specific hubs can be lost and new hubs can emerge as dictated by the change in the environment. Taken together, these findings demonstrate that the transcriptional regulatory networks are very dynamic within the time-scale of an organism and evolve rapidly across organism within an evolutionary time-scale in order to efficiently respond to changing external or internal environments.

Acknowledgements. I thank the Medical Research Council, Darwin College and Schlumberger for generous support. I thank Aswin for reading this manuscript. The work described here was presented as an invited talk at the Biowire 2007 conference held in Cambridge, UK.

References

1. Barabasi, A.L., Oltvai, Z.N.: Network biology: understanding the cell's functional organization. Nat. Rev. Genet. 5(2), 101–113 (2004)
2. Babu, M.M., Luscombe, N.M., Aravind, L., Gerstein, M., Teichmann, S.A.: Structure and evolution of transcriptional regulatory networks. Curr. Opin. Struct. Biol. 14(3), 283–291 (2004)
3. Lee, T.I., Rinaldi, N.J., Robert, F., Odom, D.T., Bar-Joseph, Z., Gerber, G.K., Hannett, N.M., Harbison, C.T., Thompson, C.M., Simon, I., Zeitlinger, J., Jennings, E.G., Murray, H.L., Gordon, D.B., Ren, B., Wyrick, J.J., Tagne, J.B., Volkert, T.L., Fraenkel, E., Gifford, D.K., Young, R.A.: Transcriptional regulatory networks in *Saccharomyces cerevisiae*. Science 298(5594), 799–804 (2002)
4. Shen-Orr, S.S., Milo, R., Mangan, S., Alon, U.: Network motifs in the transcriptional regulation network of *Escherichia coli*. Nat. Genet. 31(1), 64–68 (2002)
5. Alon, U.: Network motifs: theory and experimental approaches. Nat. Rev. Genet. 8(6), 450–461 (2007)
6. Barabasi, A.L., Albert, R.: Emergence of scaling in random networks. Science 286(5439), 509–512 (1999)
7. Albert, R., Jeong, H., Barabasi, A.L.: Error and attack tolerance of complex networks. Nature 406(6794), 378–382 (2000)
8. Albert, R.: Scale-free networks in cell biology. J. Cell Sci. 118(pt 21), 4947–4957 (2005)
9. Kitano, H.: Biological robustness. Nat. Rev. Genet. 5(11), 826–837 (2004)
10. Thieffry, D., Huerta, A.M., Perez-Rueda, E., Collado-Vides, J.: From specific gene regulation to genomic networks: a global analysis of transcriptional regulation in *Escherichia coli*. Bioessays 20(5), 433–440 (1998)

11. Guelzim, N., Bottani, S., Bourgine, P., Kepes, F.: Topological and causal structure of the yeast transcriptional regulatory network. Nat. Genet. 31(1), 60–63 (2002)
12. Teichmann, S.A., Babu, M.M.: Gene regulatory network growth by duplication. Nat. Genet. 36(5), 492–496 (2004)
13. Luscombe, N.M., Babu, M.M., Yu, H., Snyder, M., Teichmann, S.A., Gerstein, M.: Genomic analysis of regulatory network dynamics reveals large topological changes. Nature 431(7006), 308–312 (2004)
14. Madan Babu, M., Teichmann, S.A., Aravind, L.: Evolutionary dynamics of prokaryotic transcriptional regulatory networks. J. Mol. Biol. 358(2), 614–633 (2006)
15. Wuchty, S., Oltvai, Z.N., Barabasi, A.L.: Evolutionary conservation of motif constituents in the yeast protein interaction network. Nat. Genet. 35(2), 176–179 (2003)
16. Jeong, H., Mason, S.P., Barabasi, A.L., Oltvai, Z.N.: Lethality and centrality in protein networks. Nature 411(6833), 41–42 (2001)

Phase Patterns of Coupled Oscillators with Application to Wireless Communication

Albert Díaz-Guilera[1] and Alex Arenas[2]

[1] Departament de Física Fonamental, Universitat de Barcelona, 08028 Barcelona,
Spain
albert.diaz@ub.edu
[2] Departament d'Enginyeria Informàtica i Matemàtiques, Universitat Rovira i Virgili,
43007 Tarragona, Spain
alexandre.arenas@urv.cat

Abstract. Here we study the plausibility of a phase oscillators dynamical model for time division for multiple access in wireless communication networks. We show that emerging patterns of phase locking states between oscillators can eventually oscillate in a round-robin schedule, in a similar way to models of pulse coupled oscillators designed to this end. The results open the door for new communication protocols in a continuous interacting networks of wireless communication devices.

Keywords: time division for multiple access, phase oscillators, round-robin schedule.

1 Introduction

Nowadays, wireless communications have become pervasive. This form of telecommunication between elements forming a network has technically evolved to the third generation of wireless systems, that incorporates the features provided by broadband. With this evolution, wireless networks become a plausible candidate for the main telecommunication mechanism in the next future. At the same time, this technical advance comes along with new problems which requires the use of innovative ideas to solve them. One of the problems we are aware, is that of maintaining decongestion in single-hop networks, where time division for multiple access (TDMA) strategies have been shown to be a good scheme for message transmission [1]. TDMA is a channel access method for shared medium (usually radio) networks. It allows several users to share the same frequency channel by dividing the signal into different timeslots. The users transmit in rapid succession, one after the other, each using his own timeslot. This allows multiple stations to share the same transmission medium (e.g. radio frequency channel) while using only the part of the bandwidth they require.

Between the algorithms that have been proposed to solve this problem, a bio-inspired solution called *desynchronization*, has attracted our attention [2,3]. The idea is to mimic some synchronization processes in biological systems, modeled by pulse-coupled oscillators in networks. Within this scenario a mapping between

P. Liò et al. (Eds.): BIOWIRE 2007, LNCS 5151, pp. 184–191, 2008.

wireless nodes in a network and pulse-coupled oscillators is possible, providing a simple and elegant protocol for TDMA communication. Here, desynchronization refers specifically to a state where nodes perfectly interleave periodic events to occur in a round-robin schedule, in contrast with synchronization where all the oscillators collapse their phase behavior.

The model presented in Ref. [2] recalls some results in lattices of coupled oscillators where spatio-temporal pattern form in a ring of oscillators with inhibitory unidirectional pulse-like interactions [4,5] inspired in the behavior of elementary neural systems. The attractors of the dynamics are limit cycles where each oscillator fires once and only once in a cycle, and some of them correspond to the desired behavior of round-robin schedule, that maintain order in the firing succession. The limit cycle structure of the attractors of pulse-coupled oscillators system shown in [4,5], is akin to those limit cycle emergent in coupled phase oscillators, in particular in the Kuramoto model [6,7]. Using this similarity on the final states, between both descriptions, here we study the plausibility of a self-organized algorithm between nodes communicating in a wireless network, using the dynamics of phase oscillators. The results show that an equivalent desynchronized state is obtained for a finite set of initial conditions, in a continuous interacting model. A reseting mechanism is proposed to account for the entry and exit of nodes of the network, while maintaining the desynchronized state.

The paper is organized as follows. In Sect. II we review some of the results known for a ring of pulse-coupled oscillators. In Sect. III we introduce the general model of phase-coupled oscillators and in the next section we present our proposal of phase oscillators continuously coupled through a function that depends on the sinus of the phase difference of the two interacting (neighboring) units. In Sect. V we analyze the stability of the fixed points of the collective dynamics, whereas in Sect. VI we study the effect of adding or removing nodes to our system, maintaining the prescribed schedule. Finally, in the last section, we present a brief discussion of the results and provide some lines of future research.

2   Patterns in Pulse-Coupled Oscillators in a Ring

Generally speaking, coupled oscillators interact via mutual adjustment of their amplitudes and phases. When coupling is weak, amplitudes are relatively constant and the interactions could be described by phase models. In particular, pulse-coupled oscillators account for some biological processes like heart pacemaker cells, integrate and fire neurons, and other systems made of excitable units. The instantaneous interactions that take place in a very specific moment of its period makes the treatment of these systems more complicated from a theoretical point of view. In any case, the richness of behaviors os these pulse-coupled oscillatory systems include synchronization phenomena, spatio-temporal pattern formation (traveling waves, chess-board structures, and periodic waves), rhythm annihilation, self-organized criticality, etc. The reader is pointed to Ref. [5] for references on this subject.

In these models, the phase of each oscillator evolves linearly in time – usually all units having the same period. When reaching some precise value of the phase the oscillator fires emitting a pulse that is received instantaneously by its set of neighbors. At this point the neighbors change their phases according to some specific function, called phase response curve. One should notice that this response function plays the crucial role in the dynamics of the population. Since two different time scales are in play, a continuous description makes no sense and the usual way to describe mathematically the system is by means of maps. A map represents the total evolution of driving (independent linear evolution in the slow time scale) and firing (interaction between units through a pulse) processes and the change in state after a complete map reflects the nature of the dynamical behavior. Thus, we can observe the evolution towards the attractors and analyze the stability of the fixed points.

In particular, in a set of works by the authors of the current paper, it was theoretically analyzed the behavior of rings of oscillators subjected to a linear phase response curve. In the first work [4] we dealt with unidirectional couplings in the ring obtaining exact values of the fixed points of the dynamics. As we said before, the stability of the fixed points is given by the return maps of the driving plus firing process. We computed the bounds of the eigenvalues of the matrix that describes the map and showed that any excitatory coupling (positive linear phase response curve) has unstable fixed points and the only solution is a synchronized state in which the oscillators collapse one by one. On the other hand, for an inhibitory coupling (negative linear phase response curve) the fixed points become stable, giving rise to spatio-temporal patterns where a constant phase-difference between oscillators is achieved. In a second work [8] we extended the previous result to a population of bidirectionally coupled units. Finally, in a third work [5] we analyze in much more detail the patterns that appear for inhibitory couplings (negative phase response curve). In particular, we were able to find the probability of selecting a given pattern under arbitrary initial conditions. In a ring of N oscillators there are $(N-1)!$ possible permutations of the firing sequence, by keeping one of the oscillators as the initial firing one. But all these possible sequences can give rise a smaller number of fixed points, which is $N-1$. Then these fixed points or patterns have some degeneracy that can be computed analytically. From this degeneracy, it can be computed the probability of pattern selection, that depends also on the coupling strength. For instance, it can be easily found that, in the case of small coupling, the most probable state is that with the maximum phase difference between neighbors, i.e. the phase-opposition (antisynchronization) state, and as we increase the number of oscillators the patterns distribution gets sharpened around this value. There is an additional effect in the pattern selection for this construction. When the coupling strength increases there are some fixed points that disappear, i.e. there are no longer part of the available configuration space. Depending on the number of oscillators and on the periodicity of the patterns, we could estimate the critical value of the coupling strength for which the pattern disappears. This effect is, of course, very important since it alters the distribution of the pattern selection.

3   Coupled Phase Oscillators

In contrast to pulse -coupled oscillators, phase coupled oscillators are described in a single time scale by a driving term plus an interaction between their (usually relatives) phases

$$\dot{\varphi}_i(t) = f_i(\varphi_i(t)) + \sigma \sum_j g(\varphi_i(t), \varphi_j(t)) \tag{1}$$

where φ stands for the phases, σ for the coupling strength, and f and g for general functions of the specified arguments. The sum runs over the neighboring units of oscillator i.

The behavior of 1D lattices of phase models is considerably complex, even for nearest neighbor coupling. In the case of chains of oscillators, for example, when coupling is local, oscillators at the ends get different inputs from those in the middle so that phase locking may not even exist. As long as the differences in the frequencies are small enough, there will be a phase-locked solution. Interestingly, nearest neighbor interaction chains can support very small gradients when the coupling term has the form of the sinus of the phase difference (and, in fact, any odd periodic function). However, if the coupling function contains even components (that is, replace $\sin(\varphi)$ with $\sin(\varphi+\delta)$), then frequency gradients as that are can be supported in nearest neighbor chains of coupled phase oscillators [9,10].

One of the most useful connections between the description of pulse-coupled oscillators and phase oscillators is exploited in the so-called transformation to phase models [11]. Fulfilling the condition of weak coupling, and autonomous oscillatory behavior of the pulse oscillators, the entire network can be transformed into a simpler phase model by a piece-wise continuous change of variables. The interest of this mathematical equivalence is that many pulse-coupled systems can be viewed as phase-coupled systems whose continuous description is more amenable. Driven by this analogy we explore the performance of a simple set of phase oscillators in a ring compared to the use of pulse-coupled oscillators described in [2], for TDMA on wireless networks.

4   The Kuramoto Model in a Locally-Coupled Ring

We consider here a particular model of phase oscillators that was introduced by Kuramoto [6]. In the original paper, Kuramoto analyzed a population of oscillators with an all-to-all pattern of connectivity. In principle, each oscillator has its own frequency drawn from a random distribution and is coupled via a sine function to the rest of the population

$$\frac{d\varphi_i}{dt} = \omega_i + \sigma \sum_j \sin(\varphi_j - \varphi_i) \quad i = 1, ..., N \tag{2}$$

In most of the analysis the interesting issue has been the transition to the synchronized state that appears above some critical value of the coupling strength

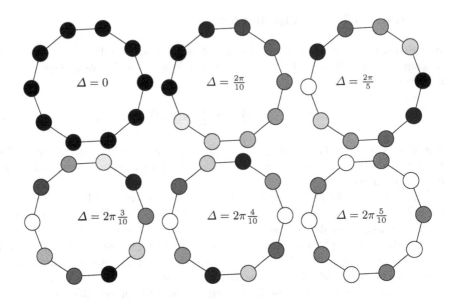

Fig. 1. Stable configurations for a ring of 10 oscillators. Top: for $\delta = 0$. Bottom: for $\delta = \pi$. The color of the node stands for the phases (white for $\varphi = 0$ to black for $\varphi = 2\pi$), but only the phase difference between neighboring nodes is important. The label at the center of each graph stands for the phase difference between neighboring nodes.

σ [7]. However, our goal here is to analyze the stationary states that can appear for a particular topology of the connections when all oscillators are driven by the same frequency (that can be taken as 0 without loss of generality). Hence, as the simples configuration, we consider a 1D ring of N oscillators, where each unit is connected to its nearest neighbors, and have zero inner frequency

$$\dot{\varphi}_i = \sin(\varphi_{i+1} - \varphi_i + \delta) + \sin(\varphi_{i-1} - \varphi_i + \delta) \quad \forall i = 1, ...N. \tag{3}$$

Here we have also introduced an arbitrary phase shift δ that will play a key role when considering the symmetries of the final stationary state.

A stationary solution $\varphi_{i+1} - \varphi_i = \Delta$, $\forall i$ exists provided that $\Delta = 2\pi m/N$, being $m \in \mathbb{N}$.[1] In this case we should have for all the oscillators

$$\dot{\varphi}_i = 2 \sin \delta \cos \Delta \tag{4}$$

i.e. the oscillators rotate at this effective frequency, all with the same value but a fixed phase difference between neighbors is kept. In Fig. 1 we plot a ring of 10 oscillators for a stationary phase difference corresponding to the cases $m = 0, 1, 2, 3, 4, 5$ (the remaining cases correspond to the complementary ones to these, because all results have to be understood as mod 2π).

---

[1] Notice that the cases m and $N - m$ are equivalent since it is a positive or a negative phase difference and all phase differences are to be understood mod 2π.

5   Linear Stability of the Attractors

Let us assume a small instantaneous perturbation to one of the nodes: $\varphi_i \to \varphi_i + \varepsilon$. Then the equation of motion for this oscillator becomes

$$\dot{\varphi}_i = \sin(\Delta + \delta - \varepsilon) + \sin(-\Delta + \delta - \varepsilon). \qquad (5)$$

Expanding the sinus functions we get up to linear order in ε

$$\dot{\varphi}_i = 2\sin\delta\cos\Delta - 2\varepsilon\cos\delta\cos\Delta. \qquad (6)$$

The derivative of the frequency with respect to the perturbation is $d\dot{\varphi}_i/d\epsilon = -2\cos\delta\cos\Delta$, providing the stability of the stationary solutions of the system. Let us now look in detail to the different combinations of these terms, keeping in mind that δ is a prescribed phase that breaks the symmetry of the problem. We will consider only two cases ($\delta = 0$ and $\delta = \pi$), any case in between these values only affects the effective frequency. Notice, however, that $\delta = \pi/2$ is a very particular case and the stability analysis requires a specific study that is beyond the scope of the current work.

For the case $\delta = 0$ all states with $\cos(\Delta) > 0$ are stable, i.e. $d\dot{\varphi}_i/d\epsilon < 0$, in particular the synchronized state $\Delta = 0$. But there are also other possibilities $0 < m < N/4$ for which the oscillators can end in a stable stationary state. Notice that the case $m = 1$, which is a stable solution whenever $N \le 4$, corresponds to the minimum phase difference between oscillators, that is the case of the round-robin schedule mentioned in the Introduction. In Fig. 1 (top) we show the three stable configurations for a ring of 10 oscillators with $\delta = 0$.

For the case $\delta = \pi$, new stable states appear, all those with $N/4 < m \le N/2$, and the synchronized state becomes unstable. For the particular case of 10 oscillators, we show the three stable configurations in the bottom of Fig. 1.

In general, we obtain a set of stable configurations where there can be subsets of nodes which are partially synchronized. For instance, if N is even, there always exists a configuration, stable for $\delta = \pi$, for which the phase difference between any two neighboring nodes is π and hence we have some sort of local *antisynchronization*, which is the maximum phase difference between neighboring oscillators. Another interesting case is that of N being a prime number; in this case all stable configurations are equivalent in the sense that there are no two synchronized oscillators, and the round-robin schedule is maintained, although not necessarily for neighboring nodes. Any stable configuration establishes a different order for the evolution of the oscillators. Although in the case of phase-coupled oscillators the firing does not make any sense, it is important to specify some value of the phase, for instance its maximum value 2π. Then any configuration stands for the time sequence of the oscillators phases reaching the value 2π.

The persistence of stable configurations where a round-robin scheduled is satisfied, opens the door for a self-organized solution to the problem of TDMA in wireless networks. However, there is still a problem concerning the inclusion of new agents (oscillators) to the system. In the next section, we investigate the effect of such new incorporations to the existing system.

6   Variation on the Number of Nodes in the Network

First of all we are going to consider how the round-robin stationary state responds to the addition of a new oscillator. Then we have, as starting configuration, a set of $N - 1$ nodes such that

$$\varphi_i = i * (2\pi)/(N - 1) \ \forall i = 1, \ldots, (N - 1) \tag{7}$$

and now we add an incoming oscillator, that we label N. The phase of the incoming oscillator is unknown. For this reason, as a first approximation, we discretize the possible values of the incoming phase φ_N in the range $[0, 2\pi]$ and count the fraction of values of this set that leads to a new round-robin configuration of N oscillators. In Fig. 2 we plot the fraction of values of the initial phase φ_n that give rise either to the round-robin state, with N oscillators, or to the synchronized state. We have not observed the emergence of other states, although at this point we can not discard their existence as spurious states.

We notice that the round-robin state is very robust in the sense that it emerges from the new configuration almost surely, although the time response is large and it increases with the number of oscillators.

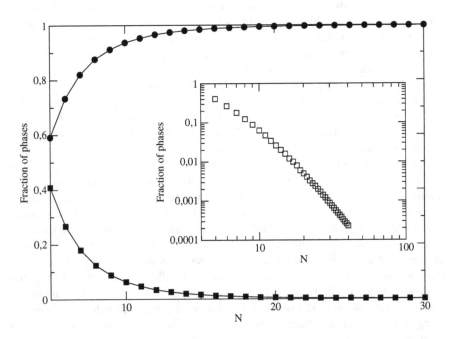

Fig. 2. Fraction of values of the initial phase of the incoming oscillator that leads to a round-robin configuration (top) and to a synchronized state (bottom), as a function of the number of final nodes. In the inset we plot the fraction of configurations that give rise to the synchronized state in log-log scale, to how fast it decreases with the system size.

On the other hand, the fact of removing one node from a stable configuration is also quite robust. The round-robin state is broken exclusively for a very small number of oscillators. This is a deterministic case in which we fix all initial phases such that $\Delta = 1/(N+1)$ and the system evolves deterministically towards $\Delta = 1/N$ for N larger than 6.

7 Discussion

We have presented a bio-inspired approach to the round-robin schedule of wireless networks, based on the synchronization of phase oscillators, particularly, Kuramoto oscillators in a ring with nearest neighbors coupling. The study of patterns of phase locking attractors shows that this continuous interaction model could be also used as an alternative protocol for TDMA, comparable to the approach of pulse coupled oscillators presented in [2], although still to be developed in deep. Upcoming experiments in complex topologies of phase oscillators show that the number of possible phase locking patterns is hugely rich. We will present a systematic study of the different stationary states, as well as their relative basis of attraction in different topologies in a future work.

References

1. Falconer, D., Adachi, F., Gudmundson, B.: Time division multiple access methods for wireless personal communications. IEEE Communications Magazine 33, 50–57 (1995)
2. Degesys, J., Rose, I., Patel, A., Nagpal, R.: Desync: self-organizing desynchronization and tdma on wireless sensor networks. In: IPSN 2007: Proceedings of the 6th international conference on Information processing in sensor networks, pp. 11–20. ACM, New York (2007)
3. Patel, A., Degesys, J., Nagpal, R.: Desynchronization: The theory of self-organizing algorithms for round-robin scheduling. In: SASO, pp. 87–96. IEEE Computer Society, Los Alamitos (2007)
4. Díaz-Guilera, A., Pérez, C.J., Arenas, A.: Mechanisms of synchronization and pattern formation in a lattice of pulse-coupled oscillators. Phys. Rev. E 57, 3820–3828 (1998)
5. Guardiola, X., Díaz-Guilera, A.: Pattern selection in a lattice of pulse-coupled oscillators. Phys. Rev. E 60, 3626–3632 (1999)
6. Kuramoto, Y.: Chemical oscillations, waves, and turbulence. Springer, New York (1984)
7. Acebrón, J.A., Bonilla, L.L., Pérez-Vicente, C.J., Ritort, F., Spigler, R.: The Kuramoto model: A simple paradigm for synchronization phenomena. Rev. Mod. Phys. 77, 137–185 (2005)
8. Díaz-Guilera, A., Pérez, C.J., Arenas, A.: Synchronization in a ring of pulsating oscillators with bidirectional couplings. Int. J. Bif. Chaos 9, 2203–2207 (1999)
9. Kopell, N., Ermentrout, G.: Symmetry and phase-locking in chains of weakly coupled oscillators. Communications on Pure and Applied Mathematics 39, 623–660 (1986)
10. Kopell, N., Ermentrout, G.B.: Phase transitions and other phenomena in chains of coupled oscillators. SIAM Journal of Applied Mathematics 50, 1014–1052 (1990)
11. Izhikevich, E.M.: Weakly pulse-coupled oscillators, fm interactions, synchronization, and oscillatory associative memory. IEEE Trans. Neural Networks 10, 508–526

Self-organizing Desynchronization and TDMA on Wireless Sensor Networks

Julius Degesys*, Ian Rose, Ankit Patel, and Radhika Nagpal

Harvard University, Cambridge MA 02138, USA
{degesys,ianrose,abpatel,rad}@eecs.harvard.edu

Abstract. Desynchronization is a recently introduced primitive for sensor networks: it implies that nodes perfectly interleave periodic events to occur in a round-robin schedule. This primitive can be used to evenly distribute sampling burden in a group of nodes, schedule sleep cycles, or organize a collision-free TDMA schedule for transmitting wireless messages. Here we present a summary[1] of DESYNC, a biologically-inspired self-maintaining algorithm for desynchronization in a single-hop network. We also describe DESYNC-TDMA, a self-adjusting TDMA protocol that addresses two weaknesses of traditional TDMA: it does not require a global clock and it automatically adjusts to the number of participating nodes, so that bandwidth is always fully utilized.

Keywords: Desynchronization, self-organization, wireless sensor networks, pulse-coupled oscillators, medium access control.

1   Introduction

It is hard to find an aspect of our lives that soon will not be affected by wireless sensor networks (WSNs). With the decreasing prices and sizes of processors and sensors, it has become possible to embed computation into almost any environment. Recent applications have ranged from volcano monitoring [16] to detecting enemy sniper gunfire [14]. The advent of these exciting new possibilities, however, has brought with it new challenges to our current programming model. In many cases, it is becoming far too complex to write centralized code that can gracefully scale while still handling the uncertain and dynamic conditions that are so prevalent in these deployments.

Despite our difficulties, nature seems to thrive in this setting. Ants and termites, with no explicit communication, are able to manage resources, construct complex structures, and forage for food. Fireflies, crickets, and frogs all synchronize their flashings, chirpings, and croakings in mating calls. Birds flock together and migrate large distances, instinctively knowing that winter's cold will soon set

---

* Corresponding author.

[1] This work was originally published in the proceedings of IPSN 2007 [1]. Here, we have removed some of the comparisons and theoretical rigor to make the document more accessible.

P. Liò et al. (Eds.): BIOWIRE 2007, LNCS 5151, pp. 192–203, 2008.
© Springer-Verlag Berlin Heidelberg 2008

in. These are all examples of self-organization, a paradigm in which individuals make simple local decisions, but collectively, produce complex global behavior. This approach can be quite powerful, as is clearly indicated by the impressive scale and robustness that these biological systems achieve. Recently several studies have begun to extend these techniques to achieve event synchronization in wireless sensor networks [6,16,3].

Here we focus on a related primitive, *desynchronization*, that is the logical opposite of synchronization; instead of nodes attempting to perform periodic tasks at the same time, nodes perform their tasks as far away as possible from all other nodes. Imagine not fireflies flashing in unison, but in a uniformly distributed, round-robin fashion.

Desynchronization is a useful primitive for periodic resource scheduling. We consider its use in implementing a time division multiple access (TDMA) medium access control (MAC) protocol in which nodes use a round-robin schedule for sending messages. In TDMA, scheduled nodes do not have to contend for the shared medium nor worry about message collisions. It is especially attractive in many settings where nodes are transmitting streams of data or there are real-time constraints on message latency, as is common in wireless sensor networks [15,5,12].

In this paper we discuss DESYNC, a recently introduced [1] biologically-inspired algorithm for achieving desynchronization in a single-hop network. Given a set of n nodes that generate events periodically with a common, fixed period T, the nodes adjust such that the events are evenly distributed throughout the time period (i.e. they are spaced at intervals of T/n). The algorithm is simple, decentralized, and requires constant memory per node regardless of network size. Furthermore, if nodes are added or removed, the system self-adjusts to re-equalize the event intervals. Thus, DESYNC implements a self-maintaining desynchronization primitive. DESYNC is then used to implement a self-organizing TDMA MAC protocol for single-hop wireless networks, DESYNC-TDMA.

DESYNC-TDMA has two improvements over classic TDMA protocols: (1) it does not require a global clock or other infrastructure overhead and (2) the schedule automatically self-adjusts to the number of participating nodes so as to fully utilize the bandwidth. Our experimental results show that DESYNC-TDMA achieves over 90% bandwidth utilization (a 25% increase from the default Telos MAC implementation) and less than 1% message loss in high traffic.

Section 2 discusses related work. Section 3 presents an overview of the DESYNC and DESYNC-TDMA algorithms. Section 4 describes some of the experimental results. Finally, we conclude in Section 5, discussing directions for future work.

2   Background and Related Work

2.1   Models of Synchronization in Biology

Many natural synchronizing systems, such as networks of neurons or swarms of fireflies, are modeled as networks of *pulse-coupled oscillators*, where each node in the network represents an adjustable oscillator that pulses at a fixed frequency. Each oscillator observes other oscillators' pulses (e.g. a neuron firing or a neighboring firefly's flash) and uses this information to adjust its own oscillator. Ultimately, all oscillators pulse synchronously.

In a seminal paper, Mirollo and Strogatz proved that a complete network of n pulse-coupled oscillators, using a simple oscillator-adjustment function, would always converge to synchrony, irrespective of the initial state [8,13]. Recently, this biological model has been extended and shown to be able to achieve decentralized time synchronization and coordinated sensor control in wireless sensor networks [6,16,3]. One of the key benefits of this model is its ability to adapt—the system adjusts automatically to nodes entering and leaving the system, even though the individual nodes are only using very simple, local rules. Thus, synchronization in this model is self-maintaining.

In some natural systems, the goal is not synchronization, but *patterned* synchronization. For example, in animal locomotion, limbs can be modeled by individual oscillators that are coupled so as to produce different gaits [13]. Similarly, in the intestines, a series of oscillators can be coupled to produce a systolic wave. In these cases, the oscillators do not first synchronize and then negotiate a schedule for the pulse pattern. Instead, they use different adjustment rules to directly generate the desired pattern, with the advantage being that these adjustment rules are also self-maintaining.

In our case of desynchronization, we are interested in the pattern in which all of the oscillators pulse at evenly spaced intervals (the oscillators are completely out of phase). We use the Mirollo and Strogatz framework to design a simple oscillator adjustment rule that causes the system as a whole to converge to desynchrony. As with the original model, the system self-adjusts to maintain desynchronization; if new nodes are introduced, or current nodes removed, the system automatically converges to a new state where the new set of nodes has evenly spaced pulses. Protocols built on top of this primitive inherit the same self-maintaining property.

2.2 Channel Sharing in Wireless Networks

In wireless networks, nodes share the medium in which they transmit messages. It is the MAC protocol's responsibility to mediate their transmissions. Any of these protocols can usually be described as being either a contention-based protocol or a schedule-based protocol [4].

In contention-based, carrier sense multiple access (CSMA) protocols, nodes check the channel before transmitting, and if the channel is busy, they randomly back off for a short time and try again. This method is simple, adaptive, and frees nodes from having to maintain complex state about their environment. As a result, CSMA is often used when the expected contention is low (i.e. few nearby nodes transmitting) or when bursty traffic is expected.

In TDMA-based protocols, nodes use a round-robin schedule to transmit messages. Time is partitioned into fixed-size slots, and each node selects a time-slot during which it may regularly send messages collision-free. Since each node gets an equally sized slot, fairness is ensured. Message latency is bounded since nodes transmit at a fixed frequency.

TDMA is especially useful when nodes are transmitting streams of data, experience periods of high contention, have a high cost for message loss (e.g. energy cost of retransmissions), or require real-time constraints on message latency. These requirements are found in many sensor network applications due to their emphasis on periodic monitoring and local, event-triggered traffic [15,5,12]. As such, several TDMA protocols have been designed specifically for these settings [4]. However, almost all traditional TDMA implementations still encounter the following difficulties:

Overhead: Nodes must know when their slots begin and end, which usually requires accurate time synchronization among nodes and a negotiation of the slot schedule. The message overhead involved in maintaining these adds to the energy consumption and implementation complexity [7].

Wasted Slots: Nodes are assigned exclusive time slots. This means that slots go unused when nodes do not have data to send or have left the network. Thus, it is important for the network to be able to reclaim this lost bandwidth.

In general, the complexity and cost of maintaining any TDMA schedule in the face of node and traffic changes can often outweigh the benefits of fairness, reliability, and high throughput. Hence, the default MAC protocols most used by sensor motes are CSMA protocols [9,17].

In reality, there is no explicit need for nodes to agree upon a global time or to maintain information about each others' identities. Rather, TDMA only requires nodes to *desynchronize the timing of their transmissions*. If nodes could self-maintain desynchronization, then both weaknesses of TDMA would be addressed simultaneously. For example, if a node does not need to transmit, it can go to sleep and the remaining nodes will adjust to fully utilize the available bandwidth without message collisions.

3   Algorithms

In this section, we first provide a description of the pulse-coupled oscillator framework, introduced by Mirollo and Strogatz [8]. We then use this framework to describe the DESYNC and DESYNC-TDMA algorithms.

3.1   Framework

Suppose there are n nodes that can communicate with each other (i.e. they are in a fully-connected network). Each node performs a task periodically with a period T. Thus, we can model each node as an oscillator with frequency $\omega = 1/T$. Let $\phi_i(t) \in [0, 1]$ denote the phase of node i at time t where the phases 0 and 1 are identical and where $0 \leq i \leq n - 1$. For example, if $\phi_i(t) = 0.75$, then node i is 75% of the way through its cycle. Upon reaching $\phi_i(t) = 1$, node i "fires" (or "pulses") indicating the termination of its cycle to the other nodes. Upon firing, the node resets its phase to $\phi_i(t^+) = 0$.

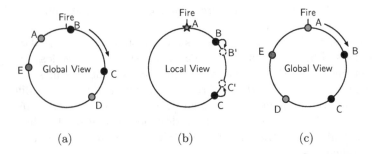

Fig. 1. DESYNC algorithm: (a) Global view of five nodes that are not yet desynchronized. (b) From node B's' local phase neighborhood view: when A fires, the node that fired immediately before it, node B, now knows both of its neighbors' positions—it heard C fire earlier and A just fired. Therefore, node B can now compute where it should have been if it were positioned ideally, B', and jump towards it. However, C has since jumped to C', unbeknownst to any other nodes. (c) The desynchronized state. All nodes are at the midpoints; thus, no node jumps and the system is stable.

We can imagine the nodes as beads moving clockwise on a ring with period T (Figure 1). When a node reaches the top, it fires. All nodes observe this firing, and can use this information to jump forwards or backwards in phase. However, nodes are otherwise oblivious of the phases of other nodes; they can not observe the current state of the ring, only the firing events.

The goal is to have each node adjust the timing (phase) of its own firing such that eventually the network is *desynchronized*. We define the desynchronized state as the state in which all of the oscillators are evenly spaced around the phase ring.

3.2 DESYNC Algorithm

A simple algorithm for achieving desynchronization in a single-hop network is DESYNC [1]. We assume that a firing event corresponds to a node broadcasting a wireless *firing message* that all other nodes can hear. Intuitively, the algorithm works as follows: each node adjusts its phase to be at the midpoint of the two nodes before and after it on the ring. In order to achieve this, a node must pay attention to the timing of the firings before and after its own. If each node can fire closer to the midpoint, then over successive periods this *jumping towards the average* will bring the system to a state in which all nodes are at the midpoints of their neighbors. This is exactly the desynchronized state.

In the algorithm, node i keeps track of the times of two events: the firing that occurs just before it fires (from node $i + 1$ (mod n)) and the firing that occurs just afterwards (from node $i - 1$ (mod n)). We call the senders of those firing messages the *phase neighbors* of node i. The firing times of the previous and next neighbors are recorded relative to node i's firing as $\tilde{\Delta}_{i+1}$ and $\tilde{\Delta}_i$, respectively. In this way, node i can approximate the phases of its previous and next phase neighbors as $\tilde{\phi}_{i+1}(t) = \phi_i(t) + \tilde{\Delta}_{i+1}$ (mod 1) and $\tilde{\phi}_{i-1}(t) = \phi_i(t) - \tilde{\Delta}_i$ (mod 1). Using this information, node i can then calculate the midpoint of its neighbors:

$$\tilde{\phi}_{\mathrm{mid}}(t) = \frac{1}{2}\left[\tilde{\phi}_{i+1}(t) + \tilde{\phi}_{i-1}(t)\right] \quad (\mathrm{mod}\ 1) \tag{1}$$

$$= \phi_i(t) + \frac{1}{2}\left(\tilde{\Delta}_{i+1} - \tilde{\Delta}_i\right) \quad (\mathrm{mod}\ 1) \tag{2}$$

Note that in calculating Equation 1, the modular arithmetic used to compute the approximations of the phase neighbors should be delayed to the end to yield the appropriate midpoint. Once the midpoint is known, node i jumps towards it:

$$\phi_i'(t) = (1 - \alpha)\phi_i(t) + \alpha\tilde{\phi}_{\mathrm{mid}}(t) \tag{3}$$

where $\alpha \in [0, 1]$ is a parameter that scales how far node i moves from its current phase towards the desired midpoint. Thus, after hearing both neighbors fire, node i instantaneously jumps from $\phi_i(t)$ to $\phi_i'(t)$. Note that if node i jumps immediately when node $i - 1$ fires, the estimate is exact, i.e. $\tilde{\Delta}_i = \Delta_i(t)$. This adjustment is not apparent to other nodes until node i fires again. Furthermore, node i's neighbors will also make adjustments without node i's knowledge.

To further illustrate this point, consider Figure 1(b) where there are three nodes: A, B, and C. First, C fired followed by B, and now, A is about to fire. In Figure 1(c), A fires; thus, B has enough information to make a jump. However, at this point, C too has heard both of its neighbors, B and D, and has already jumped to C'. Thus, by the time B makes its adjustment, it no longer knows the true distance between C and B. It is in this way that nodes continually make adjustments based on stale information. However, this system will still converge to a desynchronized state.

This algorithm has several key features:

- **Convergence to Desynchrony**: Regardless of the initial state and number of nodes, the system converges to a state in which all nodes are evenly spread out with a spacing of T/n.
- **Simple Implementation**: Nodes only record the timing of two firing events and are not concerned with the identity of the senders nor how many firings occur in a given period. Therefore, nodes use *constant memory*, regardless of network size and do not need to maintain any internal state on network composition.
- **Self-Adapting**: If the number of nodes changes (a node is added or removed) then the system is no longer desynchronized. This local imbalance causes nodes closest to the disturbance to adjust their phases, eventually leading the system back to a stable, desynchronized state. Nodes do not need to explicitly monitor the network membership. Furthermore, single-node failures are similarly accounted for in the normal operation of the algorithm. Thus, the system ensures a fair sharing of the time period, T, even when the network size changes or nodes experience faults.

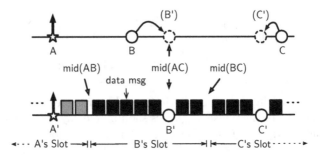

Fig. 2. DESYNC-TDMA slots: Here, we have unravelled the ring into a line segment. The nodes in on the top line represent the current state in Figure 1(c) where node A is firing. The TDMA slots associated with the next set of firings are defined by the midpoints of the firings that occurred previously. Despite the information being old, if the nodes update according to Equation 3, firings will always occur during the nodes' TDMA slots.

3.3 DESYNC-TDMA Algorithm

In this section, we describe how one can implement TDMA using DESYNC. As discussed in Section 2, TDMA-based protocols suffer from overhead and wasted slots. DESYNC allows us to design a simple low-memory TDMA protocol that automatically regulates slot sizes, fully utilizing bandwidth without incurring any collision costs.

We define node i's TDMA slot as beginning at the previously computed midpoint between node i and its previous phase neighbor. Likewise, it ends at the previously computed midpoint between node i and its next phase neighbor. Intuitively, nodes use earlier firings to compute the TDMA slots near the time of their next firing. Figure 2 illustrates this slot definition.

Defining slots in this manner also guarantee that a node will never fire outside its own slot. Note that if this were not the case, node i would be unable to send its firing message as the channel would be occupied by the current slot owner's transmissions.

DESYNC-TDMA has the following characteristics:

The algorithm fully utilizes the channel regardless of the network's state of desynchronization. The algorithm defines a set of non-overlapping slots that cover T, allowing nodes to send collision-free data, even while they are desynchronizing.

The TDMA schedule seamlessly adapts to nodes entering or leaving. When a node leaves, the neighboring nodes adjust their slot boundaries to fully utilize the bandwidth. The slot sizes equalize over time as the system approaches desynchronization, having the effect of leaving T fixed and increasing slot size. Thus, if a node does not need to transmit again for multiple periods, it can simply leave the protocol, sleep, and re-enter when it needs to send again. In the meantime, other nodes will have reaped the benefit of automatically acquiring the sleeping node's slot.

When a node enters the algorithm, it must first interrupt an existing node's data slot with its firing message. Here, the *costs of entering* for a node are the latency of one time period and the lost bandwidth that results from the one interrupted data slot. Other nodes remain oblivious to the entry and send data uninterrupted.

The algorithm is self-contained. Nodes do not need to know the network size or discover their neighbor IDs in order to create an initial schedule. The round-robin schedule order emerges as a result of the order in which nodes enter the process. Unlike other TDMA-style protocols, such as Z-MAC [11] and TRAMA [10], nodes do not need to agree on global time nor rely on a time synchronization protocol. While it is possible to write additional code to support each of these additional tasks (discovering neighbor IDs reliably, renegotiating schedule orders, electing leaders for global time consensus) this can add significant complexity to the implementation.

4   Implementation

In this section, we investigate the performance of DESYNC-TDMA. The DESYNC-TDMA algorithm was implemented on Telos wireless sensor motes.[2]

4.1   Experimental Setup

We constructed a single-hop network by placing 20 motes around a single designated base station. The base logged all messages transmitted by the other motes. As the base station did not send any messages once the experiment started, it was able to observe the algorithm without affecting its performance. For all experiments, we used the same fixed parameters: $T = 1$ sec and $\alpha = 0.95$.

4.2   Evaluation Metrics

We use two metrics to measure the performance of the system:

- **Average desync error:** We define error to be the average slot size deviation from the desired slot size (T/n) for a given round.
- **Normalized throughput:** We define *normalized throughput* as the ratio between the measured data throughput (not including bandwidth used by firing messages) and the maximum possible measured throughput of 62.8 Kbps.

4.3   Experimental Results

Figure 3 shows a single run of a fixed-size experiment for $n = 10$ motes. Plotted are the times of each mote's firing events relative to those of a single mote.

---

[2] For a full description and discussion of the experiments, we refer the reader to the original paper [1].

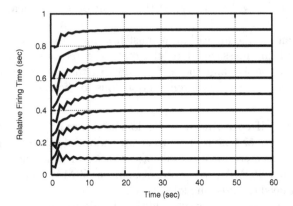

Fig. 3. Running DESYNC on 10 sensor motes: the firing times during each round are plotted relative to an arbitrarily chosen mote. The firing times stabilize to an even spacing while preserving the initial firing order.

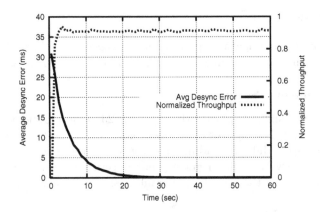

Fig. 4. The average DESYNC error and total throughput are plotted over time. The error decreases over time, but the total throughput remains high and roughly constant regardless of the state of desynchronization.

As can be seen, the motes quickly and smoothly achieved desynchronization. Figure 4 shows how the different performance metrics changed over time. The average desync error decreased exponentially with time, reaching an error of less than 1 ms within 18 rounds (note that the desired slot size is $T/n = 100$ ms). However, the total normalized throughput was high ($\sim 92\%$) and roughly constant throughout the experiment, regardless of the desynchronization error.

Figure 5 shows how the average desync error and normalized throughput varied during the mote removal and addition experiment. At $t = 135$, when a mote stopped transmitting, the resulting imbalance in slot-sizes caused a jump in error of ~ 25 ms, which decayed to less than 1 ms over the next 8 rounds. Likewise,

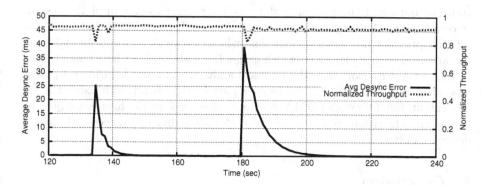

Fig. 5. Motes Arrival and Departure experiments. 8 motes were started at $t = 0$. At time $t = 135$ one mote left the system and stopped transmitting. At time $t = 180$ three motes woke up and entered the system. The system minimizes loss of throughput and rapidly re-equalizes slot sizes after addition or removal of motes.

an error of ~ 40 ms was introduced when three new motes began transmitting at $t = 180$, requiring 19 rounds to reduce to 1 ms. The total throughput was slightly impacted at each event, suffering two-round throughput losses of 12.5% and 10.6%, respectively, during the removal and addition events. Within two rounds, the total throughput had returned to normal capacity.

Overall, this experiment shows that the cost of entry and exit for a mote is low and the system is able to adapt quickly to recover bandwidth and re-equalize slot sizes. The main costs are the single round latency that an entering mote must wait before transmitting data and a temporary drop in fairness as the slot sizes are re-equalized.

From these results we can conclude general trends: for DESYNC, the average desync error decreases exponentially with time; for DESYNC-TDMA, the bandwidth utilization is consistent, and message loss is near zero, regardless of the state of desynchronization and number of transmitting motes. Thus, nodes can easily enter and leave with a limited impact on total throughput.

4.4   Summary Discussion

DESYNC-TDMA is a fundamentally new way of thinking about TDMA scheduling. Without explicit scheduling or time synchronization, DESYNC-TDMA is able to provide excellent total throughput and collision-free transmission under high loads, regardless of the state of desynchronization. Once desynchronized, it guarantees fairness and predictable (stream-like) message latencies. When nodes enter or leave, the system self-adjusts to accommodate the new nodes or to recapture the unused slots. Furthermore, unlike hybrid-TDMA methods, no contention is required for recapture.

However, DESYNC-TDMA also has some limitations and may not be appropriate for all types of traffic. One important limitation is that a node pays a "cost" when entering the system: (a) 1-round latency before being able to transmit data and (b) a smaller slot size for several rounds until the system re-converges to the desynchronized state. A second limitation is that DESYNC-TDMA provides nodes with equal slots which, although guaranteeing fairness, can also lead to inefficient bandwidth usage. If a node does not have enough data to fully utilize its slot, then the unused bandwidth is wasted. In hybrid protocols such as Z-MAC, nodes can recover part of that bandwidth using CSMA contention, but DESYNC-TDMA does not allow this. In the future, we plan to extend the algorithm to provide variable slot sizes that can reflect each node's desired bandwidth.

5   Conclusion

We have discussed a recently-introduced primitive, *desynchronization* along with a self-organizing desynchronization algorithm for single-hop networks, DESYNC. As an application of DESYNC, we also discussed DESYNC-TDMA, a self-organizing TDMA protocol. Each is a useful algorithm in its own right: DESYNC provides the ability to space out events in time, whereas DESYNC-TDMA constructs a method to share a medium fairly by simply having the events correspond to changes in ownership over the medium. There are several avenues of future work; however, a critical next step is extending DESYNC-TDMA to multi-hop networks.

Determining a slot schedule in multi-hop topologies is a much more complex problem for two reasons: nodes belong to intersecting and multiple-sized neighborhoods and overlapping broadcast regions create hidden terminals. A standard technique used in solving this problem is to color a constraint graph in which all two-hop neighborhoods in the communication graph are fully connected. Assigning each node in the graph its own color is equivalent to a global desynchronization, whereas minimal coloring constructs the fairest distribution of time amongst the nodes.

Our preliminary simulations suggest that DESYNC-TDMA converges on multi-hop topologies and produces a slot size comparable to T/k_2 (where k_2 is the size of the node's 2-hop neighborhood subgraph). However, proving that the algorithm converges on all multi-hop topologies and predicting the slot size and convergence times is currently an open question.

References

1. Degesys, J., Rose, I., Patel, A., Nagpal, R.: DESYNC: Self-Organizing Desynchronization and TDMA on Wireless Sensor Networks. In: IPSN (2007)
2. Hill, J., Szewczyk, R., Woo, A., Hollar, S., Culler, D.E., Pister, K.S.J.: System architecture directions for networked sensors. In: ASPLOS (November 2000)
3. Hong, Y., Scaglione, A.: A scalable synchronization protocol for large scale sensor networks and its applications. IEEE Journal on Selected Areas in Communication (November 2003)
4. Langendoen, K., Halkes, G.: Energy-Efficient Medium Access Control. In: Embedded System Handbook. CRC Press, Boca Raton (2005)

5. Lorincz, K., Malan, D., Fulford-Jones, T., Nawoj, A., Clavel, A., Shnayder, V., Mainland, G., Moulton, S., Welsh, M.: Sensor networks for emergency response: Challenges and opportunities. IEEE Pervasive Computing (December 2004)
6. Lucarelli, D., Wang, I.: Decentralized synchronization protocols with nearest neighbor communication. In: SenSys (2004)
7. Maróti, M., Kusy, B., Simon, G., Lédeczi, A.: The Flooding Time Synchronization Protocol. In: SenSys (2004)
8. Mirollo, R., Strogatz, S.: Synchronization of pulse-coupled biological oscillators. SIAM Journal of Applied Math. 50(6), 1645–1662 (1990)
9. Polastre, J., Hill, J., Culler, D.: Versatile low power media access for wireless sensor networks. In: SenSys (2004)
10. Rajendran, V., Obraczka, K., Garcia-Luna-Aceves, J.J.: Energy-efficient collision-free medium access control for wireless sensor networks. In: SenSys (2003)
11. Rhee, I., Warrier, A., Aia, M., Min, J.: Z-MAC: A Hybrid MAC for Wireless Sensor Networks. In: SenSys (2005)
12. Simon, G., Maróti, M., Lédeczi, A., Balogh, G., Kusy, V., Nádas, A., Pap, G., Sallai, J., Frampton, K.: Sensor network-based countersniper system. In: SenSys (2004)
13. Strogatz, S.: Sync: The Emerging Science of Spontaneous Order. Hyperion, New York (2003)
14. Simon, G., Maróti, M., Lédeczi, A., Balogh, G., Kusy, V., Nádas, A., Pap, G., Sallai, J., Frampton, K.: Sensor Network-Based Countersniper System. In: SenSys 2004: Proceedings of the 2nd International Conference on Embedded Networked Sensor Systems (2004)
15. Werner-Allen, G., Johnson, J., Ruiz, M., Lees, J., Welsh, M.: Monitoring volcanic eruptions with a wireless sensor network. In: Proc. European Workshop on Wireless Sensor Networks (EWSN) (January 2005)
16. Werner-Allen, G., Tewari, G., Patel, A., Nagpal, R., Welsh, M.: Firefly-Inspired Sensor Network Synchronicity with Realistic Radio Effects. In: SenSys (2005)
17. Ye, W., Heidemann, J., Estrin, D.: An Energy-Efficient MAC Protocol for Wireless Sensor Networks. In: INFOCOM (2002)

Bio-Inspired Multi-agent Collaboration for Urban Monitoring Applications[*]

Uichin Lee[1], Eugenio Magistretti[2], Mario Gerla[1], Paolo Bellavista[3],
Pietro Liò[4], and Kang-Won Lee[5,**]

[1] UCLA
[2] Rice University
[3] University of Bologna
[4] University of Cambridge
[5] IBM Research
{uclee,gerla}@cs.ucla.edu, emagistretti@rice.edu
pbellavista@deis.unibo.it, Pietro.Lio@cl.cam.ac.uk, kangwon@us.ibm.com

Abstract. Vehicular sensor networks (VSNs) provide a collaborative
sensing environment where mobile vehicles equipped with sensors of different nature (from chemical detectors to still/video cameras) inter-work
to implement monitoring applications such as traffic reporting, environment monitoring, and distributed surveillance. In particular, there is an
increasing interest in proactive urban monitoring where vehicles continuously sense events from streets, autonomously process sensed data (e.g.,
recognizing license plates), and possibly route messages to vehicles in
their vicinity to achieve a common goal (e.g., to permit police agents
to track the movements of specified cars). MobEyes is a middleware solution to support VSN-based proactive urban monitoring applications,
where the agents (e.g., police cars) harvest metadata from regular VSN-
enabled vehicles. Since multiple agents collaborate in a typical urban
sensing operation, it is critical to design a mechanism to effectively coordinate their operations to the area where new information is rich in a
completely decentralized and lightweight way. We present a novel agent
coordination algorithm for urban sensing environments that has been designed based on biological inspirations such as foraging, stigmergy, and
Lévy flight. The reported simulation results show that the proposed algorithm enables the agents to move to "information patches" where new
information concentration is high, and yet limits duplication of work due
to simultaneous presence of agents in the same region.

Keywords: Vehicular Ad Hoc Networks (VANET), Vehicular Sensor Networks (VSN), Bio-inspired Data Harvesting, Multi-agent Coordination.

[*] This research is supported through participation in the International Technology
Alliance sponsored by the U.S. Army Research Laboratory and the U.K. Ministry
of Defense under Agreement Number W911NF-06-3-0001, and; by ARMY MURI
under funding W911NF0510246.
[**] Corresponding author.

P. Liò et al. (Eds.): BIOWIRE 2007, LNCS 5151, pp. 204–216, 2008.

1   Introduction

Vehicular Ad Hoc Networks (VANETs) are becoming increasingly popular and relevant to the industry due to recent advances in inter-vehicular communication technologies and decreasing cost of communication devices. Unlike a typical MANET, the networking components in a vehicle have a plenty of computing and storage capacity. Thus, VANETs are considered one of the most promising forms of MANETs outside the military domain and have recently stimulated promising research ranging from safe cooperative driving to entertainment support and distributed data collection.

In this paper, we are interested in urban sensing for effective monitoring of environmental conditions and social activities in urban areas using vehicular sensor networks (VSNs). Differently from traditional wireless sensor nodes, vehicles are not typically affected by energy constraints and can easily be equipped with powerful processing units, wireless communication devices, GPS, and sensing devices such as chemical detectors, still/video cameras, and vibration/acoustic sensors. We particularly envision *proactive* urban monitoring services where vehicles continuously monitor events from urban streets, maintain sensed data in their local storage, process them (e.g. recognizing license plate numbers), and route messages to vehicles in their vicinity to achieve a common goal (e.g. to allow police agents to pursue the movements of specific cars). However, this requires the collection, storage, and retrieval of massive amounts of sensed data. In conventional sensor networks, data are dispatched to "sinks" and are processed for further use (e.g., Direct Diffusion [1]), but that is not practical in VSNs due to the sheer size of generated data. Moreover, it is impossible to filter data a priori because it is usually unknown which data will be of use for future investigations. Thus, the challenge is to find a completely decentralized VSN solution, with low interference to other services, good scalability, and tolerance to disruption caused by mobility and attacks.

To that purpose, we designed and implemented MobEyes, a novel middleware that supports VSN-based proactive urban monitoring applications [2]. In MobEyes, each sensor node performs event sensing, processing/classification of sensed data, and periodically generates data summaries with extracted features and context information tagged with timestamp and position information. Summaries are then disseminated to other regular vehicles such that mobile agents, e.g., police patrolling cars, move and opportunistically harvest summaries from neighbor vehicles. As a result, agents can create a low-cost opportunistic index which enables them to query the completely distributed sensed data storage, thus answering questions such as: which vehicles were in a given place at a given time? which route did a certain vehicle take in a given time interval?, and which vehicles collected and stored the data of interest? Unlike MobEyes, CarTel [3] utilizes opportunistic connectivity via roadside access points to send queries about sensed data and to return replies "on-demand," instead of "proactive" data collection, which should be definitely preferred in presence of constraints on query resolution latency.

Multiple agents can collaborate in harvesting relevant data, processing them, and searching for key information. It is critical to design a mechanism to effectively coordinate and geographically separate the operation of multiple agents, while allowing them to seek most productive fields in a totally distributed matter. However, multi-agent harvesting is a very challenging problem due to the dynamic nature of the target environment (e.g. continuous creation and movement of metadata) and the scale of operations (e.g. harvesting region ranging over multiple city blocks) without a priori knowledge of the location of the critical information. Incidentally, we note that social animals (ranging from bacteria to vertebrates) solve a similar problem of *foraging* to find a good food source quite efficiently using a simple communication mechanism in a fully distributed manner with lightweight and lazy coordination.

Given this observation, the primary goal of this paper is to design a novel multi-agent coordination mechanism for MobEyes harvesting agents by taking inspirations from biological systems. We realize that each species may have inched towards foraging optimality for specific tasks and various constraints (e.g., habitat niches, animal size and speed, environment, etc.). Therefore we design a mechanism by encompassing different animal foraging and behavioral ecology strategies, instead of focusing on single animal species. The natural scene examples inspiring MobEyes multi-agent coordination include: (a) Foraging behavior of *Escherichia (E.) coli* bacteria that operate in distinct modes of locomotion based on the level of nutrient concentration [4,5]; (b) Lévy walk behavior of many biological organisms and groups, e.g., albatrosses and fishing boats, to improve food search over large-scale regions [6,7]; and (c) Stigmergy found in ants and other social insects that use various types of pheromones to signal nest mates with potential conflicts, e.g., a sort of "no entry" sign [8,9].

Based on this study, we propose a novel harvesting strategy, called *datataxis* (á la chemotaxis of E. coli bacteria), that guides the agents to stay and acquire metadata on "information patches," the regions where newly created and not-harvested metadata are concentrated (based on a simple metric for metadata density estimation per road segment). MobEyes agents adapt their behavior by following a 3-state transition diagram that sometimes forces them to change their area of exploration by using Lévy walk-inspired movement patterns that are considered suitable for the large scale of the typically targeted regions. To avoid harvesting work duplication, agents exploit stigmergy-inspired techniques for conflict resolution to prevent from useless concentration of agents in the same region at the same time.

We validate the performance of our proposed data harvesting scheme via extensive simulations where we use a realistic Manhattan mobility model and compare the harvesting efficiency of our datataxis foraging (DTF) with random walk foraging (RWF), biased random walk foraging (BRWF), and an idealized preset pattern foraging (PPF). From this study, we show that the proposed DTF balances the movement of multiple agents and distributes them effectively without the need of centralized and intrusive coordination protocols.

The remainder of the paper is organized as follows: Section 2 presents a background on the MobEyes urban sensing architecture; Section 3 reviews the foraging behaviors in nature and presents our algorithm for multi-agent coordination; Section 4 presents a simulation-based performance evaluation of various agent coordination approaches; finally Section 5 concludes the paper.

2   MobEyes Vehicular Sensing Platforms

We present the MobEyes solution using one of its possible application scenarios: collecting information from MobEyes-enabled vehicles about criminals who spread poisonous chemicals in a particular section of the city (say, a subway station). We assume that the criminals use vehicles for the attack. In this scenario, MobEyes will help detect the criminal vehicles and permit tracking and capture. Here, we assume that the vehicles participating in MobEyes are equipped with cameras and chemical detection sensors. Vehicles continuously generate a huge amount of sensed data, store it locally, and periodically produce short *metadata chunks* obtained by processing sensed data, e.g., license plate numbers or aggregated chemical readings. Metadata chunks are aggregated in a summary packet that is opportunistically disseminated to neighbor vehicles, thus enabling metadata harvesting by the police to create a distributed metadata index which permits to find a set of vehicles storing data of interest for forensic purposes such as crime scene reconstruction and criminal tracking.

Any regular node periodically advertises a new summary packet with generated metadata to its current neighbors to increase the opportunities for agents to harvest the summaries. A packet header includes a packet type, generator ID, locally unique sequence number, packet generation timestamp, and generator's current position. Each packet is uniquely identified by the generator ID and its sequence number pair, and contains a set of metadata locally generated during a fixed time interval. Neighbor nodes receiving a packet store it in their local metadata databases. Therefore, depending on the mobility and the encounters of regular nodes, packets are opportunistically diffused into the network of vehicles, yet metadata diffusion is time and location sensitive. MobEyes can be configured to perform either single-hop passive diffusion (only the source advertises its packet to current single-hop neighbors) or k-hop passive diffusion (the packet travels up to k-hop as it is forwarded by j-hop neighbors with $j < k$). Figure 1 depicts the case of two sensor nodes, $C1$ and $C2$, that encounter with other sensor nodes while moving (the radio range is represented as a dotted circle). A black triangle with timestamp represents an encounter. For ease of explanation, we assume that there is only a single encounter, but in reality there may be multiple encounters with any nodes that happen to come within the dotted circles. $C1$ and $C2$ periodically advertise a new summary packet $S_{C1,1}$ and $S_{C2,1}$ respectively where the subscript denotes $\langle ID, Seq.\# \rangle$. At time $T - t_4$, $C2$ encounters $C1$, and thus they exchange those packets. As a result, $C1$ carries $S_{C2,1}$ and $C2$ carries $S_{C1,1}$.

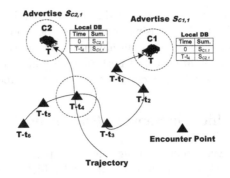

Fig. 1. MobEyes single-hop passive diffusion

In parallel with diffusion, MobEyes metadata harvesting may take place. The MobEyes police agent collects summary packets from regular nodes by periodically querying its neighbors. The goal is to collect all the summary packets generated in a specific region. Ideally, a police node should harvest only those summary packets that it has not collected so far. To focus only on missing packets, a MobEyes authority node compares its list of summary packets with that of each neighbor (i.e., a set difference problem), by exploiting a space-efficient data structure for membership checking, i.e., a Bloom filter [10]. A MobEyes police agent uses a Bloom filter to represent its set of already harvested and still valid summary packets and includes this filter when broadcasting a harvest request message [2]. Given this, each neighbor node prepares a list of missing packets. After random back-off, one of the neighbors returns those missing packets to the agent. The agent sends back an acknowledgment with a piggybacked list of returned packets and, upon listening to or overhearing this, neighbors update their lists of missing packets.

Note that each vehicle can piggyback the current position into its summary advertisement, and thus, Last Encounter Routing (LER) can be supported at no extra cost [11]. Enhanced LER with the carry-and-forward to address intermittent connectivity plays a key role in MobEyes when an agent tries to retrieve the actual data, or to send a dump request to the target vehicle.

3   Multi-agent Information Harvesting

Multiple agents can collaboratively search a given area of interest to collect desired information more rapidly. We design an algorithm to coordinate and control multiple agents to harvest target data as efficiently as possible. In particular, we are interested in designing a simple algorithm that does not involve a tight, close range control of agents' movement, since the latter would incur heavy communication overhead. At the same time, we want the algorithm to be efficient; ideally, we want our algorithm to perform similarly to a centralized coordination algorithm, in terms of data harvesting efficiency (i.e., how fast can we collect all of the interested data) and the control efficiency of agents' movement (i.e., how much redundant data was collected by multiple agents). In addition, we want the algorithm to be able to be self-organizing and adaptive to the dynamics of the environment, such as the changes in the movement patterns, the densities, and the data carried by VSN vehicles. Also some part of the network may exhibit intermittent connectivity; hence, we require our algorithm to be delay tolerant and robust to temporary disconnections.

3.1 Biological Inspirations for Data Harvesting

The main reason for us to look at biological inspiration comes from the observation that the animals and insects encounter a similar problem: they often coordinate their efforts to effectively collect food without prior knowledge of food sources; yet they are known to solve the problem quite effectively, if not optimally [12]. Accordingly to the foraging theory, animals are presumed to search for nutrients and obtain them in a way to maximize the ratio of energy intake over the time spent for foraging. Foraging constraints also shape division of labor in animal societies. This applies to both vertebrate societies where foraging tends to be associated with hunting and is based on individual recognition, and invertebrates (insect) societies which are characterized by a great deal of redundancy. In this section, we review key foraging behaviors in nature that are applied to tackle our problem.

Stigmergy: MobEyes data harvesting is directly related to the food foraging problem solved by stigmergy [9]. Ants need to find routes to possibly ephemeral food sources in an effective manner. Since it is not immediately obvious how long the current site will remain as a valid foraging site, they have to solve a dynamic problem of remembering a rewarding source while exploiting newly discovered food sites. In many cases, the nutrients are distributed in *patches*, and the main issue of foraging is finding such patches, deciding how long it will take before depleting and leaving food sources. The foraging patterns in ants change with increasing prey/food size, showing all stages intermediate between an individual and a mass exploitation of food resources. This suggests that social insects process information and solve problems in a complex environment, while keeping some parsimony at the level of the individuals' decision rules [8]. It has been known that ants can optimize their foraging by selecting the most rewarding source via the following methods. Physical contacts and other forms of direct communication, e.g., via sound or vibrations, are limited both spatially and temporally; only neighbors in the vicinity can receive the signal. On the contrary, *pheromone trails* are long lasting and can be considered a wide broadcast that slowly dissipates in time. Different types of pheromones have evolved in ants. First, there are long-lasting pheromones, used to maintain the spatial organization of ant networks, and volatile pheromones, used to quickly mark routes leading to current food sources. For instance, the pygidial gland of the Ponerine Army Ant *Leptogenys distinguenda* produces a long-lasting trail pheromone (that lasts about 25 minutes), which guides the ants back to the trail or the colony when they are detached from the trail network [13]. Second, there is a short-live *repellent* pheromone, which effectively serves as a no-entry signal.

Chemotaxis of *E. coli*: Another biological foraging behavior that we consider in the context of information harvesting is the chemotactic (foraging) behavior of many bacteria, for example *E. coli* [4]. *E. coli* is representative of a large, widespread class of bacteria, and is present everywhere in the environment and also in the lower intestines of mammals including humans. *E. coli* gets its locomotion from a set of rigid flagella that enables the bacteria to swim. When their flagella turn clockwise, bacteria tumble and do not move to any particular direction. On the other hand, when flagella turn counter-clockwise, the bacteria will swim in a directional movement. The sensors of E. coli are receptor proteins that are stimulated by the binding of molecules in the environment. Based on the level of nutrients (or attractants) a bacterium will move in different modes. More specifically, when an *E. coli* is in some substance without food or noxious substances, its flagella will alternate between moving clockwise and counter-clockwise so that the bacterium will alternate between tumbling and swimming. This alternation will move the bacterium in random directions. We can consider this movement mode a *search* for food. If the nutrients have homogenous concentration, the bacteria will exhibit a search behavior but with increased run length of swimming and decreased tumble time. In effect, they will search for nutrients more aggressively when they are in a nutrient environment. Finally, when the bacteria detect a change in the concentration level of nutrition, they will swim along the gradient of concentration toward the most nutrition rich area, and spend less time tumbling. If somehow, an *E. coli* encounters a region where nutrient gradient does not increase after the swim, it will return to the baseline search mode to look for higher concentrations.

Lévy Walk: There is a growing agreement that foraging and movement patterns of some biological organisms may have so-called "Lévy-flight" characteristics. Lévy random walks, named after the French mathematician Paul Pierre Lévy [6], are known to outperform Brownian random walks when the precise location of the targets is not known a priori but their spatial distribution is uniform. A Lévy flight is comprised of random sequences of movement segments, with lengths l, drawn from a probability distribution function having a power-law tail, $p(l) \sim \ell^{-a}$ where $1 < a < 3$. Such a distribution is said to have a "heavy" tail because large-length values are more prevalent than within other random distributions, such as Poisson or Gaussian. Viswanathan et al. demonstrated that $a = 2$ constitutes an optimal Lévy-flight search strategy for locating targets that are distributed randomly and sparsely [14]. Under such conditions, the Lévy search strategy minimizes the average distance traveled and presumably the average energy expended before encountering a target. The strategy is optimal and results in space filling paths, if the searcher is exclusively engaged in searching, has no prior knowledge of target locations, and if the average spacing between successive targets greatly exceeds the searcher's perceptual range.

3.2   Bio-inspired Multi-agent Coordination in MobEyes

In MobEyes, vehicle mobility is exploited for effective and inexpensive metadata dissemination, i.e., regular cars carry-and-forward metadata to harvesting agents. Therefore, metadata are likely located where the number of vehicles is greater. As an indicator of information concentration, we define the *information density* as the number of metadata carriers, i.e., regular cars actually transporting metadata, in a road segment. We note that our algorithm does not need to depend on this specific metric and can work with any information density metric that can be profitably measured. Like *E. coli* bacteria, our goal is to find a *patch* that contains a large number of "useful" metadata carriers with information not yet harvested by either the same or a cooperating harvesting agent. As a first level approximation, a promising solution for agents is to mimic the foraging behavior of *E. coli* by estimating the gradient of *information density* and moving to a direction where this gradient increases (á la the swim of *E. coli* in a solution with nutrient gradient), while performing a random search when there is no specific gradient (á la the tumble of *E. coli* in a homogeneous environment). We name this bio-inspired behavior of harvesting agents as *datataxis* (inspired by the chemotaxis of *E. coli*).

The key for effective datataxis is to estimate vehicle density in a decentralized way with minimum overhead. To achieve this goal, we propose to divide any road into a set of uniquely identifiable unit distance segments (or "road segments"). Any urban area can be represented as a set of road segments. While MobEyes regular nodes are in a specific road segment, they estimate density of that segment by simply counting the number of their neighbors: this per-segment density estimation is advertised by the vehicles on that road segment via the regular MobEyes summary broadcast process. Each vehicle only advertises the density information for the road segment it is currently on. In that way, the density information is locally computed and updated. Agents can collect per-road segment density samples, by exploiting the regular MobEyes protocol for summary harvesting, with no additional communication overhead.

However, the model of a simple *E. coli* behavior for all cooperating agents is insufficient to realize effective harvesting of monitoring metadata in urban environments. We have extensively explored bio-inspired coordination behaviors to identify, evaluate, and adopt the most suitable differentiated working modes to obtain high harvesting coverage with minimum overhead. In our design, MobEyes agents operate in one of the following three modes: (a) the Lévy Jump (LJ) mode, (b) the Biased Jump (BJ) mode, and (c) the Constrained Walk (CW) mode. The LJ/BJ modes are considered as the exploration stage to find the best possible location to start a more focused search, whereas the CW mode can be considered as the exploitation stage where agents try to harvest as much as possible by carefully and finely controlling their movements. Figure 2 presents a transition diagram consisting of the three possible states of operation by MobEyes harvesting agents.

First of all, a MobEyes agent starts with the LJ mode and searches for dense areas with vehicles. In the Lévy jump literature, it is known that the jump distance following a power law distribution with the exponent of 2 is known to be optimal for non-destructive foraging, i.e., a foraging scheme where agent can "productively" visit the same place many times [15]. Recall that since vehicles

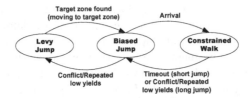

Fig. 2. Agent state diagram

move in the urban grid, it may be very possible that after a while the same area may become "productive" again. The key idea of the LJ mode is that agents can choose a long distance with some probability, due to the heavy tail of the power law distribution. Thanks to the long jumps, the area covered by the agents will be much larger than the area that would have been covered by only random walk movement patterns [15]. Since the network size is finite in our model, we use a truncated Lévy jump distribution: $f(d) = \frac{d_{max}d_{min}}{d_{max}-d_{min}} \frac{1}{x^2}$ where we set the d_{max} as the network diameter and d_{min} as the communication range. The angle of a jump from the current location is selected randomly. For each jump, the agent steers its movement towards the road segment that minimizes the distance to the new jump location. However, for a given location, it may not be feasible to jump toward a certain direction. For instance, if an agent is located at the bottom left corner of the network, a jump is feasible toward the first quadrant. The key idea of a Lévy jump is to have a long jump with some probability for efficient exploration. Thus, we modify the angle selection such that we only consider the region that can span a chosen distance. In the previous example, the jump direction is chosen from the first quadrant.

Once the agent finds a dense area above a certain threshold, the agent changes its operation state to the BJ mode so that it can move toward that location. The target location is the mid-point of the densest road segment, which is also set as the reference point of the CW mode that will be used by the agent as described below. The agent steers its movement towards the road segment that minimizes the distance to the determined reference point (i.e., a simple greedy movement).

When entering the CW region (the circular area with center the reference point and radius R), the agent switches its mode to the CW mode and starts harvesting metadata within that region. The default choice in MobEyes is to automatically set the distance parameter R as a function of the number of agents and the size of the overall search area. MobEyes supports two operating sub-modes for an agent in the CW state. First, the agent follows the road segment that maximizes the positive per-segment density change. In this case, since we exclude the current road segment from the candidate road segment for the next movement, it is possible that the rate change may be negative. If this occurs, the harvesting agent chooses the road segment that minimizes the change. Second, the agent can follow a biased random walk along a set of road segments in the vicinity; the set consists of the segments with density greater than a configurable threshold. If the explored urban area has the shape of a long strip, staying within a CW region could be inefficient. For this reason, the MobEyes agent periodically performs short range jumps to explore the nearby area after CW duration T_{cw}, thus changing its reference point. To avoid the worst case of continuous jumping around a region where there is not much gain, after a configurable threshold, the agent performs a long jump to a random direction, and switches its mode to the LJ mode to collect the density information again as in the initial phase (i.e., repeated low yield case). This behavior is repeated until the harvesting procedure has ended.

One crucial issue in multi-agent harvesting is to coordinate the movements of cooperating agents. Ideally, we want the agents to direct themselves in the richest information areas while not stepping other agents' toes. In other words, each agent coverage area should be non-overlapping with the others and, when agents encounter each other, one of them should be able to quickly move to a different non-overlapping region. To this end, similar to the pheromone trail left by ants, a harvesting agent leaves a trail on the regular vehicles while collecting metadata. The trail information will contain the ID of the collecting agent and the timestamp of data collection. Thus, agents can detect a conflict via meta-data harvesting. For conflict resolution, an agent with lower ID will perform a long jump to a random location that is outside the CW region of the conflicting agents. If it finds an information patch, the constrained random walk begins; otherwise, the LJ mode will be initiated, and the overall process starts over.

4   Evaluation

We evaluate the proposed metadata harvesting algorithm by simulation using ns-2.[1] Mobile nodes communicate using IEEE 802.11 with fixed bandwidth of 11Mbps and nominal radio range of 250m. Vehicles move in a fixed region of size $2400m \times 2400m$ according to the Manhattan mobility model (MT) from [16]. In MT, nodes are moving on the streets defined by a map (Figure 3). At each intersection, vehicles make independent decisions about the next direction; the choice of direction (straight, left, right) is equally probable. We use 7x7 grids (each grid segment is set to 300m to avoid interference between nearby streets). We populate two horizontal streets, Street 2 and Street 6, with vehicles by controlling transition probability (i.e., make left or right turns with probability 0.1, and go straight with probability 0.8). When nodes reach the boundary of the simulated region, they bounce back by inverting their direction (modeled by forcing U-turn with probability 1). If this happens, we reset the node and treat it as a new incoming node that carries no meta-data. We consider the number of nodes $N = 200$, and the maximum speed $v = 20m/s$. We fix the speed of harvesting agents to a constant (10m/s).

We evaluate the following foraging schemes by agents: (a) Random Walk Foraging (RWF), (b) Biased Random Walk Foraging (BRWF), (c) Preset Pattern Foraging (PPF), and (d) Datataxis Foraging (DTF). Agents in RWF randomly

---

[1] http://www.isi.edu/nsnam/ns

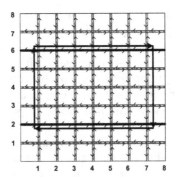

Fig. 3. Street map used for the Manhattan mobility model: Horizontal streets 2 and 6 (marked with thick solid lines) are initially populated (dense streets). The regular mobility pattern (clockwise directional cycle marked by thick areas) is traveled by agents in the PPF strategy.

choose road segments and harvest metadata from encountered vehicles. Agents in BRWF operate similarly except that they choose road segments based on a defined transition probability that is biased by knowledge about "food sources" (i.e., Street 2 or 6). In PPF, we define a preset mobility pattern representing that the agents are fully aware of the movement patterns of others, and thus, we configured the agents to move around the rectangular path that includes Streets 2 and 6. The PPF foraging strategy represents the optimal agent movement in our scenario since the agents will cover the most popular streets using this mobility pattern. DTF implements our proposed scheme for agent movement while harvesting metadata. An agent explores a region while in the Lévy Jump mode to estimate meta-data density per road segment, and switches its mode to the CW mode. After moving to an information patch, an agent stays there for 300s (in CW mode). An agent performs short jumps within a CW region, where the radius of the CW region is set to 600m. If it finds another region after the jump, a conflict is detected, and it performs a long jump, where the maximum jump range is set to 900m. The number of agents used in the simulation varies from 1 to 4 nodes.

We set the summary packet advertisement period of regular nodes and the harvesting request period to 3s in all the simulations. A new summary is generated based on a Poisson process with rate $\lambda = 1/10$ (i.e., on average it is generated every 10s). We measure the performance in terms of the total number of harvested summary packets. If multiple agents are used, we also calculate aggregate harvesting rate. When calculating the harvesting rate, we only count the number of distinct summaries harvested. For a given scenario, we report the average values of 30 runs, each of which takes 1500s (i.e., 25 minutes).

Simulation Results: Figure 4 reports the impact of the number of foragers on harvesting performance, by showing the number of harvested summaries per agent. In general, the graph shows that the value decreases as the number of

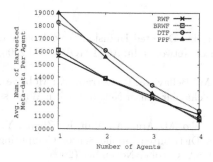
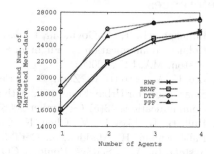

Fig. 4. Number of harvested summaries per agent

Fig. 5. Aggregated number of harvested summaries

agents increases because we only count unique summaries. We note that BRWF shows only a slight improvement over RWF. This stems from the fact that once the agents in BRWF deviate from the popular streets, it takes a long time for them to return to productive areas. The performance of DTF is consistently better than RWF and BRWF, and quite close to PPF. Recall that PPF is a foraging strategy specifically and statically optimized to our target deployment scenario and is expected to represent a quasi-optimal solution. Thus, we find that our DTF algorithm is efficient, not far from the performance achievable via static knowledge of the characteristics of the considered deployment environment. Figure 5 shows the total number of distinct summaries harvested by all the agents. In this plot we also find that the aggregate harvesting ratio of DTF is much better than both RWF and BRWF, and very close to PPF.

5 Conclusion

In this paper, we presented a novel data-harvesting algorithm for urban monitoring applications. The proposed algorithm has been designed based on biological inspirations such as (a) foraging behavior of *E. coli* bacteria, (b) stigmergy found in ants and other social insects, and (c) Lévy flights found in foraging and general movement patterns. The proposed algorithm called *datataxis* enables the MobEyes agents to move to "information patches" where new information concentration is high. This algorithm is guided by a practical metric based on local efficient estimates of information density per road segment. In our data foraging strategy, an agent starts with a random walk until it encounters an information patch; then it performs a constrained walk to move toward a higher density region. When an agent encounters some other agents in the same region it moves to another region by using a conflict resolution algorithm that has been inspired by Lévy jump, so that harvesting agents' work is not duplicated. Our simulation results showed that datataxis effectively balances the movement of agents and distributes them appropriately, performing better than other commonly used harvesting strategies.

References

1. Intanagonwiwat, C., Govindan, R., Estrin, D.: Directed Diffusion: a Scalable and Robust Communication Paradigm for Sensor Networks. In: ACM MOBICOM 2000, Boston, MA, USA (2000)
2. Lee, U., Magistretti, E., Zhou, B., Gerla, M., Bellavista, P., Corradi, A.: MobEyes: Smart Mobs for Urban Monitoring with a Vehicular Sensor Network. IEEE Wireless Communications 13(5) (Septmber 2006)
3. Hull, B., Bychkovsky, V., Chen, K., Goraczko, M., Miu, A., Shih, E., Zhang, Y., Balakrishnan, H., Madden, S.: CarTel: A Distributed Mobile Sensor Computing System. In: ACM SenSys, Boulder, CO, USA, October-November (2006)
4. Passino, K.M.: Biomimicry for Optimization, Control and Automation. Springer, Heidelberg (2005)
5. Kim, T., Jung, S., Cho, K.: Investigations into the Design Principles in the Chemotactic Behavior of Escherichia Coli. Biosystems (2007)
6. Levy, P.: Theorie de l'addition des variables aleatoires. Gauthier-Villars, Paris (1954)
7. Reynolds, A.M., Frye, M.A.: Free-Flight Odor Tracking in Drosophila Is Consistent with an Optimal Intermittent Scale-Free Search. PLoS ONE 2(4) (2007)
8. Detrain, C., Deneubourg, J.L.: Complexity of Environment and Parsimony of Decision Rules in Insect Societies. Biol. Bull. 202(3), 268–274 (2002)
9. Theraulaz, G., Bonbeau, E.: A Brief History of Stigmergy. Artificial Life 5(2), 97–116 (1999)
10. Bloom, B.H.: Space/Time Trade-offs in Hash Coding with Allowable Errors. CACM 13(7), 422–426 (1970)
11. Grossglauser, M., Vetterli, M.: Locating Nodes with EASE: Mobility Diffusion of Last Encounters in Ad Hoc Networks. In: IEEE INFOCOM, San Francisco, CA, USA, March-April (2003)
12. Stephens, D., Krebs, K.: Foraging Theory. Princeton University Press, NJ (1986)
13. Jackson, D.E., Ratnieks, F.L.: Primer: Communications in Ants. Current Biology 16(15), 570–574 (2006)
14. Bartumeus, F., da Luz, M.G.E., Viswanathan, G.M., Catalan, J.: Animal search strategies: a quantitative random-walk analysis. Ecology 86, 3078–3087 (2005)
15. Viswanathan, G., Buldyrev, S.V., Havlin, S., da Luz, M.G.E., Raposo, E.P., Stanley, H.E.: Optimizing the success of random searches. Nature 401, 911–914 (1999)
16. Bai, F., Sadagopan, N., Helmy, A.: The IMPORTANT Framework for Analyzing the Impact of Mobility on Performance of Routing for Ad Hoc Networks. Ad Hoc Networks Journal - Elsevier 1, 383–403 (2005)

Bio-Inspired Approaches for Autonomic Pervasive Computing Systems

Daniele Miorandi[1], Iacopo Carreras[1], Eitan Altman[2], Lidia Yamamoto[3], and Imrich Chlamtac[1]

[1] CREATE-NET, via alla Cascata, 56/C, 38100 – Povo, Trento Italy
`name.surname@create-net.org`
[2] INRIA, 2004 Route des Lucioles - BP 93, 06902 – Sophia Antipolis France
`eitan.altman@sophia.inria.fr`
[3] Computer Science Department, University of Basel, Bernoullistrasse 16,
4056 – Basel Switzerland
`lidia.yamamoto@unibas.ch`

Abstract. In this chapter, we present some of the biologically-inspired approaches, developed within the context of the European project BIONETS for enabling autonomic pervasive computing environments. The set of problems addressed include networking as well as service management issues. The approach pursued is based on the use of evolutionary techniques — properly embedded in the system components — as a means to achieve fully autonomic behaviour.

Keywords: pervasive computing, biologically-inspiread design paradigms, protocol evolution, chemical computing.

1 Introduction

The Future Internet will be characterized by scale, heterogeneity, complexity and dynamicity figures that call for novel approaches to the design and the management of computing and communication systems. This will enforce a shift from conventional "top-down" engineering approaches, in which systems' blueprints are designed, optimized and engineered to perform a well-defined task, to "bottom-up" approaches, in which systems will be provided with the necessary means for growing and evolving in an unsupervised manner. The final goal is to enable autonomic systems, which are able to self-manage themselves, requiring human intervention only in the definition of the high-level goals to be pursued.

The BIONETS project (Biologically-inspired Autonomic Networks and Services, `www.bionets.eu`), funded in the framework of the EC-FET initiative on Situated and Autonomic Communications, targets the introduction of biologically-inspired approaches for designing and managing pervasive computing and communication environments. The conceptual basis of the project is rooted in the observation that nature has been successfully tackling the aforementioned problems (scale, heterogeneity, complexity and dynamicity), leading to complex ecosystems which are able to self-sustain and to reach efficient equilibria in the absence of a central control.

P. Liò et al. (Eds.): BIOWIRE 2007, LNCS 5151, pp. 217–228, 2008.

In this chapter, we review some paradigms, inspired by natural (and in particular biological) systems' functioning, which could be successfully applied to architect pervasive computing/communications environments. We also discuss some of the issues to be faced when trying to engineer such paradigms into technological artifacts, presenting some examples drawn from the research activities carried out within the BIONETS project framework.

The remainder of the chapter is organized as follows. In Sec. 2 we discuss the challenges stemming from Future Internet scenarios, highlighting the need for embedding autonomic properties in computing/communication systems. In Sec. 3 we survey some potentially useful paradigms, inspired by the operations and functioning of various natural (mostly biological) systems. In Sec. 4 we discuss issues related to the use of such paradigms for designing computing/communication systems, presenting two examples taken from project's activities. Sec. 5 concludes the paper discussing promising applications of the presented paradigms.

2   Scenarios for Future Internet and the Need of Autonomicity

If in the 80s and 90s the Internet was still conceived as an "information highway", i.e., a set of physical links where myriads of data and packets were flowing, carrying the most disparate information, things are changing rapidly. The change concerns not just what users are doing with those data/packets, but the nature of the system itself. On the one hand, the Internet has become a vital ganglion of the globalised economy and society. On the other one, a lot is happening at the edges of the Internet as we have known it. Progresses in microelectronics and nanotechnologies are about to lead to situations in which electronics (including ability to communicate and to compute) gets embedded in a variety of common objects. The net results is expected to be an invisible digital halo, surrounding users and supporting them in their daily life. While this phenomenon is referred to in various ways, depending from the viewpoint (from pervasive computing environments [1] to smart spaces up to "Internet of things"), there is a general consensus that this is the direction we are moving to. This trend is bringing with it a set of challenges to current information and communication technologies and system architectures, requiring radical changes in the approaches conventionally pursued.

The first and most apparent problem is *scale*. Scale in the number of potentially connected devices as well as in the number of users and of services to be supported. Are currently solutions adequate? Can we rely on Moore laws for bandwidth/computing power/storage capacity to ensure we will be able to cope with scale? The answer is probably negative. The reason lies in the fundamentals of the current IP network architecture (which represents the backbone of almost all networked systems), which rely on an address-oriented end-to-end paradigm, assuming always-on connectivity. This incurs limitations in both the finiteness of the address space as well as problems related to the scalability of connected large-scale wireless networks.

The second problem is *heterogeneity*. Again, this is rooted in the hourglass model at the basis of the Internet TCP/IP protocol suite, where the IP acts as a "glue" between various subsystems. It is probably necessary now to rethink it, in such a way to accommodate heterogeneity and pluralism in the system, in terms of both nodes [2] and network architecture [3].

The third problem is *complexity*. Network and service management and maintenance is becoming a harder and harder job, requiring an extremely large amount of operators' time, with the consequent negative fallouts on the economic side. And with increase in scale and heterogeneity this is probably going to become the real ceiling limiting the ability to produce innovation in the ICT field.

The fourth problem is *dynamicity*. The ICT world is experiencing innovations at an extremely high rate. New technologies and services are created and disappear continuously, and new ways of profiting from the Internet are been envisaged and introduced seamlessly. There is an increasing need to design ICT systems which are able to be, in some sense, future-proof, i.e., able to plastically adapt to changing environments, services and users needs.

These four challenges motivated the research on methods and tools for building autonomic communication systems [4], in much the same way it had been proposed by IBM in the computing field [5]. The basic problem lies in the fact that current ICT systems are conceived as static ones. The ability to adapt is — in most cases — present, but in a limited form, and it is decided *a priori* in the design phase. In some sense, adaptability is "hardwired" into the system's blueprint, which is, however, unable to adapt and change. Traditional "direct engineering" approaches have the great advantage of engineering "by design" the desired system behaviour, but, at the same time, they limit the ability of systems to adapt or optimize to unforeseen scenarios.

The BIONETS project, on whose activities this chapter is mostly centred, aims at addressing such issues by looking at how nature (and biology in particular) has led to the arising of complex ecosystems, able to achieve the self-CHOP features of IBM's autonomic computing manifesto (self-configuration, self-healing, self-optimization, self-protection) [5] through open-ended evolution.

3   Nature-Inspired System Design Paradigms

Nature presents a variety of examples of systems that are able to successfully deal with scale, heterogeneity and complexity figures similar to those expected for pervasive computing environments. As far as dynamicity is concerned, things are slightly different, in that many natural phenomena (e.g., evolution) take effect over long time periods. On the other hand, in an artificial system the rate at which such phenomena can be emulated depends heavily on the available computing power.

In the early phase of the BIONETS project, a set of paradigms inspired by biological, physical and social phenomena were identified and studied [6]. We present in the following a short review of three of the most relevant paradigms identified: chemical computing, embryology and evolutionary game theory. With respect to the aforementioned issues, chemical computing provides insight into the design of computing systems able to continuously change to self-optimize to current system conditions (dynamicity). Embryology may provide means for applying evolutionary computation methods to extremely complex systems (scalability, complexity), enabling at the same time run-time optimization (dynamicity). Evolutionary game theory provides means to analyze and design systems able to work unsupervised on the basis of local interactions (scalability, complexity).

3.1   Chemical Computing

The term *Chemical computing* [7,8] refers to two distinct areas: *(i) real chemical computing*: computing with real molecules, such as DNA computing; *(ii) artificial chemical computing*: hardware and software architectures inspired by chemistry. In the case of software, it refers to computation models following a chemical metaphor, which run on regular von Neumann computers. The scope of this section is restricted to the latter.

Chemical computing can be regarded as a branch of *Artificial Chemistry* [8], the subfield of Artificial Life devoted to modelling the dynamics of chemical phenomena in order to understand the origin and evolution of organizations in general, and life in particular. The term *artificial chemistry* also refers to the specific chemical model used, defined by the molecular species involved, the reaction rules, and the algorithm for the reaction vessel.

Numerous artificial chemistry models have been proposed. A model in which molecules are λ-calculus expressions is presented in [9], showing conditions for the emergence of self-maintaining organizations out of an initial "soup" of random molecules. The chemical reaction model is catalytic (i.e. reactants are conserved after the reaction), and mass conservation is ensured by a dilution flux. This line of research led to a theory of chemical organizations [10], with several applications in biology and computer science.

We believe that chemical models have a great potential for on-line evolution in autonomic systems. This can be illustrated by the following example: In [11] Genetic Programming (GP) is applied to an algorithmic chemistry in which instructions are drawn from a multiset and executed in random order. Starting from a nearly unpredictable system, a few generations later programs exhibit highly reproducible results. The system is therefore able to evolve programs that are robust to random execution order. The authors point out the importance of the concentration of instructions, rather than their sequence. Indeed, in the solutions evolved, the concentrations of the instructions that are the most crucial to the solution are higher, and instructions that are not relevant end up with low or no concentration at all.

Chemical models raise many new questions as well. Methodologies for engineering, programming or evolving chemical reaction networks are only now emerging, for natural or artificial systems. Most of the models have a stochastic nature, which makes them inherently non-deterministic, and also more complex than traditional top-down, human-made solutions. On the other hand, we believe that making progress in this area will greatly improve our understanding of computational models close to biology, and of life in general.

3.2 Artificial Embryogenies

The application of ideas from embryology to artificial systems has been following two main research directions. The first one is *embryonics* (embryology plus electronics), an approach to improve fault tolerance in evolvable hardware by using a cellular architecture presenting dynamic self-repair and reproduction properties [12]. Approaches in this area have mostly focused on the use of *artificial stem cells*, i.e., cells which are able to differentiate into any specific kind of cell required for the organism to work. The systems devised in such way are based on the following two principles:

- Each cell contains the whole genome, i.e., the complete set of rules necessary for the organism to work and is totipotent, i.e., can differentiate into any specific function.
- The system presents self-organizing properties. Each cell monitors its neighborhood and may return to the stem cell state and differentiate into another type of cell to repair a fault detected.

The flexibility to switch functionality adds another level of robustness with respect to conventional approaches, as now not only cells with identical functionality can be used as backup or template to repair a failure, but also other cells with differentiated functionality can be used to recreate a lost one.

The second one is *artificial embryogeny* [13], which aims at extending evolutionary computing with a developmental process inspired by embryo growth and cell differentiation, such that relatively simple genotypes with a compact representation may express a wide range of phenotypes or behaviors. Indeed, researchers have recognized that "conventional" EC techniques (like GA, GP, Evolutionary Strategies, etc.) present scalability problems when dealing with problems of relevant complexity. In artificial embryogenies the genotype does not code the solution itself, but it codes recipes for building solutions (i.e., phenotypes). This can lead to a non-linear genotype-to-phenotype mapping, which may also be affected by environmental variables. In this way, a genotype change does not imply a direct change in the solution, but in the way solutions are decoded from the genotype and further grown from an initial "seed" (the embryo).

3.3 Evolutionary Games

In autonomic pervasive computing systems, users compete for resources; decisions related to flow control, to routing, to accessing common channels etc. are not under control of a central entity. Non-cooperative Game Theory [14] has naturally become a very popular paradigm for modeling the decentralized decision making. The high complexity of dynamic computing, information and communication systems suggests that one should search for models and concepts that involve large populations, dynamics, adaptation and evolution.

Equilibria concepts (the Nash and the correlated equilibrium, the Wardrop equilibrium in road traffic engineering, and the Evolutionary Stable Strategy in evolutionary games) describe relatively "static" situations, in which those involved in the game are relatively satisfied: they cannot be better off by unilaterally deviating. We may expect however that many situations of competition in autonomous complex networks are dynamic ones. Tools are needed to understand the evolution of competition, the way one converges to, or diverges from, equilibria. Such understanding can then be very useful in designing mechanisms for evolution of services related to autonomic pervasive computing systems. The evolutionary game paradigm was created by J. Maynard Smith [15,16] in a context of conflicts and competition among populations in biological complex systems. It provides tools to describe the competition dynamics between populations through differential equations that are called replicator dynamics and which relate fitness of species with their growth rate [17].

The TCP congestion control protocol is an example of a distributed network mechanism that allows flows to adjust their transmission rate in a completely decentralized way. It has had many variants that differ from each other by the degree of aggressiveness. The first protocols were the most aggressive ones and have caused catastrophic events known as congestion collapse in the Internet. These protocols have disappeared, replaced by the very gentle, non-aggressive protocol, Tahoe, that has disappeared too. New TCP protocols keep appearing. In [18], an elementary model from Evolutionary game theory, called the "Hawk and Dove game" is used to predict both equilibrium behavior between aggressive and friendly TCP versions, as well as non equilibrium behavior. It is shown that depending on system's parameters, one of two types of equilibrium can emerge: (i) One in which the two types of TCP coexist (the actual percentage of each one are given there) or (ii) One in which only the aggressive TCP survives.

A stability condition is then derived. An oscillatory behavior is identified when the system is unstable. The oscillations are perceived as instability since they do not allow the system to attain the equilibrium point that may well exist under the same set of parameters. Oscillations in population sizes are also found in the context of competition between species in biology. The difference with biology is, however, that in networks we can use our understanding of evolutionary behaviors to achieve stability and suppress oscillatory behavior. Guidelines for achieving stable behavior have been proposed in [18] in terms of delays in the system as well as some gain parameter that controls the rate of adaptation. More sophisticated stability analysis can be found in [19,20].

4   Embedding Nature-Inspired Strategies

The paradigms presented in the previous section represent promising models for building autonomic computing/communications systems. Nonetheless, for some of them (e.g., embryology), the current status of research does not allow direct application to real-world problems. On the other hand, some of the applications of nature-inspired techniques developed in BIONETS have relied on variants of tools and techniques from evolutionary computation [21].

In this section, we report two case studies related to evolution of network protocols. The aim of this section is to present some of the most relevant issues to be faced when moving from the paradigm to the application, together with a series of lessons learned from the experiments performed.

4.1   Case Study: Evolving Protocols with the Fraglets Language

In this section we report our experience with evolving network protocols using the Fraglets language [22], a programming inspired by chemical computing, and targeted at network protocols.

A *fraglet* [22] is a "virtual molecule", that represents a computation fragment as a string of symbols $[s_1 \ s_2 \ \ldots \ s_n]$. A fraglet may contain data, reaction rules involving two fraglets, or transformations of a single fraglet. Fraglets are injected for execution into a virtual reaction vessel which contains a multiset of fraglets.

The fraglets language has been designed for network protocols. For this purpose the rule processing engine is based on the "tag matching" principle: fraglets are processed according to their head symbol, which is consumed in the process, similar to protocol header processing in network packet streams.

The fraglet instruction set includes a few reaction rules and several transformations. As an example, the match reaction rule has the form [match s $tail_1$]. It reacts with any fraglet of the form [s $tail_2$] (i.e. which starts with the matching symbol s), and produces [$tail_1$ $tail_2$], i.e. the concatenation of the two tails. Other examples of transformation rules are: dup which duplicates a symbol, and exch which swaps two symbols.

The fraglets language had been originally proposed for the automatic synthesis and evolution of protocol implementations [22]. However, no actual results showing automatic evolution had been shown. We therefore performed some first GP experiments to evolve reliable transmission protocols in fraglets [23]. After several unsuccessful trials, feasible programs could finally be obtained via a primitive form of homologous recombination, which consisted in inserting pre-defined markers by hand at given points in the code, where recombination was allowed to occur. With this simple technique the system was able to find the optimum protocol for a given environment by combining existing marked modules.

The main weakness of [23] was its inability to actually produce any viable novelty: only individuals that were recombinations of existing successful ones were also successful. New individuals obtained by random mutation were either infeasible or had poor performance. We then started investigating the reasons for such difficulties in evolving fraglet programs, in contrast with the undeniable success of standard GP techniques based on trees or linear programs. Fraglets are also linear programs, therefore we had expected that it would be easy to evolve fraglet programs as it is to evolve linear ones. However, the header matching pattern turns out to be extremely constraining to the evolution process: a random mutation in a program could easily lead to tags never being matched. And in many cases, one such tag would suffice to block the whole program.

Further obstacles to evolution were later discovered: for instance, when trying to evolve programs out of a "primordial soup" of randomly generated fraglets, the system most of the times ended up clogged by a mass of instructions that were not executable, because a corresponding matching tag could never be found: that was the case of a [match exch ...] or [match dup ...]. The fraglet interpreter does not allow matching on rule keywords, therefore such fraglets can neither match nor be eliminated (since a [match match] is not allowed either). So they pollute the system forever.

Moreover, the original fraglets reaction algorithm contained a non-standard "smallest" criterion: the probability of choosing a given reaction was proportional to the smallest concentration among each of its reactants. The side effect of this rule was that the balance of molecule concentrations did not have the desired effect on the reaction probabilities that would be expected from a chemical model. As a consequence, well-known techniques from systems biology (e.g., based on stoichiometric analysis or differential equations) could not be applied. As a result, it was difficult to implement an effective code regulation mechanism [24] aimed at enforcing good genotypes and eliminating bad ones, based on the control of molecule concentrations. Some modest results on code regulation were achieved [24], but the extension of that model to more complex cases clearly showed limitations.

In spite of all these difficulties, we still believe that there is a potential in chemical models for on-line program evolution, which remains unexplored: Chemical programs are inherently parallel, robust to random execution order, and can naturally support multiple alternative execution flows.

We are now in the process of fully redesigning the fraglets language: its molecule format, instruction set and reaction algorithm, with three aims. First, programs should be fully self-modifiable: it should only be possible to generate rules that may be later eliminated. Second, the language should become plainly suited to GP, by maximizing the chance of obtaining valid programs, both syntactically and semantically. Third, it should be possible to control the concentration of instructions using closed feedback loops which monitor the system, promote good programs and eliminate bad ones.

With respect to the third goal above, we have enhanced fraglets with variants of the Gillespie algorithm, a well-known algorithm for simulating the dynamics of a real-world chemical reactor tank. Experiments with the new algorithm confirm that concentration dynamics now match the expected chemistry patterns.

Concerning the first and second goals above (self-modification and GP orientation), the modifications needed are so radical that the outcome might be an entirely new language. The basic principles however stay the same: A programming language based on the chemical reaction metaphor, in which programs are expressed as a set of virtual molecules in the form of strings that can react with each other. Virtual molecules express code and data in a uniform way. Since molecules may be long strings, it is important to choose reactants based on short keys that can be quickly looked up in a hash table, making the implementation of the chemical reaction algorithm feasible in terms of computation time. Moreover, each chemical reaction should take a short time to execute, like a thread that is scheduled for execution for a short time and then preempted.

In the new language [25], random access to any position in a fraglet will be allowed, and stack-based operations will be possible: one fraglet will be able to act as a data repository for another, accessible as a vector or as a stack, upon reaction. Since code and data are represented in a uniform way, the program can produce further code by writing on the data structure, which can then be executed by simply removing the head symbol. Rules will be able to act on other rules, such that they can always be created and eliminated, leading to fully self-modifiable code. A semantic will be assigned to each possible reaction rule, such that it can always be executed in a valid way, no matter the amount of parameters available and their types. This should enable smoother GP runs.

Given the key length restriction described above, the tag matching problem will somehow remain: the alternative would be to consider every different string as a separate molecular species, which would not be scalable. This problem is intrinsic to any artificial polymer chemistry, in which chemical species may be arbitrarily long molecular chains, and which assumes a well-stirred solution (such that any molecule can potentially collide with any other). However the problem might be significantly reduced if: first, any symbol may form a valid key, including reserved keywords of the language, numbers, etc.: this enhances the space of valid programs; second, the same key may be reused several times during sequential operations: it is easier to automatically generate a program sequence based on a single key, rather than a chain of interconnected keys as currently required in fraglets.

4.2 Case Study: Evolutionary Epidemic Dissemination Mechanisms

Epidemic-style forwarding [26] has been proposed as an approach for achieving message delivery in intermittently connected wireless ad hoc networks, also referred to as Delay-Tolerant Networks (DTNs) [27]. Such environments are characterized by a high degree of dynamism and by the unpredictable nature of the contact patterns, which make standard ad hoc wireless networks routing protocols unsuitable. Epidemic-style forwarding in DTNs is based on a "store-carry-forward" paradigm: a node receiving a message buffers it and carries it around, passing it on to new nodes upon encounter. Each time a node encounters a peer not having a copy of one message it carries, the carrier may decide to *infect* this new node by passing on a message copy. The message gets delivered when the destination first meets an infected node.

In a DTN scenario, the choice of a forwarding scheme and of its set of running parameters depend on a set of factors (mobility, traffic patterns etc.) which are — in general — not known at design time, and may significantly change over time and space. Standard adaptive techniques are limited in that they require an *a priori* definition of the actions to be taken in response to some external stimuli. The mechanism proposed in [28], on the other hand, is based on the use of concepts and tools from the Genetic Algorithms field.

In the proposed implementation, each node in the system employs its own forwarding policy, determining the actions to be taken upon the reception of a message destined to another node. The genotype in such system is represented by an array of parameters, defining the system's behavior. In the considered case-study, we considered as relevant parameters the probability of forwarding a message upon a contact with a susceptible node, as well as the maximum number of hops traversed by a message. (In general, however, there is no limitation on the number of parameters that can encoded.) Each genotype is associated with a fitness level which describes its ability to contribute to the general system's functioning. Upon encountering, two nodes may exchange information on the genotype current in use and the respective fitness level. Each node maintains a pool which contains all available information on genotypes and associated fitness levels. Such set of genotypes is used to generate new ones periodically, applying standard GA operators (crossover and mutation) to two genotypes selected with probability proportional to their fitness level.

Message delivery can be regarded as a *distributed* service, which require multiple entities to cooperate in order to achieve the desired goal. As a direct consequence, there are two main difficulties to be handled. The first one is related to the estimation of the fitness level associated to each genotype in use, which needs to be performed locally relying on partial information only. The second one is related to the fact that, in general, there is no way of controlling the behavior of the (other) nodes contributing to the delivery process. There is also a further subtlety which is worth being considered. In this case indeed, the evolutionary process is deeply intertwined with the mechanisms used for achieving communications (which represent a key component of a distributed evolutionary framework). This coupling introduces a bias in the fitness estimation process, which turns out to have a notable influence on the mechanism behaviour.

While the proposed mechanism has been shown, through extensive numerical simulations, to perform well over a wide range of operating conditions, some problems were encountered which represent as many lessons learned on the engineering of distributed evolutionary mechanisms. The main issue to be faced is the problem of mapping a global optimization problem to a distributed localized one, in which many local entities perform each its own optimization process, based on information available locally only. While in the proposed solution this was done in a rather ad hoc way, there are good chances that results from game theory can be used to provide a framework for this mapping from global optimization to local decisions. Another issue of considerable impact is the definition of suitable mechanisms for cascade fitness estimation in distributed services. In the considered service, indeed, all intermediate nodes along the path followed by the message from the source to the destination contribute to the performance of the global service (end-to-end delivery). The fitness of the overall service is computed by the two end-points, but the problem is then to decide how to reward nodes along the delivery path. A general framework for such problem is still missing. Last, the proposed solution suffer from a bias in the estimation of the delay. While this is a general sampling problem (due to the use of a finite observation window), it turns out to have a rather considerable impact on the behavior of the mechanism.

5 Conclusions

In this chapter, we have presented an overview of the bio-inspired lines of research developed within the framework of the BIONETS project for building autonomic pervasive communications/computing environments. Three of the most promising paradigms identified have been briefly presented, together with two examples reporting the difficulties to be faced when moving to the application phase.

In general, much remains to be done in order to engineer autonomic pervasive computing systems by applying bio-inspired approaches. The main limitations appear to be related to the problems encountered when trying to reproduce in a computable medium natural phenomena. While living eco-systems can indeed be regarded as a special form of distributed computing, our ability to reproduce such processes (with all their positive features) in a computing system is still limited. Attempts to bridge research communities working on biology, ecology and computer sciences appear to bring great potential for architecting the key technologies to build the Future Internet.

Acknowledgments

This work has been partially supported by the European Commission within the framework of the BIONETS project, IST-FET-SAC-FP6-027748, www.bionets. eu. The authors are grateful to S. Alouf, G. Neglia and A. Fialho for the discussions on the evolutionary forwarding case-study and to T. Meyer for the joint work on reaction algorithms for fraglets.

References

1. Weiser, M.: The computer for the 21st century. SIGMOBILE Mob. Comput. Commun. Rev. 3(3), 3–11 (1999)
2. Carreras, I., Chlamtac, I., Pellegrini, F.D., Miorandi, D.: BIONETS: Bio-inspired networking for pervasive communication environments. IEEE Trans. Veh. Tech. 56, 218–229 (2007)
3. Crowcroft, J., Hand, S., Mortier, R., Roscoe, T., Warfield, A.: Plutarch: an argument for network pluralism. In: Proc. of ACM SIGCOMM, Karlsruhe, DE (2003)
4. Dobson, S., Denazis, S., Fernandez, A., Gaiti, D., Gelenbe, E., Massacci, F., Nixon, P., Saffre, F., Schmidt, N., Zambonelli, F.: A survey of autonomic communications. ACM Trans. Aut. Adapt. Syst. 1, 223–259 (2006)
5. Kephart, J.O., Chess, D.M.: The vision of autonomic computing. IEEE Comp. Mag. 36(1), 41–50 (2003)
6. Altman, E., Dini, P., Miorandi, D., Schreckling, D.: Paradigms and foundations of BIONETS research, http://www.bionets.eu/docs/BIONETS_D2_1_1.pdf

7. Dittrich, P.: Chemical Computing. In: Banâtre, J.-P., Fradet, P., Giavitto, J.-L., Michel, O. (eds.) UPP 2004. LNCS, vol. 3566, pp. 19–32. Springer, Heidelberg (2005)
8. Dittrich, P., Ziegler, J., Banzhaf, W.: Artificial Chemistries – A Review. Artificial Life 7(3), 225–275 (2001)
9. Fontana, W., Buss, L.W.: The Arrival of the Fittest: Toward a Theory of Biological Organization. Bulletin of Mathematical Biology 56, 1–64 (1994)
10. Dittrich, P., di Fenizio, P.S.: Chemical organization theory: towards a theory of constructive dynamical systems. Bulletin of Mathematical Biology 69(4), 1199–1231 (2005)
11. Banzhaf, W., Lasarczyk, C.: Genetic Programming of an Algorithmic Chemistry. In: O'Reilly, et al. (eds.) Genetic Programming Theory and Practice II, vol. 8, pp. 175–190. Kluwer/Springer (2004)
12. Prodan, L., Tempesti, G., Mange, D., Stauffer, A.: Embryonics: artificial stem cells. In: Proc. of ALife VIII, pp. 101–105 (2002)
13. Stanley, K.O., Miikkulainen, R.: A taxonomy for artificial embryogeny. Artif. Life 9, 93–130 (2003)
14. Fudenberg, D., Tirole, J.: Game Theory. MIT Press, Cambridge (1991)
15. Maynard Smith, J.: Game theory and the evolution of fighting. In: Maynard Smith, J. (ed.) On Evolution, pp. 8–28. Edinburgh University Press (1972)
16. Maynard Smith, J.: Evolution and the Theory of Games. Cambridge University Press, Cambridge (1982)
17. Hofbauer, J., Sigmund, K.: Evolutionary Games andvPopulation Dynamics. Cambridge University Press, Cambridge (1998)
18. Tembine, H., Altman, E., El-Azouzi, R., Hayel, Y.: Evolutionary games for predicting the evolution and adaptation of wireless protocols submitted
19. Tembine, H., Altman, E., El-Azouzi, R.: Delayed evolutionary game dynamics applied to the medium access control. In: Proc. of IEEE BioNetworks, Pisa, IT (2007)
20. Tembine, H., Altman, E., El-Azouzi, R.: Asymmetric delay in evolutionary games. In: Proc. of ValueTools, Nantes, FR (October 2007)
21. Foster, J.A.: Evolutionary computation. Nature 2, 428–436 (2001)
22. Tschudin, C.: Fraglets - a metabolistic execution model for communication protocols. In: Proc. of AINS, Menlo Park, USA (July 2003)
23. Yamamoto, L., Tschudin, C.: Experiments on the Automatic Evolution of Protocols using Genetic Programming. In: Stavrakakis, I., Smirnov, M. (eds.) WAC 2005. LNCS, vol. 3854, pp. 13–28. Springer, Heidelberg (2006)
24. Yamamoto, L.: Code Regulation in Open Ended Evolution. In: Ebner, M., O'Neill, M., Ekárt, A., Vanneschi, L., Esparcia-Alcázar, A.I. (eds.) EuroGP 2007. LNCS, vol. 4445, pp. 271–280. Springer, Heidelberg (2007)
25. Yamamoto, L.: PlasmidPL: A plasmid-inspired language for genetic programming. In: Proc. of EuroGP, Napoli, IT (2008)
26. Vahdat, A., Becker, D.: Epidemic routing for partially connected ad hoc networks. Technical Report CS-200006, Duke University (April 2000)
27. Fall, K.: A delay-tolerant network architecture for challenged Internets. In: Proc. of ACM SIGCOMM, Karlsruhe, Germany, pp. 27–34 (2003)
28. Alouf, S., Carreras, I., Miorandi, D., Neglia, G.: Embedding evolution in epidemic-style forwarding. In: Proc. of IEEE BioNetworks, Pisa, IT (2007)

Biologically Inspired Self Selective Routing with Preferred Path Selection*

Boleslaw K. Szymanski, Christopher Morrell,
Sahin Cem Geyik, and Thomas Babbitt

Department of Computer Science
Center for Pervasive Computing and Networking
Rensselaer Polytechnic Institute, 110 8th Street, Troy, NY 12180
{szymansk,morrec,geyiks,babbit}@cs.rpi.edu

Abstract. This paper presents a biologically inspired routing protocol called Self Selective Routing with preferred path selection (SSRP). Its operation resembles the behavior of a biological ant that finds a food source by following the strongest pheromone scent left by scout ants at each fork of a path. Likewise, at each hop of a multi-hop path, a packet using the Self Selective Routing (SSR) protocol moves to the node with the shortest hop distance to the destination. Each intermediate node on a route to the destination uses a transmission back-off delay to select a path to follow for each packet of a flow. Neither an ant nor a packet knows in advance the route that each will follow as it is decided at each step. Therefore, when a route becomes severed by a failure, they can dynamically and locally adjust their routing to traverse the shortest surviving path. Preferred path selection reduces transmission delay by essentially removing back-off delay for the node that carried the previous packet of the same flow. The results reported here for both simulation and execution of a MicaZ mote implementation, show that this is an efficient and fault-tolerant protocol with small transmission delay, high reliability and high delivery rate.

Keywords: routing, wireless sensor networks, route repair, ant colony paradigm, link failure.

1 Introduction

Wireless sensors networks are composed of a large number of nodes equipped with radios for wireless communication, sensors for sensing the environment

---

* Research was sponsored by US Army Research laboratory and the UK Ministry of Defence and was accomplished under Agreement Number W911NF-06-3-0001. The views and conclusions contained in this document are those of the authors, and should not be interpreted as representing the official policies, either expressed or implied, of the US Army Research Laboratory, the U.S. Government, the UK Ministry of Defense, or the UK Government. The US and UK Governments are authorized to reproduce and distribute reprints for Government purposes notwithstanding any copyright notation hereon.

P. Liò et al. (Eds.): BIOWIRE 2007, LNCS 5151, pp. 229–240, 2008.
© Springer-Verlag Berlin Heidelberg 2008

and CPU's for processing applications and protocols. A significant number of wireless sensor networks consist of battery-powered nodes able to operate unattended. Such networks require autonomy of management (self-management), fault-tolerance, and energy-efficiency in all aspects of their operation. These properties are especially important for routing, since multi-hop communication is a primitive wireless sensor network operation that is fault-prone as well as energy-intensive. For instance, commonly observed in such networks are faulty (or, potentially subverted) nodes and transient and asymmetric links caused by wildly oscillating packet reception quality. Faulty nodes and transient links cause severe packet loss and spontaneous network topology changes[1,2]. In terms of energy usage by sensor network node components, radio operation is typically the most costly, as evidenced by a study in [3] and typical hardware specifications given in [4].

A traditional approach to multi-hop routing is to use routing tables that indicate the neighbor to which a packet should be forwarded to reach a destination; prominent examples include AODV[5] and Directed Diffusion[6]. This fundamental approach, which emulates traditional wired network communication, naturally requires nodes to constantly maintain individual neighbors states (e.g., active or sleeping) to support routing decisions. In operating conditions typical for wireless sensor networks, such maintenance often requires significant overhead, especially if fault-tolerance is to be supported. Hence, providing efficient routing protocols that naturally accommodate and perform well in fault-prone conditions is still an open and formidable challenge and is therefore the subject of this paper.

This paper presents the biologically inspired family of Self Selective Routing (SSR) protocols[7], which has been extended with preferred path selection, introduced in this paper. In SSR, after a node currently possessing a packet transmits it, all nodes that receive it decide which one will forward it. This decision is made autonomously by each receiver based on their respective hop distances to the destination using a transmission back-off delay to resolve potential ties.

In this paper, we discuss two novel mechanisms used by SRP, also called SSR(v3), introduced here as compared to SHR [7], also called SSR(v2): (i) an efficient and local repair of severed routes and (ii) preferred path selection. The first mechanism allows a node that detected no responders to its transmission broadcast to increase its hop distance to the destination. This increase enables the currently traveling packet to retrace a part of its path. In an effort to make the protocol more tunable, we have enabled the user to choose whether route repair occurs in each packet, or in each node. Repairing the packet increases the hop count only in the individual packet, and provides a temporary alternate route that is desired in the case of transient failures. This method of repair maintains the established topology of the network. Repairing the node increases the hopcount in the node, and provides a permanent change to the network's topology that is desired in the case of permanent failures. The second method introduced in this paper, allows the node that forwards the current packet to select itself for forwarding the next packet in the flow with essentially no delay.

This creates a protocol that is both delay efficient (minimal delay to forward a packet in a normal case) and robust (another node will forward a packet if the preferred node is down or has lost its link to the sender) at the same time.

There are other protocols that, like SSR, route on the premise of avoiding neighbor state maintenance and letting receivers contend for forwarding packets. However, they all require geographical location information, which SSR does not. Three such protocols, GRAd[8], GRAB[9], and BLR[10] are not capable of a route repair. Other protocols, GeRaF[11], IGF[12], PSGR[13] and SIF[14] define eligibility regions for packet forwarding and therefore require detailed knowledge of geographical placement of currently active nodes which is difficult to obtain and maintain in wireless sensor networks.

2   Self Selective Routing

The SSR protocol has been inspired by the use of pheromones by the biological ants to mark paths to guide other ants to food sources without memorizing or prescribing a path explicitly[15,16]. Accordingly, the SSR protocol consists of three phases: (i) an initial destination request flooding that finds the destination node, (ii) a destination reply flooding that establishes hop distances between each node and the given flow's destination, and (iii) data transmission proper.

The destination request phase corresponds to the initial search for food in which ant scouts randomly explore the environment. In the process, they mark the branching paths with pheromones, which will later guide the ant scout back to the home colony (retracing the path, an ant will follow the strongest marks as they were most recently visited on the way out). Packets sent in this stage are referred to as DREQ (Data Request) packets. The destination reply phase corresponds to a walk back to the colony by an ant that found a food source. Walking back, an ant will mark branches on the path home with pheromones to distinguish the return path from other, unused paths. Packets sent in that stage are called DREP (Data Reply) packets. This initial flooding is done once at the sensor network deployment to all potential destinations (in wireless sensor networks there is often only one destination, the base station, making the initial two stages particularly simple). We used for this purpose the signal-strength aware flooding technique described in [17] which also provides more details on the these two initial stages. This paper focuses on data transmission stage itself.

2.1   Data Transmission in SSR

As shown in figure 1, the data transmission stage can be represented by a Finite State Automaton (FSA) that defines the input, actions and output generated in each state of a node in the network as it routes data (similar FSAs can be defined for the destination request and reply stages). For example, when a node receives a packet that it has not seen before, it immediately moves into the *NEW* state, and depending on its input and status (e.g. data packet received by the destination, data packet received by a node closer to the destination then the

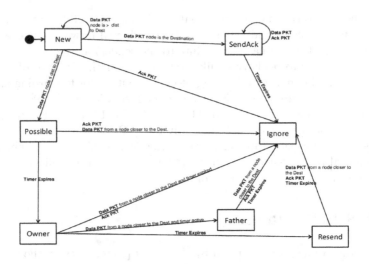

Fig. 1. State diagram for SSR(v3)

sender, acknowledgment packet received, etc) the node transitions itself into the corresponding state and executes the associated actions (for clarity, not shown in the figure).

When the source transmits a DATA packet, only neighbors that are closer to the destination than the sender will react. Depending on the reacting nodes proximity to the destination in relation to the sending node, it selects a transmission back-off delay. That delay is uniformly distributed between 0 and $\lambda/2$ if the reacting node is one hop closer to the destination. If the reacting node is more than one hop closer, the back-off delay is selected between $3\lambda/4$ and λ. This difference in back-offs ensures that the more reliable single hop closer neighbors have priority over the less reliable multiple hop closer neighbors. λ is a scaling factor that allows us to tune the probability of collision of the nodes' responses. If, during the back-off delay, a DATA packet is received from a node that is closer to the destination, the receiving node cancels the forwarding of the DATA packet and moves to the Ignore state. When the transmission back-off time expires, the node increments the packet's actual hop count by one, sets the expected hop count to its hop distance to the destination and then transmits the packet.

After forwarding the packet, the node monitors the carrier to determine if the packet has been forwarded. Lack of forwarding causes retransmissions, and finally route repair which is accomplished by increasing the node or packet's hop distance to the destination by 2 and retransmitting.

To promote reliable links, we introduced a preferred path selection, in which a node which forwarded the current packet will respond almost immediately to a transmission of a new packet in the same flow. To simplify processing, these nodes calculate their delay by dividing the regularly selected back-off delay by 625, while ensuring that it remains larger than the radio transition time. This results in a back-off delay between 20 and 160μs, given λ is 100ms. This minimizing of back-off delay ensures the node future self-selections, thereby stabilizing repeatedly traversed paths. In the ant pheromone model, as ants move over different paths, and the once strongly scented but now less used paths begin to fade, ants shift their routes to the paths that are most frequently used. In reference to the slow fading of the pheromone, we have chosen to not follow the biological inspiration literally. Instead, we restore the full range back off delay immediately after the preferred node fails to self-select, as such failure indicates that the recently used node is no longer reliable. Despite its simplicity, the effect of using the preferred path selection in SSR(v3) is very positive, as demonstrated in the section below.

3   Performance Evaluation

Using both the SENSE wireless network simulator [18] and MicaZ sensor motes [4], we performed a series of experiments to compare the performance of SSR(v2) with the newly designed SSR(v3). Additionally, in the case of simulations, both protocols were compared with a traditional routing protocol, AODV [5].

3.1   Simulations

We tested three different scenarios. The first one involved a single sink (base station) collecting data from many sources, which is a typical sensor network setting. The second scenario investigated transient failures, while the third one evaluated the performance of the protocols under permanent failures. In failure simulations, faults occurred with varying probabilities, while the sink network simulation evaluated the performance with a varying number of sources.

The simulation topography consists of an 8 unit by 8 unit terrain populated with 500 nodes placed randomly. Each node is stationary and has a single unit nominal transmission range. The wireless medium is simulated with the free space propagation model[20], and the radio modeled operation at 914 MHz with 1 Mb/s of bandwidth. Packet sizes were uniformly distributed around a mean of 1000 bytes and were sent at uniformly distributed intervals with a mean of 40 seconds. MAC broadcast was used in which a node senses the carrier and broadcasts only if no other transmissions are detected. The average hop distance between sources and their respective destinations is 7.8 hops.

Each simulation was executed eleven times, each time with a different random number seed for a simulation time of 3,000 seconds per seed. The same 11 seeds were used for all simulation sets. λ was set to 100ms for all simulations.

Single Sink Network. In a wireless sensor network, using a single sink is common. For example, any network that contains a single base station is usually configured that way. Such configuration may result in heavy traffic congestion near the sink. Such congestion has the possibility of causing massive amounts of collisions, and could possibly stop the network from functioning at all. In sink network simulations, we varied the number of sources transmitting to a single sink from 10 to 100 to test the scalability of each protocol.

As is apparent in figure 2, a single sink network is where SSR(v3) shows its worth, and where AODV breaks under its limitations. The protocols' end-to-end delays were so drastically different, that a logarithmic scale was necessary to plot them together. As the density of sources increases from 70 to 100, which is 14% to 20% of the nodes in the network transmitting, AODV required approximately 100 seconds to transfer a packet from the source to the destination. Although SSR(v3) does increase its delay slightly, it still manages to keep that delay to under 0.1 seconds, even with 100 nodes transmitting. Clearly, the preferred path selection allows packets to move across the network quickly enough that a packet reaches the destination before the following packet is transmitted, thus avoiding any significant impact from congestion.

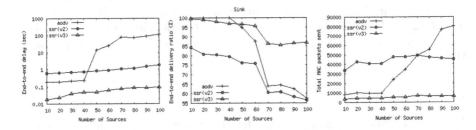

Fig. 2. Transmission delay, delivery ratio, and total MAC packets sent in the case of a single sink network for three compared protocols: AODV, SSR(v2) and SSR(v3)

SSR(v3) is also superior in terms of delivery ratio. As sources increase to 100, SSR(v3)'s delivery ratio decreases to near 90%, while AODV's drops to nearly 55%. The reasons are the same as described earlier, where AODV succumbs to the congestion around the sink node, while SSR(v3) is fast enough to avoid significant congestion. Also in total MAC packets sent, SSR(v3) manages to use less than 10% of the packets that AODV uses at 100 sources.

Failure Simulations. The failure sensitivity of SSR's route repair routine can be tuned by adjusting the number of retransmissions by the forwarding node required to invoke route repair. By increasing this value, SSR can be successfully employed in a network with a high rate of transient failures, but maintains performance in a network with a high rate of permanent failures. In our tests, two retransmissions were required to invoke route repair. Since a packet transmission interval is 40 seconds, a node failure lasting less than 80 seconds on average would not change the route from the source to the destination. As mentioned earlier, the protocol is also tunable, because route repair can be executed temporarily on individual packets, or permanently on the nodes.

Transient Failures. There are several possible causes for transient node failures, such as error-prone links, power management induced duty cycles, or excessive packet collisions. Of these, the duty cycle induced failures are the least disruptive since they may be coordinated with the networking protocol. The presented simulation results are based on a random transient failure model, so they exaggerate the effect of duty cycles on the protocols. In the transient failure simulations, each node was assigned a mean active time and a mean sleep time. The sum of these two times was fixed at 200 seconds. The time spent in each mode was distributed exponentially about the mean value.

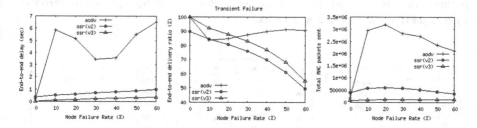

Fig. 3. Transmission delay, delivery ratio, and total MAC packets sent in the case of transient failures for three compared protocols: AODV, SSR(v2) and SSR(v3)

As seen in figure 3, AODV has the worst transmission delay that increases significantly with the transient failure rate. SSR(v3) has by far the smallest delay of the three protocols, with a factor 10 advantage over AODV for the most failure prone case. SSR(v3) has lower delays than AODV for all cases in which transient failures are present. Both SSR(v2) and SSR(v3) only slightly increase the incurred transmission delay when the transient failure rate is growing.

In terms of delivery ratio, AODV is the best, dropping from 100% in a reliable case to 90% for 60% transient failure rate. SSR(v3) delivery ratio drops from 100% to 55% over the same region while SSR(v2)'s is slightly lower, dropping from 90% to 50%. However, AODV requires a much larger number of MAC packet transmissions than either SSR(v2) or SSR(v3). This is because to find a new path, AODV's route repair algorithm initiates a new route request phase, causing a flood of packets from the point at which the route is severed. AODV uses over 30 times more packets than SSR(v3). Hence, by implementing a simple replication scheme, in which each packet in SSR(v3) is sent 3 times, we could bring the SSR(v3) delivery rate into a range that is more comparable with AODV, while still keeping the number of MAC packets 10 times lower. The impact of this huge difference in packets required will show itself primarily in the energy consumption of the protocols.

Permanent Failures. In the permanent failure model, each node had a random chance of failing. Nodes that fail had their failure start time uniformly distributed over the simulation time. In this scenario, trends observed for transient failures continue but are less pronounced.

As seen in figure 4, as the number of node failures increase, the transmission delay also increases while the delivery ratio generally decreases. SSR(v3) achieves the lowest and most stable transmission delay of all three protocols. Even at 60% failure rate, its delay is only slightly increased compared with its delay in the reliable network, and is nearly 10 times better than that of AODV. Although SSR(v3) delivery ratio is not 100% as is AODV, it still shows a 16% improvement over SSR(v2), and stays at or above 96%. This improvement arises because any node that tends to get entangled in external collisions will not be able to forward packets consistently and therefore sooner or later it will be replaced in SSR(v3) by a node that can, if such a node exists.

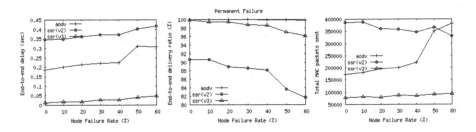

Fig. 4. Transmission delay, delivery ratio, and total MAC packets sent in the case of permanent failures for three compared protocols: AODV, SSR(v2) and SSR(v3)

Again, the most significant difference between AODV and SSR arises in MAC packet sent. As failures increase, the number of packets required for AODV to maintain 100% delivery begins to quickly increase, while SSR(v3) maintains practically the same number for all failure rates. Hence, for the same reasons as discussed in transient failure simulations, the ratio of the numbers of MAC packets used increases from an initial factor of 2 to a factor of 5 for the 60% permanent failure rate.

SSR's approach to route repair is clearly more local and efficient, as evidenced by the plots. It should also be noted that under SSR(v2) and SSR(v3), the path lengths and number of packets per hop remain nearly constant over the range of permanent and transient failure rates. This demonstrates that priority-driven opportunistic behavior of these protocols is highly accommodative to potentially disruptive duty cycles and node failures.

3.2   Implementation on MicaZ Motes

We have implemented the new SSR(v3) protocol on MicaZ motes [4] using TinyOS version 1.1.7 to compare performance of this implementation with the implementation of SSR(v2) on the same hardware [19]. In the implementation,

we used B-MAC with acknowledgments disabled to provide link layer functionality. DATA packets of 29 bytes were sent for 12.5 min at a rate of 5sec/packet in
an indoor environment. The radio power was set to -21dBm and a distance of 1m
provided a reliable delivery rate. However, with moderate probability some long
distance transient links also formed. Both compared protocols used the same λ
of 22ms.

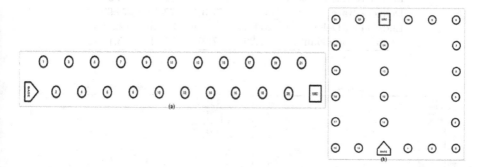

Fig. 5. (a) Double line topology, (b) Route repair topology. Nodes have reliable connections with their closest neighbors and transient connections with others. The base
indicates the direction in which all motes are oriented.

SSR(v2) was compared to SSR(v3) on two topologies. Double line topology,
shown in figure 5(a), has two motes at each hop eligible to forward the packet.
Route repair topology from figure 5(b), contains three unequal length and disjoint paths: a short, medium and long one. With these topologies, we tested
the repair capabilities of each protocol. During testing we blocked motes 12 and
13 in the network by placing a metal container over the motes after the first 5
minutes of the test.

As shown in table 1, in double line topology experiments, SSR(v3) provided
a large improvement in delivery rate, more than halving the percentage of lost
packets in SSR(v2). It also achieved a modest improvement in the end-to-end
delay compared to SSR(v2). On route repair topology both protocols performed
equally well.

To better understand these results, we plotted the time versus delay of each
successfully transmitted packet in both topologies for SSR(v3) (see figure 6).
Initially, packets frequently followed different length paths showing transient
nature of links in the experiment and therefore decreasing the effectiveness of
the preferred path selection. However, later on, the nodes with stable link tend
to persist longer on paths used for transmission, increasing the advantage of
SSR(v3) over SSR(v2). The failure of nodes 12 and 13 in the middle of a run
(around packet 160) on route repair topology prevents this effect from occurring,
resulting in similar performance of both protocols.

In the current implementation, both SSR(v2) and SSR(v3) allowed longer but
transient links to win self-selection. On the first glance, this seems to be beneficial

Table 1. Experimental results for double line and route repair topologies

	Double line		Route repair	
	SSR(v2)	SSR(v3)	SSR(v2)	SSR(v3)
Packets Sent	246	277	110	117
Packets Received	1070	1279	304	317
Packet Ratio (rec/sent)	4.33	4.61	2.74	2.69
Delivery Rate	47.3%	77.3%	77.3%	74.9%
End-to-end Delay	209 ms	174 ms	117 ms	122 ms
Average Hop Count	7.26	7.07	5.11	5.15

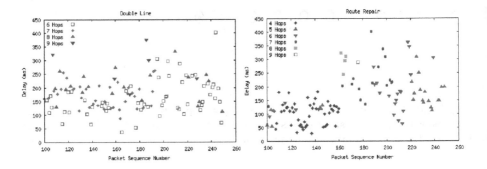

Fig. 6. Packet sequence number versus delay for SSR(v3) executed over the two topologies

as such links may decrease the number of hops needed to reach destination. However, closer inspection reveals that such links may increase the chance for retransmissions because the long links have relatively small probability of being overheard by the sender when they respond and transmit a packet towards the destination.

4   Conclusion and Future Works

In this paper, we have presented SSR(v3), which naturally accommodates fault-prone sensor network routing conditions and takes full advantage of the properties of the broadcast communication primitive of such networks. SSR provides seamless route repair in cases of permanent or transient failures of nodes or links. The preferred path selection introduced here allows the packet to traverse not only the shortest path to the destination, but also the most reliable one. It also preserves SSR(v2)'s ability to use other links if the preferred link is down. The resulting significant decrease in the transmission delay and increase in delivery ratio address the most important weaknesses of SSR(v2).

In future work, we intend to extend the SSR family of protocols to address issues of mobility and energy efficiency, both of which are common in wireless sensor network applications. While SSR(v3) may currently accommodate mobility, it is not yet explicitly optimized for it. Mobility shortens the time over which hop distance tables remain valid. To retain SSR's autonomic behavior, we are researching how to efficiently update these tables based on local observations of node movement. SSR can already accommodate topology changes caused by energy-efficient topology control algorithms, such as ESCORT [21]. However, explicitly incorporating a topology control algorithm into SSR is still a challenge, as it requires ensuring that the algorithm is not so aggressive that it overcomes SSR's ability to find eligible forwarders for every packet.

References

1. Woo, A., Tong, T., Culler, D.: Taming the underlying challenges of reliable multihop routing in sensor networks. In: Proc. ACM SenSys 2003, pp. 14–27. ACM Press, New York (2003)
2. Zhao, J., Govindan, R.: Understanding packet delivery performance in dense wireless sensor networks. In: Proc. ACM SenSys 2003, pp. 1–13. ACM Press, New York (2003)
3. Anastasi, G., Falchi, A., Passarella, A., Conti, M., Gregori, E.: Performance measurements of motes sensor networks. In: Proc. 7th ACM Intern. Symp. Modeling, Analysis and Simulation of Wireless and Mobile Systems, pp. 174–181. ACM Press, New York (2004)
4. Crossbow Technology, Inc., http://www.xbow.com
5. Perkins, C., Belding-Royer, E., Das, S.: RFC 3561-ad hoc on-demand distance vector (AODV) routing, http://www.faqs.org/rfcs/rfc3561.html
6. Intanagonwiwat, C., Govindan, R., Estrin, D.: Directed diffusion: a scalable and robust communication paradigm for sensor networks. In: Proc. ACM MobiCom, pp. 56–67. ACM Press, New York (2000)
7. Branch, J.W., Lisee, M., Szymanski, B.K.: SHR: Self-Healing Routing for wireless ad hoc sensor networks. In: Proc. Intern. Symp. Performance Evaluation of Computer and Telecommunication Systems SPECTS 2007, pp. 5–14. SCS Press, San Diego (2007)
8. Poor, R.: Gradient routing in ad hoc networks, http://www.media.mit.edu/pia/Research/ESP/texts/poorieeepaper.pdf
9. Ye, F., Zhong, G., Lu, S., Zhang, L.: Gradient broadcast: a robust data delivery protocol for large scale sensor networks. ACM Wireless Networks 11(2) (2005)
10. Heissenbüttel, M., Braun, T., Bernoulli, T., Waelchli, M.: BLR: beaconless routing algorithm for mobile ad hoc networks. Computer Communications Journal 27(11) (2004)
11. Zori, M., Rao, R.R.: Geographic Random Forwarding (GeRaF) for ad hoc and sensor networks: multihop performance. IEEE Trans. Mobile Computing 2(4), 337–348 (2003)
12. Blum, B.M., He, T., Son, S., Stankovic, J.A.: IGF: a robust state-free communication protocol for sensor networks. Technical Report CS-2003-11, University of Virginia, Charlottesville (2003)

13. Xu, Y., Lee, W.-C., Xu, J., Mitchell, G.: PSGR: priority-based stateless geo-routing in wireless sensor networks. In: Proc. IEEE Conf. Mobile Ad-hoc and Sensor Systems. IEEE Computer Society Press, Los Alamitos (2005)
14. Chen, D., Deng, J., Varshney, P.K.: A state-free data delivery protocol for multihop wireless sensor networks. In: Proc. IEEE Wireless Communications and Networking Conf. IEEE Computer Society Press, Los Alamitos (2005)
15. Cordon, O., Herrera, F., Stutzle, T.: A review on the Ant Colony Optimization Metaheurstics: Basis, Models and New Trends. Mathware & Soft Computing 9 (2002)
16. Koenig, S., Szymanski, B.K., Liu, Y.: Efficient and Inefficient Ant Coverage Methods. Annals of Mathematics and Artificial Intelligence 31(1-4), 41–76 (2001)
17. Chen, G., Branch, J., Szymanski, B.K.: Local leader election, signal strength aware flooding, and routeless routing. In: 5th IEEE Intern. Workshop Algorithms for Wireless, Mobile, Ad-Hoc Networks and Sensor Networks WMAN 2005. IEEE Computer Society Press, Los Alamitos (2005)
18. Chen, G., Branch, J.W., Pflug, M., Zhu, L., Szymanski, B.K.: SENSE: a wireless sensor network simulator. Advances in Pervasive Computing and Networking, pp. 249–267. Springer, Heidelberg (2004)
19. Wasilewski, K., Branch, J., Lisee, M., Szymanski, B.K.: Self-healing routing: a study in efficiency and resiliency of data delivery in wireless sensor networks. In: Proc. Conference on unattended Ground, Sea, and Air Sensor Technologies and Applications, SPIE Symposium on Defense & Security, April, Orlando, FL (2007)
20. Rappaport, T.S.: Wireless Communications: Principles and Practice. Prentice Hall, Englewood Cliffs (1996)
21. Branch, J.W., Chen, G., Szymanski, B.K.: ESCORT: Energy-efficient Sensor network Communal Routing Topology using signal quality metrics. In: Lorenz, P., Dini, P. (eds.) ICN 2005. LNCS, vol. 3420, pp. 438–448. Springer, Heidelberg (2005)

Biologically Inspired Approaches to Networks: The Bio-Networking Architecture and the Molecular Communication

Tatsuya Suda, Tadashi Nakano, Michael Moore, Akhiro Enomoto, and Keita Fujii

Information and Computer Science University of California, Irvine
Irvine, CA 92697-3425, USA
{suda,tnakano,mikemo,kfujii,enomoto}@ics.uci.edu

Abstract. This article describes two branches of biologically inspired approaches to networks; biologically inspired computer networks and biologically inspired nanoscale biological networks. The first branch, biologically inspired computer networks, applies techniques and algorithms from biological systems to design computer networks. The second branch, biologically inspired nanoscale biological networks, applies techniques and algorithms from biological systems to design nanoscale biological networks. This paper describes these two branches of approaches proposed by the authors of this paper; biologically inspired computer networks (i.e., the Bio-Networking Architecture) and biologically inspired nanoscale biological networks (i.e., the Molecular Communication).

Keywords: biological inspiration, computer networks, nano-scale biological networks, bio-networking architecture, molecular communication.

1 Introduction

Information processing systems today are composed of a larger number of devices gathering, exchanging and processing information. In addition, system components (e.g., computational and sensing devices) are becoming smaller, potentially less reliable, and integrated directly into the environment, (e.g., sensor networks for environmental monitoring, sensor networks imbedded in a human body). As a result, information processing systems today face challenges of scaling to a larger number of system components and adapting to dynamical changes in the environment and failures of system components.

Observation of the biological world shows that a biological system is composed of a massive number of nano/micro-scale biological components that adapt to changes in the environment and failures of biological components. In biological systems, biological components self-organize into complex systems in a hierarchical manner, where a higher level structure is composed of a number of lower level components. For example, an organism is composed of multiple organs, each of which is composed of multiple cells. Components at each level communicate and coordinate through a variety of mechanisms (e.g. electric signals, molecule diffusion), allowing biological systems to adapt to environmental changes and component failures.

P. Liò et al. (Eds.): BIOWIRE 2007, LNCS 5151, pp. 241–254, 2008.
© Springer-Verlag Berlin Heidelberg 2008

Design principles from biological systems may be, thus, applicable to designing today's information processing systems that need to integrate large number of components and that need to adapt to environmental changes and component failures. This paper describes two branches of biologically inspired approaches to network systems; (1) biologically inspired computer networks and (2) biologically inspired nanoscale biological networks. The first branch, biologically inspired computer networks, applies the principles found in biological systems to design computer networks. The second branch, biologically inspired nanoscale biological networks, applies the principles to design nano to microscale biological networks.

The remainder of this paper is organized as follows. Section 2 describes an example of biologically inspired computer networks (i.e., the Bio-Networking Architecture), and Section 3 describes an example of biologically inspired nanoscale networks, (i.e., Molecular Communication). Section 4 concludes this paper.

2 Biologically Inspired Computer Networks: The Bio-Networking Architecture

Future computer networks are expected to be autonomous, scalable and adaptive. They autonomously operate with minimal human intervention. They scale to and support billions of nodes and users. They adapt to diverse user demands and dynamically changing network conditions such as network failures.

Key features of future networks mentioned above, i.e., autonomy, scalability and adaptability, have already been achieved in biological systems. For example, scalability is observed in large-scale social insect colonies (e.g., ant colonies) that contain millions of entities (e.g., worker ants), yet exhibit highly sophisticated and coordinated behaviors (e.g., division of labor in foraging and nest building). Social insect colonies also adapt to dynamically changing conditions (e.g., change in the amount of food in the ant colony) through local interaction of entities that adjust their behavior based on environmental conditions (e.g., more ants go out of a colony and gather food when the food level in the colony becomes low.). Based on the observation that biological systems exhibit desirable features that are required for future networks, a number of researchers are currently investigating the feasibility of applying biological concepts and principles to computer networks design [1-10].

The authors of this paper believe if a network and network applications are modeled after biological concepts and principles, they satisfy key requirements such as autonomy, scalability and adaptability. Thus, the authors of this paper have designed a biologically.inspired network, called the Bio-Networking Architecture [11, 12] where key biological concepts and principles (e.g., self-organization, emergence, redundancy, natural selection, and diversity) are applied to design of networks and network applications. The following describes biological principles applied to the Bio-Networking Architecture, and then describes the design of a middleware framework for the Bio-Networking Architecture.

2.1 Biological Concepts and Principles Applied in the Bio-Networking Architecture

A key concept in biological systems is emergent behavior. In biological systems, useful behavior often emerges through the collective, simple and autonomous behaviors of individual biological entities. For example, when a bee colony needs more food, a large number of bees will leave the hive and go to the flower patches in the area to gather nectar. When the bee colony is near its food storage capacity, only a few bees will leave the hive to gather nectar. This adaptive food gathering function emerges from the relatively simple and local interactions among individual bees. If a returning food gathering bee can quickly unload its nectar to a food storing bee, it means that the food storing bees are not busy and that there is little food in the hive. This food gathering bee then encourages other nearby bees to leave the hive and collect nectar by doing the well-known "waggle dance." If a returning food gathering bee must wait a long time to unload its nectar, it means that the food storing bees are busy and that there is plenty of food in the hive. This food gathering bee then remains in the hive and rests. Since the food gathering bee performs a localized interaction only between itself and the food storing bee, the interaction can scale to support the large or growing nutritional needs of a bee colony. The bee colony also exhibits other types of emergent behavior, such as self-organization, evolution, and survivability. Thus, emergent behavior is the formation of complex behaviors or characteristics through the collective, simple and autonomous behaviors or characteristics of individual entities.

The Bio-Networking Architecture applies the concept of emergent behavior by implementing network applications as a group of autonomous entities called the cyber-entities (see Figure 1). This is analogous to a bee colony (an application) consisting of multiple bees (cyber-entities). Each cyber-entity implements a functional component related to the application and also follows simple behavior rules similar to biological entities (such as reproduction, death, migration, relationship establishment with other cyber-entities). In the Bio-Networking Architecture, useful application functionality emerges from the collaborative execution of application components (or services) carried by cyber-entities, and useful system behaviors and characteristics (e.g., self-organization, adaptation, evolution, security, survivability) arise from simple behaviors and interaction of individual cyber-entities.

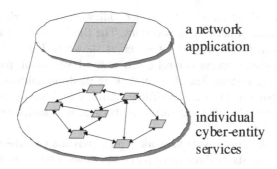

a network
application

individual
cyber-entity
services

Fig. 1. Cyber-entities and Network Applications

Another key biological concept applied in the Bio-Networking Architecture is evolution. Evolution in biological systems occurs as a result of natural selection from diverse behavioral characteristics of individual biological entities. Through many successive generations, beneficial features are retained while detrimental behaviors become dormant or extinct, and the biological system specializes and improves itself according to long-term environmental changes,

Within a biological system, diversity of behaviors is necessary for the system to adapt and evolve to suit a wide variety of environmental conditions. In the Bio-Networking Architecture, different cyber-entities may implement the same behavior with different behavior policies to ensure a sufficient degree of diversity. For instance, different cyber-entities may implement the migration behavior with different behavior policies (e.g., one cyber-entity with a migration policy of moving towards a user, and another cyber-entity with a migration policy of moving towards a cheap resource cost node). In the Bio-Networking Architecture, when a cyber-entity reproduces, mutation and crossover may occur in its behavior policy. This also ensures a sufficient degree of diversity.

Within a biological system, food (or energy) serves as a natural selection mechanism. Biological entities naturally strive to gain energy by seeking and consuming food. Similarly to an entity in the biological world, each cyber-entity in the Bio-Networking Architecture stores and expends energy for living. Cyber-entities gain energy in exchange for performing a service, and they pay energy to use network and computing resources. The abundance or scarcity of stored energy affects various behaviors and contributes to the natural selection process in evolution in the Bio-Networking Architecture. For example, an abundance of stored energy is an indication of higher demand for the cyber-entity; thus the cyber-entity may be designed to favor reproduction in response to higher levels of stored energy. A scarcity of stored energy (an indication of lack of demand or ineffective behaviors) may eventually cause the cyber-entity's death.

Reproduction to create diverse cyber-entity behaviors along with a mechanism for natural selection will result in the emergence of evolution to allow applications to adapt to long-term environmental change in the Bio-Networking Architecture.

2.2   Middleware Framework of the Bio-Networking Architecture

The Bio-Networking Architecture is a middleware framework that supports design and development of network applications. The Bio-Networking Architecture consists of interconnected Bio-Networking nodes as shown in Figure 2. A virtual machine capable of resource access control (such as the Java virtual machine) runs atop the native operating system. The Bio-Networking platform software, which provides an execution environment and supporting facilities for cyber-entities, runs using the virtual machine. The platform software also manages underlying resources such as CPU and memory.

Cyber-entities that provide network applications run atop the platform software. A cyber-entity consists of three main parts: attributes, body, and behaviors. Attributes

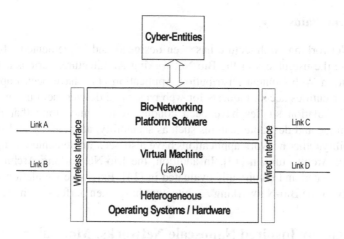

Fig. 2. A Bio-Networking Node

carry information regarding the cyber-entity (e.g., cyber-entity ID, description of a service it provides, cyber-entity type, stored energy level, and age). The body implements a service that a cyber-entity provides and contains materials relevant to the service (such as data, application code, or user profiles). For instance, a cyber-entity's body may implement control software for a device, while another cyber-entity's body may implement a hotel reservation service. A cyber-entity's body that implements a web service may contain a web page data. Cyber-entity behaviors implement non-service related actions that are inherent to all cyber-entities. Major behaviors of cyber-entities are listed below:

- Energy exchange and storage: Cyber-entities may gain/expend and store energy as described earlier.
- Communication: Cyber-entities communicate with other cyber-entities to request a service, to provide a service, or to forward messages (e.g. discovery messages) to other cyber-entities.
- Migration: Cyber-entities may migrate from bionet platform to platform.
- Replication and reproduction: Cyber-entities may make copies of themselves (replication). Two parent cyber-entities may create a child cyber-entity (reproduction), possibly with mutation and crossover in child cyber-entity's behavior policy.
- Death: Cyber-entities may die because of old age or energy starvation.
- Relationship establishment: Cyber-entities may establish relationships with other cyber-entities.
- Discovery: Cyber-entities may discover other cyber-entities by sending a discovery message through their relationships.
- Resource sensing: Cyber-entities may sense the type, amount, and unit cost of resources (e.g. CPU cycles and memory space) available on both a local and neighboring platforms. Each platform determines the unit cost of resources provided on that platform based on their availability.

2.3 Current Status

The Bio-Networking Architecture has been designed and implemented. In order to demonstrate the usefulness of the Bio-Networking Architecture, various applications (e.g., Aphid, a Web Content Distribution Application [11]) have been implemented. Aphid cyber-entities accept requests for web pages and deliver them using the HTTP protocol. Simulation studies have been conducted to demonstrate that the Aphid exhibits unique and desirable features such as scalability, adaptability and survivability/availability.Other network applications have also been implemented using the Bio-Networking Architecture in [13]. In addition, the Bio-Networking Architecture has been implemented and empirically evaluated in [14]. Extensive simulation studies on evolvability of the Bio-Networking Architecture have been performed in [15].

3 Biologically Inspired Nanoscale Networks: Molecular Communication

With the current bionanotechnology, it is feasible to engineer biological systems (e.g., receptors, nano-scale reactions), as demonstrated through modification of DNA to produce new cell functionality. Techniques to engineer biological systems is now applicable to design nano-scale or micro scale biological network systems.

The authors of this paper have proposed and are currently designing molecular communication [16, 17]. Molecular communication is a new paradigm for communication between biological nanomachines over a short-range (a nano- and micro-scale range). Biological nanomachines are nano- and micro-scale devices that either exist in the biological world or are artificially created from biological materials and that perform simple functions such as sensing, logic, and actuation. As biological nanomachines (or nanomachines in short) are too small and simple to communicate through traditional communication mechanisms (e.g. through sending and receiving of radio or infrared signals), molecular communication provides a mechanism for nanomachines to encode information onto molecules and communicate information by propagating the information encoded molecules.

Molecular communication represents a direct and precise mechanism for interacting with biological nanomachines, and it may be used, for instance, in implant devices [18] to directly interact with human cells for human health monitoring. Molecular communication may also be useful in molecular computing. In the molecular computing area, researchers are currently creating molecular logic gates (e.g., an inverter and a NAND gate) and memory [19-24] using components from biological systems. Those molecular computing components may be interfaced through molecular communication to perform more complex computing functionality.

Section 3.1 introduces communication mechanisms that biological entities use, using two examples of how communication is performed within and between cells. Section 3.2 describes a molecular communication architecture. Section 3.3 presents a design of two molecular communication systems based on specific biological communication mechanisms.

3.1 Biological Communication

Biological nanomachines (e.g., cells) exhibit a wide variety of mechanisms for exchanging information at the nano and micron scales. Some specific biological mechanisms include mechanisms for intracellular communication and intercellular communication [25].

In intracellular communication, communication occurs between components within a single biological cell. For example, acetylcholine is transported within a neuron along the axon and then released, causing a response in the adjacent neurons that have a receptor for acetylcholine. A component of the cell emits molecules (e.g., vesicles containing acetylcholine), and molecules are carried by molecular motors. A molecular motor (e.g., kinesins and dyneins) is a protein or protein complex that transforms chemical energy (e.g., ATP hydrolysis) to mechanical movement of the molecular motor along a cytoskeletal track (e.g. microtubules).

In intercellular communication, communication may occur through cell-cell channels called gap junctions. Gap junctions allow connected cells to share small molecules such as Ca^{2+} (calcium ions) and inositol 1,4,5-trisphosphate (IP_3), and therefore, enable coordinated actions among adjacent cells. For example, ciliated airway epithelial cells communicate with each other through the diffusion of IP_3 via gap junctions. A stimulated cell first increases the intracellular IP_3 concentration, and this results in the release of Ca^{2+} from the intracellular calcium store. IP_3 diffuses through gap junctions to adjacent cells, resulting in the release of Ca^{2+} from intracellular stores in each of the adjacent cells. Diffusion of IP_3 continues and thus propagates calcium waves over a number of cells.

3.2 Molecular Communication Architecture

Molecular communication may be described using a generic communication architecture that consists of system components (information molecules, sender nanomachines, receiver nanomachies, environment) and communication processes of a molecular communication: encoding, sending, propagation, receiving and decoding (see Figure 3).

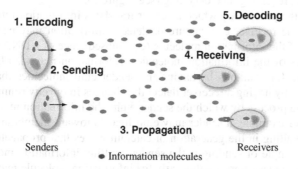

Fig. 3. Molecular Communication Architecture

Molecular Communication System Components

The generic molecular communication system consists of *information molecules* (i.e., molecules such as proteins or ions) that represent the information to be transmitted, *sender nanomachines* that emit the information molecules, *receiver nanomachines* that receive and react to information molecules, and the *environment* in which the information molecules propagate from the sender nanomachine to the receiver nanomachine (see Figure 3). It may also include transport molecules (e.g., myosin and dynein molecular motors) that transport information molecules through the environment using chemical energy, guide molecules (e.g. microtubules) that direct the movement of transport molecules, and interface molecules that encapsulate information molecules and allow a variety of information molecules to bind to the same transport molecule (e.g. a vesicle that can store many types of molecules and bind to a transport molecule).

The sender and receiver nanomachines may be, for instance, biological cells that use peptides, ions, or phosphates such as inositol-triphosphate as information molecules (e.g. cells that communicate through calcium ions). The sender stores information molecules and releases the information molecules according to some event (e.g. high concentration of an external signal molecule). For instance, calcium ions (information molecules) may be pumped into a sender cell from the external environment and stored in the endoplasmic reticulum of a sender cell. A sender cell may then release the stored calcium ions (information molecules) when a ligand binds to a receptor of a sender cell. The receiver cell includes calcium sensitive components to detect increased calcium concentration and reacts to the increased calcium concentration. For example, calcium ions bind to receptors of the receiver cell, resulting in neuron action potential or a cellular immune response such as inflammation.

The information molecules propagate from the sender to the receiver in the environment. The environment of molecular communication may be an aqueous medium with various ions and molecules dissolved in solution.

Molecular Communication Processes

The communication processes of molecular communication include encoding, sending, propagation, receiving, and decoding (see Figure 3).

Encoding is the process by which a sender translates information into information molecules that the receiver can capture or detect. Information may be encoded in a subcomponent of the information molecule (e.g. subsequence of a DNA sequence), or in characteristics of the information molecules. Information may also be encoded in the environment, for example, by having the sender emit molecules that modify the environment and by having a receiver detect the changes in the environment.

Sending is the process by which the sender emits the information molecule into the environment. For example, a sender may emit ligands toward membrane receptors of nearby cells, resulting in the generation of calcium waves that propagate from cell to cell. Another example of sending is the sender emitting information molecules using peptide translation machinery. In this case, the information molecule may be a peptide sequence that is encapsulated into a vesicle, transported by molecular motor machinery of the cell from the endoplasmic reticulum (a site of vesicle encapsulation) to the cellular membrane (a site of vesicle exocytosis), and emitted outside of the sender cell using vesicle exocytosis.

Propagation is the process by which information molecules move through the environment from a sender to a receiver. Propagation may occur through simple passive propagation (e.g. Brownian motion) in which the information molecules do not actively use energy to move through the environment. Propagation may also be controlled by constraining the volume of the environment in which information molecules can move. For example, in propagation through gap junctions, propagation is limited to inside the cell and the gap junction and, thus, molecules do not propagate in all directions. Another example of controlled propagation is molecular motors that walk over rail molecules to transport information molecules.

Receiving is the process by which the receiver captures carrier molecules propagating in the environment. The receiver may contain a selective receptor (e.g. a receptor that is sensitive to calcium ions or specific peptides) to capture the informatoin molecule. The receiver may contain gap junctions that allow molecules (e.g. calcium ions) to flow into the cell without using receptors. Another option for receiving is to use fusion of vesicles (observed in vesicle transport) containing information molecules into the membrane of receivers.

Decoding is the process by which the receiver, after receiving information molecules, decodes the received information molecules into a reaction. The design of a reaction is dependent on the application. If biological cells are used as receivers, potential reactions include enzyme-mediated reactions or protein synthesis. For instance, to report a detected information molecule, the receiver may express GFP (Green Fluorescent Protein) in response to the received information molecules.

3.3 Molecular Communication Systems

The authors of this paper have designed two instances of molecular communication; one using intracellular communication mechanisms, and the other using intercellular communication mechanisms. The following illustrates the design of these two molecular communication systems.

A Molecular Communication System Using Intracellular Communication Mechanisms

In the molecular communication system using intracellular communication mechanisms (see Figure 4) [26, 27], the sender nanomachine emits information molecules that are contained in an interface molecule. The interface molecule is a container that stores various information molecules. The interface molecule isolates information molecules inside from the propagation environment, and thus, it reduces interference from environmental noise. Transport molecules (e.g. molecular motors such as dynein or kinesin) propagate along guide molecules (e.g., microtubule) and transport interface molecules (containing information molecules) from the sender to the receiver nanomachines. The network topology of guide molecules determines the direction that molecular motors move. Communication processes in this molecular communication system are described below.

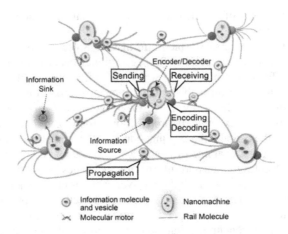

Fig. 4. Molecular Communication Using Intracellular Communication Mechanisms

Encoding: Sender nanomachines encode information on information molecules (e.g., DNA molecules, proteins, peptides). For example, nanomachines encode information on sequences of peptides and inject the peptides into vesicles. Vesicles can be loaded on molecular motors, and thus a variety of encoded molecules can be transmnitted.

Sending: Sender nanomachines emit information molecules. The information molecules are, then, attached to molecular motors.

Propagation: Propagation is performed through molecular motors that move along rail molecules from sender nanomachines to receiver nanomachines in a directed manner.

Receiving: Receiver nanomachines receive interface molecules from molecular motors using protein tags. When molecular motors approach receiver nanomachines, interface molecules (such as vesicles) may be fused into receiver nanomachines, releasing information molecules into the receiver.

Decoding: In decoding, receiver nanomachines invoke reactions in response to information molecules. For example, peptides (e.g. neurotransmitters) transported through molecular motors in a neuron cause receiver neurons to generate an action potential.

A Molecular Communication System Using Intercellular Communication Mechanisms

In the molecular communication system using intercellular communication mechanisms, a sender nanomachine (i.e., a cell) communicates with a receiver nanomachine (i.e., a cell) through a network of cells that passively propagate information molecules from a sender nanomachine to a receiver nanomachine (see Figure 5) [28]. The particular intercellular communication mechanisms used in this system is cell-cell calcium signaling that is described in section 3.1. Communication processes in this molecular communication system are described below.

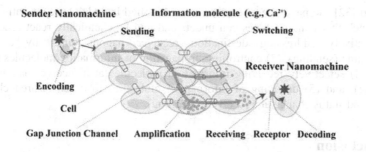

Fig. 5. Molecular Communication Using Intercellular Communication Mechanisms

Encoding: Ca^{2+} waves are used to encode various cellular information such as muscle contraction, chemical secretion in biological systems. Similarly, in this molecular communication system, a sender nanomachine encodes various information onto Ca^{2+} waves by varying the properties of Ca^{2+} waves (e.g., frequency, amplitude, and duration of Ca^{2+} waves). Encoding in this system is, thus, the process of selecting the properties of Ca^{2+} waves (e.g., release of agonistic substances in certain amounts) that represent different information at the receiver nanomachines(s).

Sending: A sender nanomachine releases chemical substances (e.g., Ca^{2+} and/or Ca^{2+} mobilizing molecules such as IP_3)) in the manner decided in the encoding process. The released chemical substances stimulate a nearby cell to initiate the following propagation process.

Propagation: A stimulated cell increases its Ca^{2+} level, and the increased Ca^{2+} level propagates from cell to cell through gap junctions. The network of cells may perform amplification of the Ca^{2+} level to increase the distance that the Ca^{2+} wave propagates. The network of cells may also perform switching by opening and closing gap junctions so that the Ca^{2+} wave propagates to a specific receiver nanomachine(s).

Receiving: Receiving is a process by which the receiver nanomachine detects cellular responses of the neighboring cell in the environment. For example, cells in the environment receiving the propagating Ca^{2+} waves may release molecules into the environment, and a receiver nanomachine detects the released molecules in the environment.

Decoding: The receiver nanomachine invokes an application specific response corresponding to the information molecule it receives. Possible decoding at a receiver nanomachine (e.g., a nanomachine based on a cell) includes differential gene expression, secretion of molecules, and generation of movement.

3.4 Current Status

The authors of this paper are currently designing and investigating the feasibility of the two molecular communication systems described in section 3.3. Other reearchers are also examining various designs of molecular communication [29, 30, 31].

Molecular communication is a new and emerging research area that is attracting a number of researchers. NSF has recently organized a workshop (in Feb, 2008) on molecular communication and discussed a number of research issues that need to be addressed [32] Some of the research issued identified include (1) representing information such that nanomachines can understand and biochemically react (encoding), (2) selectively addressing destination receiver nanomachines by a sender nanomachine (sending), (3) controlling propagation of information molecules (propagation), (4) selectively receiving information molecules at a receiver nanomachine (receiving), and (5) decoding of information molecules to cause desired chemical reaction and status change at a receiver (decoding).

4 Conclusion

This paper described biologically inspired approaches to two different classes of networks; computer networks and nanoscale biological networks. As an example of each approach, this paper described the work of the Bio-Networking Architecture and the Molecular Communication. In the Bio-Networking Architecture, biological principles such as emergence and evolution are applied to design a broad range of network applications that need to scale to a large number of network components and that need to adapt to environmental changes. In the Molecular Communication, components and mechanisms from biological communication are used to design and implement nano and micro scale biological networks. In both approaches, biological inspiration is explored to construct scalable and adaptive networking systems.

References

1. Aickelin, U., Greensmith, J., Twycross, J.: Immune System Approaches to Intrusion Detection - A Review. In: Nicosia, G., Cutello, V., Bentley, P.J., Timmis, J. (eds.) ICARIS 2004. LNCS, vol. 3239, pp. 316–329. Springer, Heidelberg (2004)
2. Di Caro, G., Dorigo, M.: AntNet: distributed stigmergetic control for communications networks. Journal of Artificial Intelligence Research (JAIR) 9, 317–365 (1998)
3. Dorigo, M., Di Caro, G., Gambardella, L.M.: Ant algorithms for discrete optimization. Artificial Life 5(3), 137–172 (1999)
4. Forrest, S., Hofmeyr, S., Somayaji, A.: Computer Immunology. Communications of the ACM 40(10), 88–96 (1997)
5. George, S., Evans, D., Marchette, S.: A Biological Programming Model for Self-Healing. In: Proceedings of the 2003 ACM workshop on Survivable and self-regenerative systems, pp. 72–81 (2002)
6. Hariri, S., Khargharia, B., Chen, H., Yang, J., Zhang, Y., Parashar, M., Liu, H.: The Autonomic Computing Paradigm. Cluster Computing: The Journal of Networks, Software Tools, and Applications 9(1), 5–17 (2006)
7. Hofmeyr, S.A., Forrest, S.: Architecture for an Artificial Immune System. Evolutionary Computation, vol. 8(4), pp. 443–473. MIT Press, Cambridge (2000)
8. Kephart, J.O.: A Biologically Inspired Immune System for Computers. In: Proceedings of the Fourth International Workshop on the Synthesis and Simulation of Living Systems, pp. 130–139 (1994)
9. Kephart, J.O., Chess, D.M.: The Vision of Autonomic Computing. IEEE computer 36(1), 41–50 (2003)

10. Montresor, A.: Anthill: a Framework for the Design and Analysis of Peer-to-Peer Systems. In: Proceedings of the 4th European Research Seminar on Advances in Distributed Systems (2001)
11. Wang, M., Suda, T.: The Bio-Networking Architecture: A Biologically Inspired Approach to the Design of Scalable, Adaptive, and Survivable/Available Network Applications. In: Proceedings of the 1st IEEE Symposium on Applications and the Internet (2001)
12. Suda, T., Itao, T., Matsuo, M.: The Bio-Networking Architecture: The Biologically Inspired Approach to the Design of Scalable, Adaptive, and Survivable/Available Network Applications. In: Park, K. (ed.) The Internet as a Large-Scale Complex System, the Santafe Institute Book Series. Oxford University Press, Oxford (2005)
13. Itao, T., Tanaka, S., Suda, T., Aoyama, T.: A framework for adaptive UbiComp applications based on the Jack-in-the-Net architecture. Kluwer/ACM Wireless Network Journal 10(3), 287–299 (2004)
14. Suzuki, J., Suda, T.: A Middleware Platform for a Biologically-inspired Network Architecture Supporting Autonomous and Adaptive Applications. IEEE Journal on Selected Areas in Communications (JSAC), Special Issue on Intelligent Services and Applications in Next Generation Networks 23(2), 249–260 (2005)
15. Nakano, T., Suda, T.: Self-Organizing Network Services with Evolutionary Adaptation. IEEE Transactions on Neural Networks 16(5), 1269–1278 (2005)
16. Hiyama, S., Moritani, Y., Suda, T., Egashira, R., Enomoto, A., Moore, M., Nakano, T.: Molecular Communication. In: Proc. of the 2005 NSTI Nanotechnology Conference (2005)
17. Moore, M., Enomoto, A., Nakano, T., Egashira, R., Suda, T., Kayasuga, A., Kojima, H., Sakakibara, H., Oiwa, K.: A Design of a Molecular Communication System for Nanomachines Using Molecular Motors. In: Fourth Annual IEEE Conference on Pervasive Computing and Communications and Workshops (March 2006)
18. Freitas Jr., R.A.: Nanomedicine. Basic Capabilities, vol. I. Landes Bioscience (1999)
19. Mao, C., Labean, T.H., Reif, J.H., Seeman, N.C.: Logical Computation Using Algorithmic Self-assembly of DNA Triple-crossover Molecules. Nature 407, 493–496 (2000)
20. Sakamoto, K., Gouzu, H., Komiya, K., Kiga, D., Yokoyama, S., Yokomori, T., Hagiya, M.: Molecular Computation by DNA Hairpin Formation. Science 288(2469), 1223–1226 (2000)
21. Weiss, R., Basu, S., Hooshangi, S., Kalmbach, A., Karig, D., Mehreja, R., Netravali, I.: Genetic Circuit Building Blocks for Cellular Computation, Communications, and Signal Processing. Natural Computing 2, 47–84 (2003)
22. Head, T., Yamamura, M., Gal, S.: Aqueous Computing - Writing on Molecules. In: The Proc. CEC 1999, pp. 1006–1010 (1999)
23. Elowitz, M.B., Leibler, S.: A synthetic oscillatory network of transcriptional regulators. Nature 403, 335–338 (2000)
24. Mao, C., LaBean, T.H., Relf, J.H., Seeman, N.C.: Logical computation using algorithmic self-assembly of DNA triple-crossover molecules. Nature 407, 493–496 (2000)
25. Alberts, B., Johnson, A., Lewis, J., Raff, M., Roberts, K., Walter, P.: Molecular Biology of the Cell, Garland Science, 4th Bk&Cdr edn (2002)
26. Enomoto, A., Moore, M., Nakano, T., Egashira, R., Suda, T., Kayasuga, A., Kojima, H., Sakibara, H., Oiwa, K.: A molecular communication system using a network of cytoskeletal filaments Communication. In: 2006 NSTI Nanotechnology Conference (May 2006)
27. Moore, M., Enomoto, A., Nakano, T., Egashira, R., Suda, T., Kayasuga, A., Kojima, H., Sakakibara, H., Oiwa, K.: A Design of a Molecular Communication System for Nanomachines Using Molecular Motors. In: Fourth Annual IEEE Conference on Pervasive Computing and Communications and Workshops (March 2006)

28. Nakano, T., Suda, T., Moore, M., Egashira, R., Enomoto, A., Arima, K.: Molecular Communication for Nanomachines Using Intercellular Calcium Signaling. In: IEEE NANO 2005 (June 2005)

29. Moritani, Y., Hiyama, S., Suda, T.: Molecular Communication among Nanomachines Using Vesicles. In: 2006 NSTI Nanotechnology Conference (May 2006)

30. Hiyama, S., Isogawa, Y., Suda, T., Moritani, Y., Suto, K.: A Design of an Autonomous Molecule Loading/Transporting/Unloading System Using DNA Hybridization and Biomolecular Linear Motors in Molecular Communication. In: European Nano Systems (December 2005)

31. Sasaki, Y., Hashizume, M., Maruo, K., Yamasaki, N., Kikuchi, J., Moritani, Y., Hiyama, S., Suda, T.: Controlled Propagation in Molecular Communication Using Tagged Liposome Containers. In: BIONETICS (Bio-Inspired mOdels of NEtwork, Information and Computing Systems) (December 2006)

32. NSF Workshop on Molecular Communication: Biological Communications Technology, http://netresearch.ics.uci.edu/mc/nsfwf08

User-Centric Mobility Models for Opportunistic Networking*

Chiara Boldrini, Marco Conti, and Andrea Passarella

IIT-CNR, Via G. Moruzzi 1, 56124 Pisa, Italy
{c.boldrini,m.conti,a.passarella}@iit.cnr.it

Abstract. In this chapter we survey the most recent proposals for modelling user mobility in mobile pervasive networks, and specifically in opportunistic networks. We identify two main families of models that have been proposed. The first modelling approach is based on the observation that people tend to visit specific places in the physical space, which therefore exert special attraction on them. The mechanics of user movements are defined based on these attractions. The second approach is based on the fact that people are social beings, and therefore they move because they want to interact and meet with each other. Movements are thus defined based on the social relationships established by users among themselves. Both modelling approaches show good match with popular traces available in the literature. However, we note that each approach misses the other's point: people actually move *both* because they are attracted by other people, and because they spend time in preferred physical places. Therefore, we describe a new mobility model (Home-cell Community-based Mobility Model, HCMM) that takes both properties into account, i.e., social relationships and attraction of physical places. HCMM matches well-known statistical features of real human mobility traces. Furthermore, it provides intuitive and easy-to-use knobs to control overall system statistical properties generated by users' movements (e.g., the average time spent by users inside or outside preferred places).

Keywords: opportunistic networks; mobility models.

1 Introduction

Modelling mobility of users is a topic witnessing renewed interest in the research community over the last few years. As discussed in more detail in Section 2, this comes hand-in-hand with the evolution of the legacy Mobile Ad hoc Networking (MANET) paradigm towards the concept of *opportunistic networking* [26]. Opportunistic networks see disconnections and partitions originating from nodes' mobility as very features of multi-hop ad hoc networks, rather than exceptions to mask (as legacy MANETs do). In these networks data addressed to nodes currently disconnected are not dropped (as in MANETs) but, according to the

---

* This work was partially funded by the IST program of the European Commission under the HAGGLE (027918) FET-SAC project.

P. Liò et al. (Eds.): BIOWIRE 2007, LNCS 5151, pp. 255–267, 2008.

store-carry-and-forward paradigm [11], node movements are exploited to bridge disconnections and bring data closer and closer to the intended destination. The simplest approach to forwarding is that of flooding the network with copies of the message, as in the case of Epidemic protocol [30]. Recently, more sophisticated solutions have been proposed, that try to detect the way nodes move and exploit these features for efficiently forward messages (see, for example, the HiBOp protocol in [35]). Whether mobility is used as an input for testing the performance of the system or is also intrinsic to the way the forwarding algorithm works, a clear understanding of user movements is in any case one of the cornerstones to design efficient solutions for opportunistic networks.

The need for an accurate knowledge of user mobility has been partly fed by a research stream looking at the statistical features of large traces either of mobile users' associations to WLANs, or to pair-wise contacts between users' mobile devices (see Section 2 for an extended reference list). Originally collected mainly to study the usage patterns of WLANs (e.g., [22]), those publicly available traces has been exploited to infer properties of the users' mobility process (e.g., [2]). Most notably, some of these studies have highlighted that legacy mobility models adopted for MANET research (e.g., the Random Waypoint Model [7]) are not realistic when compared with real traces (e.g., [17]). In turn, this has motivated the definition of novel, more realistic, mobility models (e.g., [14]).

In this chapter, we present the state-of-the-art of mobility models for opportunistic networks, after discussing in more depth the driving forces that led to renewed interest on this topic (see Section 2). We identify two main approaches along which to categorise models: i) models based on the attraction exerted by physical places on users (see Section 3), and ii) models based on the underlying social relationships between users (see Section 4). We discuss that both approaches capture very important features of realistic mobility patterns. However, we also highlight that each approach misses the other's point. Therefore, in Section 5 we discuss a recently proposed mobility model (the Home-cell Community-based Mobility Model, HCMM) that joins together both modelling approaches. To define users' movements, HCMM takes into consideration both the fact that people tend to spend time close to preferred physical locations, *and* the fact that people move also to meet each other because they have social relationships. HCMM shows the same statistical features of well-known real-world mobility traces, and provides simple handles to customise the mobility patterns. Therefore, it is a valuable tool to investigate opportunistic networking systems under realistic mobility conditions. Final remarks and open issues are discussed in Section 6.

2   Driving Forces

At the origin of the work on mobility models for opportunistic networks, we can identify two main driving forces. On the one hand, the very features of opportunistic networks themselves, i.e., the fact that mobility is a key parameter to design networking protocols. On the other hand, the availability of large traces about usage of wireless networks, that has shown that previously adopted mobility models are not able to reproduce real traces' features. These complementary aspects are discussed separately in the following sections.

2.1 Mobility in Opportunistic Networks

As sketched in Section 1, unlike in MANETs, in opportunistic networks a continuous end-to-end path has *not* to be established prior to exchange messages between a sender and a receiver. As shown in Figure 1, forwarding is generally multi-hop, and based on the *store-carry-and-forward* paradigm. Nodes store messages they have to forward and carry them until encountering another node deemed more suitable to bring the message (closer) to the eventual destination. Such paradigm is much more suitable to pervasive networking environments with respect to the legacy MANET assumptions. Mobile devices (phones, PDAs, etc.) carried by users may be just sporadically connected to a common network, e.g. because users turn them off, or they get out of reach of other nodes, or due to the intrinsic variability and instability of wireless links. Furthermore, despite the increasing penetration of 3G and WiFi networks, assuming that the core infrastructure will be so extended to seamlessly cover any mobile device users may carry on is not very realistic.

Fig. 1. Example of communication in opportunistic networks

In this scenario mobility plays a crucial role, as it permits to bridge disconnected clouds, and ultimately enable end-to-end communications despite connectivity impairments. Motivated by this remark, there has been a significant body of work focused on collecting traces of people contact patterns, in order to understand whether mobile devices carried by users could enable communications along the opportunistic networking paradigm. These works do not directly provide mobility models. However, they show distinctive features of opportunistic networking environments induced by real users' movements. Reproducing similar properties should be a target of any mobility model for such scenarios.

Works in [27,28] report about traces collected from mobile devices carried by a total of 20 students, selected from two separate classes. Students carried the devices during their daily life for two-and-a-half weeks (the first group) and eight weeks (the second group). Background software collected contacts between users by automatically establishing a Bluetooth connection when two users met. The goal of the study was to understand whether simple opportunistic routing schemes (such as Epidemic Routing) could actually work over such an opportunistic network. Therefore, authors investigated properties such as node reachability, transfer capacity, delivery probability and delay.

Similar traces have been collected in the framework of the Haggle project (http://www.haggleproject.org). The analysis of these traces [8] has mainly focused on two fundamental features, i.e., the distribution of contact duration and inter-contact times, showing that both distributions can be well approximated with a power law over a significant time frame. Assuming that the ideal distributions (i.e., without any effect due to the limited duration of the measurement period) are actually Pareto, authors have found analytical conditions on the distribution's shape for simple routing protocols (such as Epidemic) to provide finite average delays. Unfortunately, these conditions are not satisfied in the traces they have collected.

The final example we mention in this class is the Reality Mining dataset [10]. This work is particularly interesting because authors show that it is possible to derive indexes of the underlying social structure between people participating to the data gathering experiment. As we will discuss in more detail in Section 4, considering the underlying social structure to drive mobility is one of the most interesting approaches in the state of the art.

2.2 WLAN Traces Datasets

Traces discussed in the previous section require the instrumentation of mobile devices to log pair-wise contacts between users. While providing quite precise pictures of connectivity patterns, they are costly to setup. Therefore, researchers have worked to exploit another set of traces available in the literature (WLAN association traces) to infer mobility patterns of mobile users. Clearly, WLAN association is much more easy to log than real contacts between users. In the WLAN case, two nodes are assumed to be in contact if they are associated to the same access point at the same time. Sometimes, correcting techniques are used to reduce the inaccuracy of this rough assumption [20]. Despite the unavoidable approximations, this approach permits to use rich data sets to understand some feature of user mobility patterns.

The original motivation of collecting these data sets was to understand the usage of wireless LANs (see, for example, [22]). This impacts the kind of statistics that can be extracted. Typical statistics (see, e.g., [2,24]) are the number of users per AP, the number of APs visited by users, the fraction of time spent by users under each AP, etc. From a mobility modelling standpoint, the most interesting features of this work is the fact that users tend to spend a large fraction of the time under a few APs. This clearly highlights that users are attracted towards specific physical places, where they spend most of their time. This remark is at the basis of a very interesting family of mobility models that we describe in Section 3.

Several mobility models have been developed by exploiting WLAN traces. For example, authors of ModelT [17] and ModelT++ [23] exploit WLAN traces to model the registration patterns of users to APs. This is clearly not enough to build an accurate mobility model for opportunistic networks, but might be "good enough" to study other types of pervasive networking environments. Another interesting finding of these papers is the fact that popular models used for MANET research (e.g., RWP) cannot reproduce the same registration patterns observed in the real traces. These works thus confirm the need of improved models for opportunistic networking scenarios.

An interesting work in this class is presented in [15]. Authors assume that two nodes are in contact if they are connected at the same time with the same AP. Then, based on this assumption, the number of peers each mobile node encounters during the trace collection period is computed, and a corresponding graph (an edge is added if two nodes have met) is build . This graph, that shows the overall connectivity opportunity of the network, reveals small-world properties [32]. Similar properties appear also in the models used in the social networks field. This is a strong indication about exploiting social network models to define the basic mechanisms of users' movements, which is a second trend actually pursued in mobility model for opportunistic networks (see Section 4).

To conclude this section, it is worth mentioning a set of works that, starting from WLAN traces, try to define users mobility models. Specifically, we mention works in [20,33,34]. We do not describe them in detail because what they provide is actually a set of distributions that fit some statistic (e.g., the speed distribution) observed in the trace. While the fitting is usually extremely good ([34]), these papers do not propose a general mobility model that describes users movements. Therefore, their applicability outside the specific environment they have been derived from is not clear.

3 Location Driven

Models we describe in this section typically rely on WLAN traces as in the cases reported in Section 2.2 (sometimes they actually use the same traces indeed). However, they do not simply fit some distribution to model particular statistics extracted from the traces. Instead, they start by proposing a general model of how users move, and then show that the model – properly tuned – is able to match statistics extracted from the traces. This approach is more valuable, as it provides a comprehensive model of users' behaviour, that can be used to study opportunistic networks in arbitrary scenarios.

As mentioned in Section 1, the models we describe in this section capture one very significant feature of human movements, i.e., the fact that people move towards and spend time around (a few) *preferred physical places*, and do not wander in the space at random. The importance of identifying physical places where people preferentially roam is actually seen as one of the main pieces of context that should be exploited to design pervasive applications, in general (see, e.g., [18]).

The work in [16] proposes the *Weighted Waypoint Mobility Model* (WWP), as a simple extension of the random waypoint model. In WWP users select movements' destinations among a set of possible waypoints, according to a distribution that reflects the users' preference for waypoints. The probability of selecting a waypoint depends on the previous waypoint (i.e., the location a user is leaving), and the time of the day. The pause time in each waypoint is also dependent on the time of the day. To provide a concrete application example, authors tuned the waypoint transition and pause time distributions after surveying the behaviour of students in USC campus. This model captures the fact that locations are not equally preferable for users, and the fact that the time spent in different places may depend on that place and on the time of the day. The main limitations of this model are that it uses aggregate statistics to define the behaviour of *all* users, thus resulting in all users behaving (statistically) exactly the same. Furthermore, it does not consider social interactions between users, which is one of the fundamental forces driving users movements.

The fact that users tend to visit preferentially specific places is also the basis for the (richer) model in [12]. This model relies on the concept of *mobility profile*. A profile is a set of places a user visits during a day. To reduce variability across users, the model actually considers clusters of profiles, e.g., it clusters together mobility profiles that differ for just a few places. Each user can follow different representative profiles during different days. The model of their mobility is therefore represented as a mixture of profiles (see [12] for the details about how the mixtures can be evaluated). This approach clearly accounts for preferential visits of users to specific places. However, the model is a bit involved, and does not provide intuitive knobs to tune it in scenarios different from the traces it has been tuned on.

A flexible and pretty intuitive model is presented in [14]. The main idea is to assign a *community* to each user, as a physical place where users are likely to spend a significant part of their time, and re-appear periodically (e.g., their working place). Authors model single-user behaviour through two states: *local* and *roaming* epochs. In local epochs users move (following a RWP-like model) within their community, while in roaming epochs they are free to move in the whole simulation space. Transitions between epochs is governed by a simple two-state Markov chain. In the simplest version of the model, time is divided in *concentration movement periods* (CMP) and *normal movement periods* (NMP). Periods are of fixed length and regularly alternate. The parameters of the Markov chain defining transitions between local and roaming epochs change between CMPs and NMPs. The main idea is to model different periods of the day or days of the week through the alternation of CMPs and NMPs. Indeed, people behave (and move) quite differently, e.g., during working hours and leisure time. The last enrichment authors propose is defining hierarchies of communities to be used during roaming epochs: instead of wandering in the whole simulation space, users select their destination (during roaming epochs) preferentially inside regions physically close to their "home" community (in which they stay during local epochs). By exploiting the same line of reasoning (i.e., definition of local and roaming epochs, and division of time in alternating periods), the model can be further enriched. For example, authors show that the model is able to predict distinctive features of several well-known traces in the literature by using three time periods and a six-tier communities definition. A similar modelling approach (even though the model is much less intuitive), is also used in [29]. While the model in [14] is very flexible in defining user behaviour, it still does not permit much customisation on a per-user basis. Furthermore, it is basically limited to describing the users' movements, without providing any insight about the *basic mechanics* of users' behaviour that result in the described movements. In the end, this model does not shed light on the very features of people behaviour that drive their movement patterns.

The last drawback we have highlighted is actually common to all models discussed in this section and is, in our opinion, the main limitation of this body of work. This limitation is overcome by the models we discuss in the next section.

4 Social Inspired

The set of models we describe in this section aim at characterising user movements as a consequence of the social relationships people establish between each other. This capture a second very important aspect of human mobility (besides attraction towards physical places). People actually move (also) to interact with each other, and such interactions are defined by the social "links" between themselves. This remark actually opens a very interesting direction for mobility models, that is being pursued over the last few years. There has been a lot of work in the social network field to model networks of people on the basis of their social interactions ([21,32,1,9], just to mention a few well-known examples). This body of work could be highly leveraged to describe user behaviour and providing sound mobility models emanating directly from it.

To the best of our knowledge, the first work proposing such approach was [13]. The most interesting idea of this model is organising users in groups (cliques), where groups are identified according to the social network of relationships between users. Then, a physical location is assigned to each group, and users are forced to follow a predefined periodic schedule to visit all locations representing the groups they are part of. Despite resulting in a quite rigid model, this is the first work exploiting social structures to organise users in groups, and define their mobility patterns.

Much more complete and flexible models exploiting the same idea are presented, respectively, in [25], and [31]. In the following of the section we will describe the former work only, as the latter follows pretty much the same ideas, but provides less strong evidence about matching real traces.

The work in [25] describes the *Community-based Mobility Model* (CMM). In CMM every node belongs to a social community (group). Nodes that are in the same social community are called *friends*, while nodes in different communities are called *non-friends*. Relationships between nodes are modelled through social links (each link has an associated weight). At the system start-up all friends have a link to each other. Also two nodes that are not friends can have a link, according the *rewiring probability* (p_r) parameter. Specifically, for each node, each link towards a friend is rewired to a non-friend with p_r probability. This definition of social communities and social links is pretty close to the caveman model used in the social network field (see, e.g., [32]).

Social links are then used to drive node movements. Nodes move in a grid, and each community is initially randomly placed in a square of the grid. Nodes' movement is made up of two component: first, a node has to select the cell towards which to move. A node selects the target cell according to the social attraction exerted by each cell on the node. Attraction is measured as the sum of the links' weights between the node and the nodes currently moving inside or towards the cell. The target cell is selected based on the probabilities defined by cells' attraction (i.e., if a_j is the attraction of cell j, then the probability of selecting that cell is $a_j / \sum_j a_j$). After selecting the target cell, the "goal" within a cell (the precise point towards which the node will be heading) is selected according to a uniform distribution. Finally, speed is also selected accordingly to a uniform distribution within a user-specified range. CMM also allows for collective group movements. Specifically, once every *reconfiguration period* nodes of each group select a (different) cell and move to that cell. Reconfigurations are synchronous across groups, i.e., all groups start moving to the new cell at the same time. Therefore, during reconfigurations nodes of different groups may get in touch.

CMM shows several interesting features. As in [13], users are organised in groups according to social network models. However, the mechanisms driving users movements are less static and much more sensible. In CMM users movements are driven by their social relationships. Therefore, users spend most of their time together with friends, even though they go once in a while "visiting" people they have "less solid" social relationships with. Furthermore, CMM includes also collective group movements (once every reconfiguration interval), that can be exploited to model, for example, change of classes in University campuses. Simulation results [25] have shown that CMM is able to reproduce the main statistical features of pair-wise contacts between devices (specifically, authors have compared the distributions of contact and inter-contact times provided by their model with traces used by [8], showing that they are in good agreement).

To the best of our knowledge, CMM is the most flexible and complete example of mobility model inspired by social network theories. However, despite considering relationships between people, CMM does not include the other important aspect of human mobility, i.e., the attraction towards physical places. Therefore, we have proposed a new mobility model taking into account both aspects.

5   Joining Social and Physical Attraction

Main concern about models seen in Section 3 and Section 4 is that they all capture a real aspect of human mobility, but just a single one. Our belief is that location attractions and social attractions are equally important drivers of node mobility, and therefore both of them should be included in a mobility model. In fact, using one and ignoring the other could lead to mobility patterns not representative of the most frequent mobility scenarios. For example, in [36] we highlighted that in CMM nodes are prone to what we called the *gregarious behaviour*: all nodes of a community follow the movements of the first node of that community that has decided to exit the physical location where they were all roaming. The condition for the gregarious behaviour to take place is that the probability that all nodes remain in the home cell after the first node has moved out approaches zero. In [37] we proved that this condition holds for the majority (and most common) values of CMM's configuration parameter, thus revealing that the gregarious behaviour is not a mishap but a feature of the model. In real life, users can actually act like this, e.g. when all colleagues follow the first one who has suggested to go to the canteen, but, in the majority of situations, not all friends follow the first friend who decides to go out of the community. In conclusion, the gregarious behaviour does not hold in general.

Our work in [37] shows how to avoid the gregarious behaviour by joining the concepts of CMM and the concept of preferential locations of nodes. The resulting *Home-cell Community-based Mobility Model (HCMM)* maintains the social model of CMM, but introduces a different way of computing attractions and of making movement decisions. In HCMM each node is attracted by its *home cell* (i.e. the cell to which its community is assigned after a reconfiguration), based on the social attraction exerted on that node by all other nodes that are part of its community, *irrespective* of their current physical locations. In a sense, the attraction between nodes of the same group is transferred to the place where they usually roam and, hence, are expected to be found (e.g. students go to the Faculty building when they search for a professor, but here they may also find out that he has his day off). Similarly, the social attraction towards an external cell is evaluated based on the social relationships with nodes having their home in that cell. When a node is in its home cell, the cell for the next movement is selected as in CMM (but the attraction of the cells is constant within a reconfiguration period due to the new algorithm for computing attractions). When a node is outside its home cell, it will continue to roam within the external cell with a probability p_e, and it will go back home with probability $1 - p_e$. HCMM allows us to model a kind of scenario in which nodes are attracted towards a place (e.g., their office building) in which usually people of their group roam. Nodes are also attracted outside that place because of social relationships between groups, and spend some time in the foreign groups before heading back home. HCMM reproduces a world where nodes are attracted toward specific locations, but these are selected based on the friends which usually roam there, i.e in a social-aware fashion. Thus, HCMM succeeds in merging location attractions and social attractions.

When evaluating the performance of networking protocols under a specific mobility pattern, one should be able to configure the underlying mobility model for obtaining a desired behaviour. For example, the average time each node remains within its home community ($E\,[T_{in}]$) and the average time it stays outside ($E\,[T_{out}]$) are key factors that strongly impacts the performance of routing protocols for opportunistic networks. In fact, each round-trip movement is a

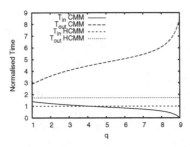

Fig. 2. Average time a node roams within and outside its home cell as functions of q

time-disjointed connection between two potentially disconnected subnetworks. Knowing $E[T_{in}]$ and $E[T_{out}]$ means knowing which is the effective connectivity level available for the routing protocols or, in other word, which is the physical upper bound on their performance. In [37] we found that, while in CMM these values change during the evolution of the network, in HCMM are constant. This means that, by initially tuning appropriately the configuration parameters, we can obtain the desired statistical behaviour for $E[T_{in}]$ and $E[T_{out}]$. In CMM this is not possible because the average time a node is within or outside its home cell varies with the state of the system, i.e, the number of nodes that are currently in each cell. Therefore, in CMM it is very hard to set model parameters to achieve the desired nodes' behaviour. These observations are confirmed by Figure 2, which plots $E[T_{in}]$ and $E[T_{out}]$ for a generic node i under CMM and HCMM as functions of q (with normalised time). q is the number of nodes of i's home cell that are currently outside and its value ranges from 1 to $n-1$, where n is the number of nodes initially assigned to the home cell (here $n = 10$).

HCMM has been shown to match realistic mobility patterns. In fact, simulation results [37] indicate that contact time and inter-contact time distributions for HCMM matches that of CMM model, which has the same pattern of the traces used by [8].

6   Future Directions

In this chapter we have surveyed and discussed research efforts aimed at refining traditional mobility models used for MANET research in the framework of opportunistic networking. Despite the results we have highlighted, we believe there are still issues that need to be satisfactorily addressed.

Collection and analysis of real traces is one of the pillars of the new wave of research on mobility models. However, recent papers [34,19,5] have clearly shown that the understanding and analysis of popular traces has still to be refined. Specifically, [5] has shown that the shape of important distributions (such as the inter-contact times between users) changes depending on the size and length in time of measurements. This is an important result that is likely to generate significant further work, because it is still not clear how much the statistical properties extracted from available traces represent very features of humans mobility, and how much they are artefacts of the measurement methodology.

Another issue related to traces exploitation is how to exploit them without sticking too much to the specific settings they have been collected in. An interesting initial work dealing with this issue has recently been published in [6]. In this paper authors propose a connectivity trace generator (CTG), that takes as input real connectivity traces, and produces traces with similar connectivity statistical properties, but with scaled parameters (e.g., the number of nodes). Tools like CTG are required to make trace-based analyses general enough by abstracting from the trace collection setup. This is required in order to achieve general understanding of opportunistic networking systems, based on realistic (but general enough) mobility patterns.

Finally, we believe that exploiting social networks results to describe users' behaviour is a very promising idea that has just started to be explored. CMM and HCMM are initial efforts in this direction. Models presented in this paper should be further extended to include more complex (and realistic) users' behaviour. For example, users community are clearly not closed, but evolve over time following the evolution of social relationships between people. Users belong to more than one community, and may change the "home" community they belong to over time (e.g., at a day scale). Grasping these features into simple and, possibly, analytically tractable mobility models is a challenge for future works.

References

1. Albert, R., Barabasi, A.L.: Statistical Mechanics of Complex Networks. Reviews of Modern Physics (2002)
2. Balazinska, M., Castro, P.: Characterizing mobility and network usage in a corporate wireless local-area network. In: Proc. of ACM/USENIX Mobysis (2003)
3. Borrel, V., Dias de Amorim, M., Fdida, S.: On natural mobility models. In: Proc. of IFIP WAC (2005)
4. Borrel, V., Dias de Amorim, M., Fdida, S.: A Preferential Attachment Gathering Mobility Model. IEEE Communications Letters 9(10) (2005)
5. Cai, H., Eun, D.Y.: Crossing Over the Bounded Domain: From Exponential To Power-law Inter-meeting Time in MANET. In: Proc. of ACM MobiCom (2007)
6. Calegari, R., Musolesi, M., Raimondi, F., Mascolo, C.: CTG: a connectivity trace generator for testing the performance of opportunistic mobile systems. In: ACM/SIGSOFT ESEC-FSE (2007)
7. Camp, T., Davies, V.: A survey of mobility models for ad hoc network research. Wireless Communication and Mobile Computing 2(5) (2002)
8. Chaintreau, A., Hui, P., Diot, C., Gass, R., Scott, J.: Impact of human mobility on opportunistic forwarding algorithms. IEEE Trans. Mob. Comp. 6(6), 606–620 (2007)
9. Dorogovtsev, S.N., Mendes, J.F.F.: Evolution of Networks. Oxford University Press, Oxford (2003)
10. Eagle, N., Pentland, A.S.: Reality mining: sensing complex social systems. Springer-Verlag Personal Ubiquitous Comput. 10(4) (2006)
11. Fall, K.: A delay-tolerant network architecture for challenged internets. In: Proc. of ACM SIGCOMM (2003)

12. Ghosh, J., Beal, M.J., Ngo, H.Q., Qiao, C.: On Profiling Mobility and Predicting Locations of Wireless Users. In: Proc. of ACM/SIGMOBILE REALMAN (2006)
13. Herrmann, K.: Modeling the Sociological Aspects of Mobility in Ad Hoc Networks. In: Proc. of ACM MSWiM (2003)
14. Hsu, W.J., Spyropoulos, T., Psounis, K., Helmy, A.: Modeling Time-Variant User Mobility in Wireless Mobile Networks. In: Proc. of IEEE Infocom (2007)
15. Hsu, W.J., Helmy, A.: On Nodal Encounter Patterns in Wireless LAN Traces. In: Proc. of WiNMee (2006)
16. Hsu, W.-J., Merchant, K., Shu, H.-W., Hsu, C.-H.: Weighted Waypoint Mobility Model and its Impact on Ad Hoc Networks. ACM SIGMOBILE Mob. Comput. Commun. Rev. (2005)
17. Jain, R., Lelescu, D., Balakrishnan, M.: Model T: an empirical model for user registration patterns in a campus wireless LAN. In: Proc. of ACM MobiCom (2005)
18. Kang, J.H., Welbourne, W., Stewart, B., Borriello, G.: Extracting places from traces of locations. ACM/SIGMOBILE Mob. Comput. Commun. Rev. 9(3) (2005)
19. Karagiannis, T., Le Boudec, J.-Y., Vojnovic, M.: Power law and exponential decay of inter contact times between mobile devices. In: Proc. of ACM MobiCom (2007)
20. Kim, M., Kotz, D., Kim, S.: Extracting a mobility model from real user traces. In: Proc. of IEEE INFOCOM (2006)
21. Kleinberg, J.: The small-world phenomenon: An algorithmic perspective. In: Proc. 32nd ACM Symposium on Theory of Computing (2000)
22. Kotz, D., Essien, K.: Analysis of a Campus-wide Wireless Network. In: Proc. of the ACM/SIGMOBILE MobiCom (2002)
23. Lelescu, D., Kozat, U.C., Jain, R., Balakrishnan, M.: Model T++: an empirical joint space-time registration model. In: Proc. of ACM MobiHoc (2006)
24. McNett, M., Voelker, G.M.: Access and mobility of wireless PDA users. ACM/SIGMOBILE Mob. Comput. Commun. Rev. 9(2) (2005)
25. Musolesi, M., Mascolo, C.: Designing Mobility Models based on Social Network Theory. ACM/SIGMOBILE Mob. Comput. Commun. Rev. 11(3) (2007)
26. Pelusi, L., Passarella, A., Conti, M.: Opportunistic Networking: data forwarding in disconnected mobile ad hoc networks. IEEE Communications Magazine 44(11) (November 2006)
27. Su, J., Chin, A., Popivanova, A., Goel, A., de Lara, E.: User Mobility for Opportunistic Ad-Hoc Networking. In: Proc. of IEEE WMCSA (2004)
28. Su, J., Goel, A., de Lara, E.: An Empirical Evaluation of the Student-Net Delay Tolerant Network. In: Proc. of Mobiquitous (2006)
29. Tuduce, C., Gross, T.: A Mobility Model Based on WLAN Traces and its Validation. In: Proc. of IEEE INFOCOM (2005)
30. Vahdat, A., Becker, D.: Epidemic Routing for Partially Connected Ad Hoc Networks. Technical Report CS-2000-06, CS. Dept. Duke Univ. (2000)
31. Venkateswaran, P., Ghosh, R., Das, A., Sanyal, S.K., Nandi, R.: An Obstacle Based Realistic Ad-Hoc Mobility Model for Social Networks. Academy Publisher Journal of Networks 1(2) (2006)
32. Watts, D.J.: Small Worlds The Dynamics of Networks between Order and Randomness. Princeton Studies on Complexity. Princeton University Press, Princeton (1999)
33. Yoon, J., Noble, B.D., Liu, M., Kim, M.: Building realistic mobility models from coarse-grained traces. In: Proc. of ACM/USENIX MobiSys (2006)
34. Zhang, X., Kurose, J., Levine, B.N., Towsley, D., Zhang, H.: Study of a Bus-based Disruption-Tolerant Network: Mobility Modeling and Impact on Routing. In: Proc. of ACM/SIGMOBILE MobiCom (2007)

35. Boldrini, C., Conti, M., Iacopini, I., Passarella, A.: HiBOp: a History Based Routing Protocol for Opportunistic Networks. In: Proc. IEEE WoWMoM (2007)
36. Boldrini, C., Conti, M., Passarella, A.: Impact of Social Mobility on Routing Protocols for Opportunistic Networks. In: Proc. IEEE WoWMoM AOC Workshop (2007)
37. Boldrini, C., Conti, M., Passarella, A.: Users Mobility Models for Opportunistic Networks: the Role of Physical Locations. In: Proc. of IEEE WRECOM (2007)

Wavelet-Domain Statistics of Packet Switching Networks Near Traffic Congestion

Pietro Liò[1], Anna T. Lawniczak[2], Shengkun Xie[2], and Jiaying Xu[2]

[1] The Computer Laboratory, University of Cambridge, 15 JJ Thomson Avenue,
Cambridge CB3 0FD, UK
[2] Department of Mathematics and Statistics, University of Guelph,
Guelph, Ont N1G 2W1, Canada

Abstract. Recent theoretical and applied works have demonstrated
the appropriateness of wavelets for analysing signals containing non-
stationarity, unsteadiness, self-similarity, and non-Markovity. We applied
wavelets to study packet traffic in a packet switching network model, focusing on the spectral properties of packet traffic near phase transition
(critical point) from free flow to congestion, and considered different
dynamic & static routing metrics. We show that "wavelet power spectra" and variance are important estimators of the changes occurring with
source load increasing from sub-critical, through critical, to super-critical
and it depends on the routing algorithm.

Keywords: OSI Network Layer; wavelet spectra, packet traffic,
congestion.

1 Introduction

The identification of the conditions of traffic congestion in the Internet and other
types of communication networks, such as wide area networks (WANs), local area
networks (LANs), wireless communication systems, ad-hoc networks, and sensors networks, is an important area for data analysis and modeling. A general
paradigm of these networks is represented by the Packet Switching Network technology. A Packet Switching Network (PSN) is a data communication network
consisting of a number of nodes (i.e., routers and hosts) that are interconnected
by communication links; see [1], [2] and the references therein; see also [3], [4],
[5], [6], [7]). In this paper, we analyse packet traffic in a data communication
network model of the packet switching type using an ensemble of wavelet-domain
statistical methods. We apply these methodology to the analysis of the number
of packets in transit (NPT) from their sources to their destinations in our PSN
model [3], [8] for various routing algorithms and network connection topologies
when source loads are close to the critical ones, i.e., the phase transition points
from free flow to congestion. We characterize the critical point by the level of
packets production at sources in the PSN model. In our model we consider dynamic and static routing algorithms. Data exploratory analysis of PSN signals
showed some statistical difficulties: the mean is not independent by the variance, signals need to be denoised or thresholded, they are non stationary, and

P. Liò et al. (Eds.): BIOWIRE 2007, LNCS 5151, pp. 268–279, 2008.

may have self similarity and multiscale properties. Many studies have indicated the importance of capturing scaling properties when analysing and modeling packet traffic [9], [10]; however, the influence of long-range dependence (LRD) and marginal statistics still remains on unsure footing. Tackling these problems will lead to improved understanding of the effects of network connection topology, routing and cost parameters on packet traffic dynamics. With these topics in mind we organise the paper as follows. In Section 2 we provide a brief description of our PSN model [3], [8]. Section 3 provides a justification for the appropriateness of using wavelets in our study; and sections 4 and 5 reports on results and conclusions.

2 Packet Switching Network Model

We study packet traffic behaviour of the PSN model, developed in [3], [8], and implemented as a C++ simulator called Netzwerk-1 [11]. The PSN model is an abstraction of the Network Layer of the ISO OSI Reference Model [1], [2] and like in real networks is concerned primarily with packets and their routings; it is scalable, distributed in space, and time discrete. We use the following naming convention for various considered PSN model set-ups. We denote the PSN model parameters by $L_\beta^\alpha(L, ecf, \lambda)$, where $\alpha = p$ (periodic) or np (non-periodic) stands for periodicity of network connection topology $L_\beta^\alpha(L)$ isomorphic to a lattice of type β (e.g., two-dimensional square or triangular lattice) with L nodes in the horizontal and vertical directions. The parameter ecf represents an edge cost function. The PSN model set-up is using $ecf = ONE$, or QS, or $QSPO$; the ecf ONE assigns a value of 'one' to each edge in the lattice L. Thus, this results in a static routing. The ecf QS assigns to each edge in the lattice L a value equal to the length of the outgoing queue at the node from which the edge originates. The ecf $QSPO$ assigns a value that is the sum of the $ecfs$ ONE and QS. The routing decisions made using ecf QS or $QSPO$ imply adaptive or dynamic routing, where packets have the ability to avoid congested nodes during the PSN model simulation. The parameter λ stands for the source load value. In the PSN model each node performs the functions of a host and a router and maintains one incoming and one outgoing queue which is of unlimited length and operates according to a first-in, first-out policy. At each node, independently of the other nodes, packets are created randomly with probability λ corresponding to the source load. In the PSN model all messages are restricted to one packet carrying only the following information: time of creation, destination address, and number of hops taken. In the PSN model time is discrete and we observe its state at the discrete times $k = 0, 1, 2, \ldots, T$, where T is the final simulation time. The set-up of the PSN model is initialized with empty queues and the routing tables are computed. At each simulation step, the time-discrete, synchronous and spatially distributed PSN model algorithm consists of the sequence of five operations: (1) *Update routing tables,* (2) *Create and route packets,* (3) *Process incoming queue,* (4) *Evaluate network state,* (5) *Update simulation time;* see [3] and [8] for details. In the PSN model, for each family of network set-ups, which differ only in the value of the source load λ, values of λ_{sub-c} for which packet traffic is congestion-free are called *sub-critical source loads*, while values λ_{sup-c} for which traffic is congested are called *super-critical source loads*. The critical source load λ_c is the largest sub-critical source load, i.e., the maximum source load, at which the PSN network traffic is free from congestion Thus, λ_c is a phase transition point from free flow to congested state of a network. In this paper we present simulation results for PSN model set-ups with a network connection topology that is isomorphic to $L_{Sq}^{p}(16)$ (i.e., a two-dimensional periodic square lattice with 16 nodes in the horizontal and vertical directions) and source load values: $SUBCSL$ (i.e., sub-critical source load $\lambda_{sub-c} = \lambda_c - 0.005$), CSL (i.e., the critical source load λ_c) and $SUPCSL$ (i.e., sup-critical source load $\lambda_{sup-c} = \lambda_c + 0.005$). Depending on the analysis type we consider simulations with the final simulation time $T \in [8000, 12800]$. For the considered PSN model set-ups the λ_c values are as follows:$\lambda_c = 0.115$ for $L_{Sq}^{p}(16, ONE, \lambda)$, $\lambda_c = 0.120$ for $L_{Sq}^{p}(16, QS, \lambda)$ and $\lambda_c = 0.120$ for $L_{Sq}^{p}(16, QS, \lambda)$.

3   Wavelet-Domain Statistics

Wavelet theory has a profound impact on signal processing as it offers a rigorous mathematical approach to the treatment of multiresolution. Benefits of carrying out an analysis in the wavelet domain, rather than in the domain of original observations, come from decorrelation and regularization considerations, as well as dimension reduction properties of the wavelet transforms [9]. There is now days a wealth of wavelets statistics that can be effective in analysing networks; for sake of space we present the most interesting methodologies. Using wavelets we study spectral properties of number of packets in transit from their sources to their destinations in our PSN model for various routing algorithms and network connection topologies when the mean flow density of packets into PSN model is closed to the phase transition point [12]. [3], [4], [5], [6], [7], [9], [10], [13], [14].

3.1   Denoising, Thresholding

In statistics, the recovery of the underlying function from a noisy signal is generally modeled using regression models; several authors have proposed wavelet estimators [15]. Consider the standard univariate regression: $y_i = f(x_i) + \epsilon_i$, where $i = 1, ..., n$, and ϵ_i are independent $N(0, \sigma^2)$ random variables; f is the "true" function. Assuming that the noise in the wavelet transform is, at each resolution level, Gaussian noise that is approximately stationary, we can reformulate the problem in terms of wavelet coefficients: $\hat{w}_{jk} = w_{jk} + \epsilon_{jk}$, where j is the level ($j = 0, ..., J - 1$), and k, the displacement ($k = 0, ..., 2^J - 1$), where $n = 2^J$ is the length of the signal. It is often reasonable to assume that only a few large coefficients contain information about the underlying function, while small coefficients can be attributed to noise. Shrinkage consists in attenuating or eliminating the smaller wavelet coefficients and reconstructing the profile using mainly the most significant wavelet coefficients and all the scaling coefficients. Several shrinkage approaches have been proposed. For example, the 'hard' threshold approach selects coefficients using a 'keep or kill' policy, while using the 'soft' thresholding, the absolute values of coefficients are shrunk by a value equal to the threshold. A meaningful approach we have challenged our data is the data-dependent change-point statistics proposed by Ogden and Parzen [16]. Data-adaptive thresholds might become very important in analyzing traffic in real networks because hypothesis testing procedures can be used to test the appropriateness of various thresholds to the data under different assumptions.

3.2   Mean - Variance Stabilization

We found that for supercritical load signals the variance of the noise increases with the mean of the signal (see Figure 1); therefore the straightforward application of wavelet shrinkage is not appropriate. A possible approach consists of transforming the problem to one where the variance of the noise is constant with respect to the mean of the signal, i.e. apply a variance stabilization transform to restore homoscedasticity. Fryzlewicz and Nason [17] proposed a Haar-Fisz (HF) transform for Gaussianising and stabilizing the variance of sequences of Poisson counts. The HF transform is performed in linear computational time as a computationally straightforward modification of the Discrete Haar Transform.

3.3   Detecting Self-similarity and Scaling Properties

Multiscale properties have been observed in packet traffic since the landmark paper [13]. Meaningful approaches to detect fractal and self-similarity and scaling property behaviours of the signal are the estimation of fBm and wavelet variance. The wavelet variance, [18], is a scale-by-scale decomposition of the variance of a signal. An estimate of the wavelet variance at a given scale is obtained by summing the squares of the wavelet coefficients (usually only those not affected by boundary conditions) and dividing by the number of them. When a bivariate signal is available, summing, at a given level, cross-products of coefficients with the same location will instead lead to an estimate of the wavelet covariance at that level. The wavelet cross-covariance at a given level and lag $\hat{o}$ can be estimated by summing cross-products of coefficients at locations whose distance from each other is equal to the given lag. Estimates of the wavelet correlation and cross-correlation are obtained by dividing the-wavelet covariance and cross-covariance by the product of the wavelet standard deviations. An approximate $100(1 - 2p)\%$ confidence interval for the MODWT (maximal overlap discrete wavelet transform) wavelet correlation at level j can be constructed, where p is a significance level, see [19]. A key property to most of multiscale systems is the self-similarity, which can be analysed by means of fractional brownian process estimators. A Gaussian stochastic process, $B_H(t)$, with mean zero, $B_H(0) = 0$, $EB_H(t)^2 = \sigma^2 t^{2H}$ for some $\sigma > 0$ and $0 < H < 1$, and with stationary increments is fractional Brownian motion (fBm) [20]. When $H = 1/2$, fBm is reduced to Brownian motion. FBm $B_H(t)$ is an example of a self-similar process

Table 1. Wavelet estimates of fractional Brownian motion of NPT signals of PSN model set-ups specified in the first column. The second column provides estimates based on second order discrete derivative (SODD). The third column provides wavelet based estimates and the fourth column provides estimates based on the linear regression in loglog plot of the variance of detail versus level.

PSN model set-up	SODD b.e.	Wavelet b.e.	Regress b.e.
$L^p_{Sq}(16, ONE, SUBCSL)$	0.5000	0.4914	0.4652
$L^p_{Sq}(16, ONE, CSL)$	0.4859	0.4869	0.4280
$L^p_{Sq}(16, ONE, SUPCSL)$	0.4871	0.4926	0.4726
$L^p_{Sq}(16, QS, SUBCSL)$	0.4979	0.4991	0.4770
$L^p_{Sq}(16, QS, CSL)$	0.4940	0.4966	0.4884
$L^p_{Sq}(16, QS, SUPCSL)$	0.5096	0.5032	0.4565
$L^p_{Sq}(16, QSPO, SUBCSL)$	0.4753	0.4694	0.4554
$L^p_{Sq}(16, QSPO, CSL)$	0.4863	0.4875	0.4928
$L^p_{Sq}(16, QSPO, SUPCSL)$	0.5222	0.5176	0.4416

with exponent H, that is for any $c \geq 0$, processes $B_H(ct)$ and $c^H B_H(t)$ have the same finite dimensional distributions and can serve as a stochastic model for nonstationary fractal data [9]. A meaningful approach to detect self-similarity is the estimation of the fBm. The main problem that occurs when using fBm as a model is to properly estimate the H parameter. Many H estimators are available and the choice of a method is a difficult issue. Among them, the wavelet based is one of the most interesting since it naturally matches the structure of the fBm process for two reasons. First, although fBm is nonstationary, its wavelet transform is stationary. Second, even if fBm is long range dependant, its wavelet coefficients are almost uncorrelated. From a practical point of view, two reasons also can motivate the use of this method: its complexity is only O(N) and it is known that the wavelet based estimator has interesting asymptotical properties. Namely, for $1/f$ processes this estimator is efficient. From now on, we will focus on properties of discrete processes denoted $BH[i]$ (i.e., fBm with a starting value $BH[0] = 0$, zero mean, Gaussian, with stationary increments and second order nonstationary). [20]. In this paper we estimated the fBm of NPT signals of various PSN model set-ups for source loads below, at and above the critical loads, see Table 1. The results are obtained using wfbmesti in Matlab.

4   Results and Discussions

In this article we use wavelets to analyse time series of PSN model with different traffic load, topology, and routing characteristics. Figure 1 shows the time plots of the NPT (number of packets in transit) signals for the PSN model set-up $L^p_{Sq}(16, QSPO, \lambda)$, respectively, for the considered $SUBCSL$, CSL and $SUPCSL$ source loads. The time plots of NPT signals for PSN model set-up $L^p_{Sq}(16, ONE, \lambda)$, for $\lambda = SUBCSL, CSL, SUPCSL$, are in [12] and for PSN model set-ups $L^{np}_{Sq}(16, ONE, \lambda)$ and $L^{np}_{Sq}(16, QS, \lambda)$, for $\lambda = SUBCSL, CSL,$

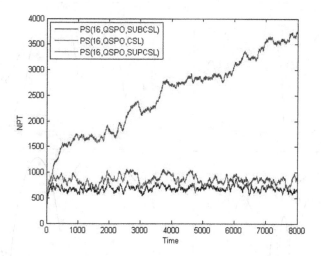

Fig. 1. Time plots of the NPT signals for the PSN model set-ups $L_{Sq}^{p}(16, QSPO, \lambda)$ for $\lambda = SUBCSL$ (blue graph), $\lambda = CSL$ (green graph) and $\lambda = SUPCSL$ (red graph) in the time window $(0, 8000)$

$SUPCSL$, are in [14]. These plots show that there are remarkable differences among the graphs of NPT signals corresponding, respectively, to $SUBCSL$ (blue graph), CSL (green graph) and $SUPCSL$ (red graph) loads for PSN model set-ups with the same type of ecf. Similar differences have been observed for other PSN model set-ups not shown here, e.g., [12].

The differences in the nature of fluctuations among these graphs are even more noticeable after detrending NPT signals. We detrend NPT signals corresponding to $SUBCSLs$ and $CSLs$, respectively, by removing from each NPT signal its sample mean. The NPT signals corresponding to $SUPCSLs$ are detrended by removing, respectively, their upward linear trends first, follow by the removal of the sample means from their residuals.

Since in congested network states mean and variance are not steady, therefore, we conduct a Haar-Fisz (HF) transform on the DNPT signals. Our results (not display here) show that the variation of calculated mean and variance has been reduced for SUB and CSL source load values. Thus, HF transform is appearing to be successful in decreasing of the range of variation for our dataset. Note that HF transform does not change the shape of the series, but it narrows down the range of variation. This fact implies that variance stabilization on the dataset is improved compared with the original data. Therefore, we may conclude that HF transform could be used as a classifier to separate free flow network states from the congested ones by using HF transform on the DNPT series. We have also carried out exploratory analysis of DNPT signals (not shown) for other PSN model set-ups.

We show in each left column of Figures 2, 3 and 4, the scaled wavelet spectra of DNPT signals of the set of PSN model parameterization considered here,

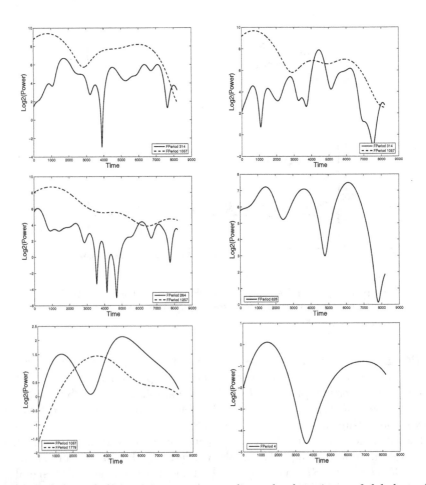

Fig. 2. Scaled wavelet power spectra corresponding to local maximum of global wavelet power spectra within cone of influence (COI) of DNPT signals (left column) and HF transformed DNPT signals (right column) of PSN model set-up $L_{Sq}^p(16, ONE, \lambda)$ for $\lambda = SUBCSL$ in the first row, $\lambda = CSL$ in the second row, $\lambda = SUPCSL$ in the third row

and in each right column of these figures the scaled wavelet spectra, respectively, of the HF transformed DNPT signals. We provide results for PSN model set-up $L_{Sq}^p(16, ONE, \lambda)$ in Figure 2, for $L_{Sq}^p(16, QS, \lambda)$ in Figure 3 and for $L_{Sq}^p(16, QSPO, \lambda)$ in Figure 4. In each of these figures the first row provides results for $\lambda = SUBCSL$, the second one for $\lambda = CSL$, and the third one for $\lambda = SUPCSL$.

We have carried out a comparative analysis of the static and adaptive routing with respect to network load and topology. Wavelet spectral properties show significant differences when load changes from $SUBCSL$ to $SUPCSL$, and these changes depend on the routing type, i.e. when static routing (ecf ONE) is replaced by adaptive routing (ecf QS or $QSPO$). Report on performances of

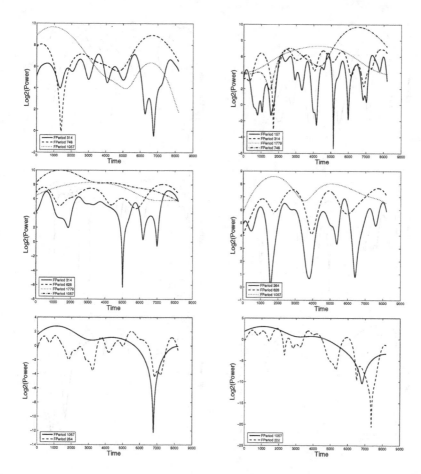

Fig. 3. Scaled wavelet power spectra corresponding to local maximum of global wavelet power spectra within cone of influence (COI) of DNPT signals (left column) and HF transformed DNPT signals (right column) of PSN model set-up $L_{Sq}^p(16, QS, \lambda)$ for $\lambda = SUBCSL$ in the first row, $\lambda = CSL$ in the second row, $\lambda = SUPCSL$ in the third row

PSN model set-ups with different types of network topologies will be provided elsewhere. As a summary, the spectral properties of number of packets in transit change with the increase of source load values from sub-critical, through critical to super-critical ones for each PSN model set-up (i.e., each selection of network connection topology and edge cost function). Additionally, for each fixed network connection topology they are dependent on the routing algorithm being used.

To determine the nature of the dependency in NPT signals we estimated the Hurst parameter of the increment of DNPT signals using the discrete wavelet transform method. The estimates were obtained using waveletFit in R and they are listed in Table 2. Based on these estimates and their confidence intervals we do not detect LRD in the increment of DNPT signals for PSN model set-ups

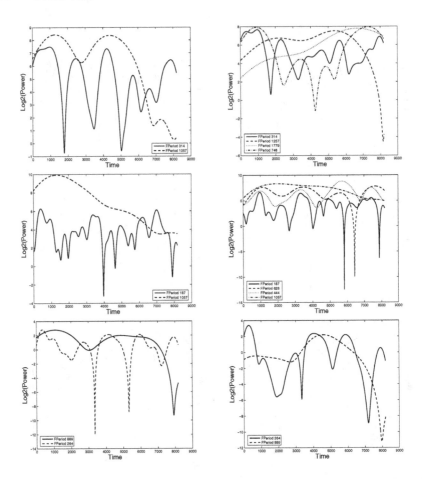

Fig. 4. Scaled wavelet power spectra corresponding to local maximum of global wavelet power spectra within cone of influence (COI) of DNPT signals (left column) and HF transformed DNPT signals (right column) of PSN model set-up $L^p_{Sq}(16, QSPO, \lambda)$ for $\lambda = SUBCSL$ the first row, $\lambda = CSL$ the second row, $\lambda = SUPCSL$ the third row

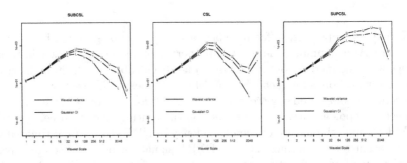

Fig. 5. Wavelet variance of DNPT signals of $L^p_{Sq}(16, QSPO, \lambda)$ for $\lambda = SUBCSL$ in the left plot, $\lambda = CSL$ in the middle plot and $\lambda = SUPCSL$ in the right plot

Table 2. Hurst exponents estimates (second column) with confidence intervals (third column) of increment process of DNPT signals (going from 4096 to 12800) of PSN model set-ups specified in the first column. The estimates were obtained using waveletFit in R function.

PSN model set-up	Hurst Exp. Est.	Confidence Interval
$L^p_{Sq}(16, ONE, SUBCSL)$	0.46	(0.41, 0.5)
$L^p_{Sq}(16, ONE, CSL)$	0.45	(0.42, 0.48)
$L^p_{Sq}(16, ONE, SUPCSL)$	0.52	(0.47, 0.56)
$L^p_{Sq}(16, QS, SUBCSL)$	0.46	(0.38, 0.53)
$L^p_{Sq}(16, QS, CSL)$	0.44	(0.34, 0.54)
$L^p_{Sq}(16, QS, SUPCSL)$	0.51	(0.48, 0.54)
$L^p_{Sq}(16, QSPO, SUBCSL)$	0.40	(0.3, 0.49)
$L^p_{Sq}(16, QSPO, CSL)$	0.44	(0.34, 0.55)
$L^p_{Sq}(16, QSPO, SUPCSL)$	0.51	(0.48, 0.54)

$L^p_{Sq}(16, ONE, SUBCSL)$, $L^p_{Sq}(16, ONE, CSL)$ and $L^p_{Sq}(16, QSPO, SUBCSL)$. However, the results for $L^p_{Sq}(16, ecf, SUBCSL)$ for $\lambda = ONE$, QS and $QSPO$ suggest possibility of LRD in our data. Since the normality test of D'Agostino of increment NPT (INPT) signals showed that INPT signals are Gaussian, the Augmented Dickey-Fuller unit root test showed that INPT signals are stationary (and DNPT non-stationary) and the 'variance plot method' showed that INPT signals are self-similar we modeled NPT signals using fBm, Table 1. The Hurst parameter estimates in Table 2, seems to be consistent (with some exceptions) with wavelet estimates of fractional Brownian motion of NPT signals listed in Table 1.

Finally, we computed the wavelet variance, covariance and correlation of DNPT signals. In Figure 5 we show the wavelet variance of DNPT signals of $L^p_{Sq}(16, QSPO, \lambda)$ for $\lambda = SUBCSL$ in the left plot, for $\lambda = CSL$ in the middle plot and for $\lambda = SUPCSL$ in the right plot. The wavelet variance of $SUPCSL$ signals (and the covariance of $SUPCSL$ versus CSL or $SUBCSL$) moves to higher scales for $QSPO$. We found a confirmation of this behaviour with a large data set of $SUBCSL$, CSL, $SUPCSL$. This suggests that wavelet variance analysis may become an insightful approach to detect LRD and phase transition associated to abrupt changes in the PSNs due to increase in load.

5   Conclusions and Future Directions

This paper addresses a few methodological approaches to analysing performance of packet traffic of PSNs. It demonstrates the applicability of these methodologies in a case study of simulation data of PSN model. In particular, it proposes a comparison of packet traffic dynamics under different conditions, from free flow to congestion. We study how this dynamics is affected by the coupling of network connection topology with routing algorithms. We have shown that wavelets power spectra are important estimators of the desiderable characteristics of a PSN traffic. Clearly wavelet-based statistics opens a new direction in the analysis of networks signals. Future work will focus on multiscale properties of wavelet analysis, and on dynamical properties of networks generated from combinations of motifs, i.e. basic patterns of interconnections (small connected subgraphs). We believe that this approach may also lead to insights on designing modular and self-organized networks and identify hidden network communities and unstable nodes.

Acknowledgement

A.T.L. acknowledges partial financial support from Sharcnet and NSERC of Canada, S. X. from the Univ. of Guelph, and J.X. from Sharcnet and the Univ. of Guelph. The authors thank B. Di Stefano and X. Tang for helpful discussions.

References

1. Bertsekas, P.D., Gallager, R.G.: Data Networks. Prentice Hall, Upper Saddle River (1992)
2. Leon-Garcia, L., Widjaja, I.: Communication Networks. McGraw-Hill, Boston (2000)
3. Lawniczak, A.T., Gerisch, A., Di Stefano, B.: OSI Network-layer Abstraction: Analysis of Simulation Dynamics and Performance Indicators. In: Mendes, J.F.F., Dorogovtsev, S.N., Povolotsky, A., Abreu, F.V., Oliveira, J.G. (eds.) Science of Complex Networks. AIP Conference Proceedings, vol. 776, pp. 166–200 (2005)
4. Tretyakov, A.Y., Takayasu, H., Takayasu, M.: Phase Transition in a Computer Network Model. Physica A 253, 315–322 (1998)
5. Fukś, H., Lawniczak, A.T.: Performance of data networks with random links. Mathematics and Computers in Simulation 51, 101–117 (1999)
6. Woolf, M., Arrowsmith, D.K., Mondragón-C, R.J., Pitts, J.M.: Optimization and phase transitions in a chaotic model of data traffic. Physical Reviews. E 66, 046106 (2002)
7. Lawniczak, A.T., Tang, X.: Packet Traffic Dynamics Near Onset of Congestion in Data Communication Network Model. Acta Physica Polonica B 37 (5), 1579–1604 (2006)
8. Lawniczak, A.T., Gerisch, A., Di Stefano, B.: Development and Performance of Cellular Automaton Model of OSI Network Layer of Packet-Switching Networks. In: Proceedings of the 16th IEEE Canadian Conference on Electrical and Computer Engineering, IEEE CCECE 2003, Montreal, Quebec, Canada, May 04-07, 4 pages (2003)
9. Abry, P., Baraniuk, R., Flandrin, P., Riedi, R., Veitch, D.: The Multiscale Nature of Network Traffic. IEEE Signal Processing Mag. 19(3), 28–46 (2002)
10. Park, K., Kim, G., Crovella, M.E.: On the Relationship Between File Sizes Transport Protocols, and Self-Similar Network Traffic. In: Int'l. Conf. Network Protocols, pp. 171–180. IEEE CS Press, Los Alamitos (1996)
11. Gerisch, A., Lawniczak, A.T., Di Stefano, B.: Building Blocks of a Simulation Environment of the OSI Network Layer of Packet Switching Networks. In: Proc. IEEE CCECE 2003-CCGEI 2003, Montreal, Quebec, Canada, pp. 001–004 (May/mai 2003)

12. Lawniczak, A.T., Lio, P., Xie, S., Xu, J.: Wavelet Spectral Analysis of Packet Traffic near Phase Transition Point from Free Flow to Congestion in Data Network Model. In: Proc. 20th IEEE CCECE 2007-CCGEI 2007, Vancouver, BC, Canada, April 22-26, p. 4 (2007)
13. Leland, W., Taqqu, M., Willinger, W., Wilson, D.: On the self-similar nature of Ethernet traffic (extended version). IEEE/ACM Trans. Networking, 1–15 (1994)
14. Lawniczak, A.T., Lio, P., Xie, S., Xu, J.: Study of Packet Traffic Fluctuations near Phase Transition Point from Free Flow to Congestion in Data Network. In: Proc. 20th IEEE CCECE 2007-CCGEI 2007, Vancouver, BC, Canada, April 22-26, p. 4 (2007)
15. Donoho, D., Johnstone, I., Kerkyacharian, G., Picard, D.: Wavelet shrinkage: Asymptopia? J. R. Statist. Soc. B 57, 301–369 (1995)
16. Ogden, R.T., Parzen, E.: Data dependent wavelet thresholding in nonparametric regression with change-point applications. Comput. Stat. Data Anal. 22, 53–70 (1996a)
17. Fryzlewicz, P., Nason, G.P.: A Haar-Fisz algorithm for Poisson intensity estimation. J. Comput. Graph. Stat. 13, 621–638 (2004)
18. Percival, D.B.: On estimation of the wavelet variance. Biometrika 82, 619–631 (1995)
19. Whitcher, B., Guttorp, P., Percival, D.B.: Wavelet Analysis of Covariance with Application to Atmospheric Time Series. Journal of Geophysical Research - Atmospheres 105, 14941–14962 (2000)
20. Mandelbrot, B.B., Van Ness, J.W.: Fractional Brownian motions, fractional noises and applications. SIAM Review 10, 422–437 (1968)

A Circulatory System Approach for Wireless Sensor Networks*

Vasileios Pappas[1], Dinesh Verma[2], and Ananthram Swami[3]

[1] IBM Research, T.J. Watson
vpappas@us.ibm.com
[2] IBM Research, T.J. Watson
dverma@us.ibm.com
[3] Army Research Lab
aswami@arl.army.mil

Abstract. One of the challenges in a military wireless sensor network is the determination of an information collection infrastructure which minimizes battery power consumption. The problem of determining the right information collection infrastructure can be viewed as a variation of the network design problem, with the additional constraints related to battery power minimization and redundancy. The problem in its generality is NP-hard and various heuristics have been developed over time to address various issues associated with it. In this paper, we propose a heuristic based on the mammalian circulatory system, which results in a better solution to the design problem than the state of the art alternatives.

1   Introduction

Wireless sensor networks are an important aspect in military networks for tracking and defending military installations. An array of sensors of many different modalities (including acoustics, chemical, thermal, and video) are used in several military contexts, including but not limited to defending sensitive installations like military bases or strategic objectives, detect and track movements of enemy forces, monitor movements of vehicles, etc. Sensors may also be mounted on airborne blimps or surveillance aircraft to collect information while flying over an area under observation.

In the common and prevalent deployments of present day US military sensor networks, the sensors themselves tend be relatively static. Sensors may be mounted on pre-selected locations, e.g. on the top of a set of buildings in an urban area, and connect via wireless or wired links to a processing location, usually in a relatively safe location. The sensors situated in locations that are relatively inaccessible may have wireless

---

* Research was sponsored by US Army Research laboratory and the UK Ministry of Defence and was accomplished under Agreement Number W911NF-06-3-0001. The views and conclusions contained in this document are those of the authors and should not be interpreted as representing the official policies, either expressed or implied, of the US Army Research Laboratory, the U.S. Government, the UK Ministry of Defense, or the UK Government. The US and UK Governments are authorized to reproduce and distribute reprints for Government purposes notwithstanding any copyright notation hereon.

P. Liò et al. (Eds.): BIOWIRE 2007, LNCS 5151, pp. 280–294, 2008.
© Springer-Verlag Berlin Heidelberg 2008

connectivity to the processing location either directly or indirectly through other sensor nodes acting as relays. The sensors in such locations also tend to be running on batteries, and need to maximize the battery longevity. On occasions, soldiers may reposition the sensors to improve coverage, but such repositioning is relatively infrequent.

Another issue facing the military wireless sensor networks is that of reliability. Due to the remote location of some of the sensors, especially those located at the perimeter of the sensor field, or others that may be relatively easy to access by outsiders, a sensor can be easily destroyed. Thus, the sensor nodes are susceptible to a high rate of failures.

While such a present day environment is very different than the more futuristic scenarios of smart dust [13] or fully mobile ah-hoc wireless networks, it still poses many interesting technical challenges that remain unsolved. When a sensor network needs to be deployed, the location of the sensor nodes is often determined by criteria such as accessibility, availability of power, location.s suitability for monitoring etc. The location of each sensor can be determined relatively accurately using GPS technology. Once the locations are determined, the team planning the layout of the sensor field needs to determine the set of links among the different sensors that need to be enabled. The set of links are chosen so as to get the information from the sensors to the central processing location. The choice of the links determines the configuration of the sensor nodes for communication purposes. Making the choices to have the best sensor network is the wireless sensor network design problem.

The network design problem needs to be solved before the actual deployment of the sensor network. During the actual deployment, soldiers install the sensors, and reconfigure them so that the right links are active. The sensor network can then operate to send the monitored information back to the processing locations, or to the intermediate processing units.

In this paper, we consider the possible solutions to network design, and propose a biologically inspired solution for the same. The biological approach of network design is inspired by the blood circulation system in humans and other mammals. We also demonstrate several advantages of the biologically inspired solution over the traditional approaches for network design when used in the context of sensor networks.

We begin first with a formal statement of the problem in Section 2, followed by a review of the state of the art techniques for network design in Section 3. This is followed by a brief description of the circulation system in biology in Section 4. In Section 5, we present the biologically inspired technique for network design and discuss its advantages and disadvantages when compared to the traditional approaches. Finally, we present our conclusions and identify avenues for further work.

2 Problem Formulation

Figure 1 demonstrates the problem of wireless sensor network design in a hypothetical context. It considers a police station in the middle of a city in Afghanistan which needs to be kept safe against potential insurgent attacks. Because the police station is located in the middle of a crowded and commercially active area, the choices of locations at which surveillance sensors can be placed for proper functional operation are limited. For the problem formulation that we are considering, the modalities of the sensors are

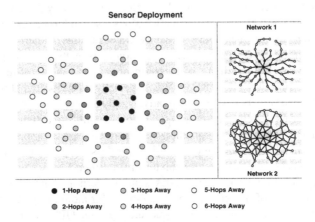

Fig. 1. Hypothetical Sensor Deployment

not significant, although they would be for the processing and fusion of information received from the different sensors. Sensors placed at the locations shown in the figure are the least likely to be vandalized, destroyed or disrupted, and are usually located on top of houses overlooking commercial streets. The information from the various sensors needs to be fed back to a processing center at the police station. Because of the poor and unreliable electric infrastructure in the city, sensors run on batteries. This provides a constraint on the inbound receiving degree of any sensor node acting as a relay. Reception of signals requires power expenditure because the receiver has to be tuned to the frequency of the transmission. Due to the limited power budget, each of the sensor nodes may be limited to be listening to one or two of the neighbors. As yet another part of power saving cycle, the transmitters and receivers may need to sleep periodically (duty-cycle) so that the battery life can be maximized. The processing center can be assumed to be active all the time, and be running off an electric generator. While the security forces can go out and reposition sensors and replace expired batteries, such activities are time-consuming and dangerous.

In a graph theoretic formulation of the problem, we can consider the problem of connecting the various nodes as shown in Figure 1. The figure also shows the possible wireless links that can potentially exist between the different sensor nodes. The connectivity links are shown as directed edges, since intervening buildings, differences in building heights, and direction of the sensors could often cause situations where information can be transmitted in one direction but not in the other. The goal is to select a set of transmission edges so that each of the nodes is connected to the processing center without violating the inbound degree constraints of any relay nodes.

The wireless sensor network design problem can be formulated in graph theoretical terms. Given the location of N nodes marking the sensor locations in a graph, and a set P of potential possible links between the different nodes, and an upper bound on the inbound degree of nodes, select a subset E of P such that the graph characterized by the set of edges E and N nodes is connected, and the inbound degree constraints of all of the nodes in the system are maintained.

Two additional attributes are important in the design of the military sensor networks, namely reliability and energy efficiency. Since many elements of the sensor network are running on batteries, one needs to minimize the number of transmissions and retransmissions needed to get the information over to a site for processing. The other constraint is that of availability . since any node can be destroyed by hostile action, e.g. tampering by an insurgent, the network topology must be able to support node failures. The requirements on constraints, power consumption and availability are often competing with each other and all can not be optimized together.

Network design problems, even with many simplifying assumptions, are known to be NP-hard [10], and this problem is no exception. As a result, the goal of network design has always been to develop appropriate heuristics to solve the design problem. In the next section, we look at some heuristics for network design that are known in the current literature.

3 Current Approaches to the Solution

Network design has been an established problem in the broader networking domain with several heuristics used in different schemes to address the topic of creating a suitable graph. In its basic formulation, the solutions approaches were initially formulated in the domain of wired t elecommunication networks.

The oldest method known to authors to connect nodes into a tree is the Esau-Williams [3] algorithm. Other commonly used heuristics involve the use of a spanning tree algorithm like Prim or Kruskal as an approximation and to add the constraints of capacity or inbound degree as a filtering mechanism during the formation of the spanning tree. A unified heuristic for tree formation is provided by Kershenbaum [9] as a general case of the above. Connecting nodes into a general mesh network rather than a tree by provided by the heuristics of clustering [2] and Cut-Set Saturation [5], with most other schemes being a variations or combination of these two. The clustering approach combines nearby nodes into a single super-node, and then finds a spanning tree that can connect those clusters together while incorporating capacity and other constraints as links are selected. The cut-set saturation approach calculates the utilization of links based on anticipated traffic, and removes links with low utilizations and high costs in the minimum cut-set till a mesh satisfying all constraints are obtained.

Another set of heuristics for designing the topologies on overlays in Internet [12] consists of two basic approaches . The add approach and the drop approach. The add approach starts with an empty graph and progressively adds edges chosen according to some weighing function assigned to each link. Links are added until no more links can be added without violating a constraint. The drop approach starts with all set of feasible links and drops edges according to some weighing function assigned to each link. Links are dropped until dropping a link results in a disconnection.

The above heuristics were developed for the design of telecommunication links, were the cost of acquiring long distance transmission links was the dominating aspect of the design process. In comparison, the cost of a link in the military wireless sensor network is effectively zero, since the links basically communicate at zero cost for the life of the network. This fact renders the basic driving principle behind the telecommunication network design ineffective.

Within the realm of sensor networks themselves, the focus of issue of network design has largely focused on maximization of coverage using mobile sensors. Several approaches for network design using mobile sensors can be found, including approaches based on mathematical programming [8], geometric properties [1], hierarchical structures [11] and flat structures [6]. However, in the specific domain that is examined in the context of this paper, based on static sensors with a predetermined location and coverage, the algorithms are not really applicable.

Another related aspect considered in the literature is that of the routing protocols developed for ad-hoc wireless networks [7]. The routes created by the protocols can be viewed as selecting specific subset of feasible links for the purpose of communication. Usually, the paths selected for routing assume bidirectional links, and focus on maintaining connectivity in the presence of mobility.

Although the various results available in prior literature can be adapted into heuristics for design the wireless sensor network, they all suffer from various drawbacks in the context of the military network deployments. As mentioned earlier, the telecommunication network designs are geared towards minimizing link costs. The mobile networks are looking at coverage, an important role but something constrained significantly by the topology of the network, and the routing protocols are optimized for a mobile environment rather than trying to optimize the transmissions in a static network.

The problem of collecting information from a set of static points in the network, with a high degree of resiliency and minimizing the number of translations can be viewed as an analogue of the blood circulation system in the human body. The circulation system provides oxygen (new configuration information) via arteries and brings back the results via veins for oxygenation (information fusion) at the lungs. The movement is synchronized by the pulsation of a single source (the heart), and such synchronization can be used to minimize the number of message transmissions and save power due to synchronization of duty-cycles of different nodes.

4   Overview of the Circulatory System

Before describing how to use the circulatory system approach to the wireless sensor network design problem, let us take a quick look at the operation of the circulation system.

The main components of the human (or mammal) circulatory system are the heart, the blood, and the blood vessels. The blood vessels consist of arteries, capillaries and veins. Arteries bring oxygenated blood to the tissues, and veins bring deoxygenated blood back to the heart. The pulmonary arteries and veins are an exception since they provide blood to the lungs for oxygenation. Blood passes from arteries to veins through capillaries, which also provide oxygen and nutrition to the different cells.

The systems of fish, amphibians, reptiles, and birds show various stages of the evolution of the circulatory system. In fish, the system has only one circuit, with the blood being pumped through the capillaries of the gills and on to the capillaries of the body tissues. In amphibians and most reptiles, a double circulatory system is used. In the first

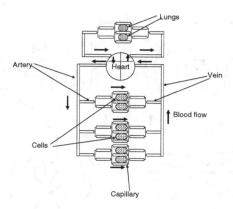

Fig. 2. Circulation System in Mammals

circuit, the blood is pumped to the lungs, where it acquires oxygen. It then returns to the heart and enters the second circuit, going to the rest of the body, eventually returning to the heart. Figure 3 illustrates a simplified model of the way the circulatory system works in mammals.

The circulatory system is powered by the heart whose contractions and expansion synchronize the pressure experienced by the entire system throughout the body. The synchronization is illustrated by the fact that the heart rate can be measured by monitoring the pulse at any of the arteries or sub-arteries in the body.

From a mathematical perspective, the circulatory system can be seen as a composition of two graphs, one characterizing the pulmonary system which goes through the lungs, and the other characterizing the circulation through the rest of the body. Each graph is a directed graph and consists of several cycles that all go through the heart. Any edge in the graph is part of a cycle going through the heart.

The circulatory system has many key characteristics that are desirable in the context of military wireless sensor networks. The circulation system can be operated in a semi-synchronized manner thereby reducing the number of transmissions that are required to collect information in any cycle. The circulation system can be created with a branching degree that is compatible with the constraints of the various sensor nodes. Furthermore, one can augment the circulation system with backup cycles to account for failures of nodes in the system.

In the next section, we describe the scheme for developing a circulation system analogue for a given set of sensor node locations and analyze its properties.

5   The Circulatory System Approach

As mentioned earlier, the primary problem in the sensor network would be to find a set of links that would create a system analogous to a circulation system in the sensor network. The circulation system comprises of a series of cycles which all include the processing center and branch at appropriately selected points. In this section, we will discuss the following topics: *(i)* how to determine the interconnection topology

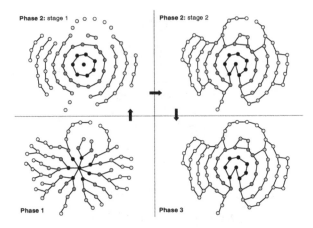

Fig. 3. Circulatory topology construction for the example sensor network of Figure 1

modeled after the circulatory system, *(ii)* how to use the discovered topology to collect information from all of the sensors, *(iii)* schemes to improve reliability and robustness to failure, *(iv)* schemes for reducing the energy consumption of the network, and *(iv)* analysis of the scheme in terms of its node degree, path length and energy savings characteristics.

5.1 Constructing the Topology

Next we describe a distributed algorithm that each node runs in order to construct the circular network topology. The algorithm has three phases. In the fist phase each node determines its distance to the sink (the collecting node), measured in hops. This distance corresponds to distance on the shortest path. There are many distributed protocols that can be used in order to compute this distance. For instance, the sink can broadcast a special message with a filed that indicates a hop count of zero. When a node receives such a message it either drops it, if it has recently seen another message with a smaller or equal hop count, or it rebroadcasts the message after increasing the message hop count by one. The node's distance from the sink is equal to the smallest hop count message that the node has received. This process is repeated periodically in order to accommodate changes in the network topology, such as the arrival of new nodes or the departure of old ones. The frequency of this process is a faction of the network dynamics. In a wireless sensor network where dynamics are minimal we expect that it is repeated quite infrequently.

The actual topology is created in the second phase of the algorithm. Links are constructed in the following two stages: In the first stage, links are created between neighboring nodes if they are in the same distance from the sink. In the case that three nodes are connected in a clique the one additional link is dropped. This can be achieved by running a localized algorithm that selects only the two links that form the Gabriel graph [4] of the three nodes, that is dropping the less energy efficient link. As a result at the end of this first step nodes that are in equal distance from the sink form a chain network. Note, that it is not necessary true that all equal-distance nodes form just one chain. Thus,

Table 1. Circulatory System Construction Algorithm

```
// Variables
sink : indicates if the node is the sink
links : the set of neighbors that it is connected to
nexthop : the next hop to the sink on the shortest path
neighbors : the set of all current neighbors
distance : distance from the sink in hops

// Initializations
links ← ∅
nexthop ← ∅
distance ← ∞
neighbors ← getNeighbors()
if sink then
    distance ← 0
end if

// Phase 1 : fi nd the shortest distance to the sink
for each nbr in neighbors do
    if nbr.distance < distance − 1 then
        distance ← nbr.distance + 1
        nexthop ← nbr
    end if
end for

// Phase 2 : add selected neighbors to the set of active links
for each nbr in neighbors do
    if nbr.distance = distance then
        links ← links ∪ nbr
    end if
end for
if ‖ links ‖> 2 then
    links ← links\ getRedundantLinks(links)
else
    if ‖ links ‖= 2 and isSelectedAsTrunk() then
        for each nbr in neighbors do
            if nbr.isSelectedAsTrunk() then
                links ← links \ nbr
            end if
        end for
        links ← links ∪ nexthop
    end if
    if ‖ links ‖< 2 then
        links ← links ∪ nexthop
    end if
end if

// Phase 3 : merge degenerated paths if possible
if ‖ links ‖= 1 then
    for each nbr in neighbors do
        if nbr.distance ≠ distance and nbr ∉ links then
            links ← links ∪ nbr
            break
        end if
    end for
end if
```

---

the network may have multiple chains of equal-distance nodes. Furthermore, a chain of equal-distance nodes is not always closed, meaning that the chain may have a starting and ending node.

The second stage of of the second phase of the construction algorithm connects all the chains of equal-distance nodes with the minimum number of links. This happens as follow: If the chain is open, then the starting and the ending nodes create a link with their neighbor that is one hop closer to the sink. In this way they become the uplink and the downlink of the chain. We call a node that is either an uplink or downlink a trunk node. If the chain is closed then two adjacent nodes are selected in order to become the trunk nodes. The selection can be random, by running a leader election algorithms between the nodes that form the chain, or it can be deterministic. In our construction algorithm, the deterministic selection happens as follow. A node becomes an uplink if its neighboring node that is one hop closer to the sink is an uplink, and if one of its neighboring nodes that is in equal-distance to the sink can become a downlink. In a similar way a node can become a downlink. Ties are solved in a localized fashion (node with the higher ID wins). Also, the equal-distance trunk nodes is drop the link that exists between them.

In a degenerated case, a node becomes both uplink and downlink when it does not have any neighbors that are in equal distance to the collecting node. The third phase of the construction algorithm is an optimization step that minimizes the number of degenerated cycles. The optimization is based on the following observation: Two degenerated cycles can be merged if their edge nodes, i.e. the ones that do not have a link to a neighbor that is in a longer distance to the sink, can be connected. Note that these two nodes cannot be equal-distance nodes because in such a case they had already been connected at the first stage of the second phase of the construction algorithm. Note that the same optimization applies in the case of just one degenerated cycle. If the edge node of such cycle happens to have a neighbor that does not have a link with it is because this neighbor is one hop closer or further from the sink. In such a case the edge node will create a link to that neighbor in order to eliminate the degenerated cycle.

The final outcome of the above algorithm is a network topology that mimics the circulatory system. The trunk links correspond to arteries and veins and the chains that connect equal distance nodes correspond to the capillaries. Table 1 provides the sketch of the construction algorithm, while Figure 3 depicts the different phases of the construction. Note that the only dashed line at the lower left corner graph corresponds to the additional link created during the optimization phase.

5.2   Information Flow

While the construction algorithm determines the final topology of the network, it does not specify the exact way that information flows into the network. The circulatory graph is a directional graph, meaning that information flows only in one way in each link[1]. Next we present a distributed algorithm that determines the direction of each each link at the flow level. The main property of the resulting flow graph is that it does not contain any loops, when the sink node is removed from the graph. In other words, the information flow algorithm creates a directed acyclic graph (DAG) on top of the network. That

---

[1] This does not necessarily mean that transmission happen only in one direction. Most of the MAC protocols require bidirectional links, so at the MAC level links are bidirectional in the circulatory graph.

is, information generated from any node in the network flows at most once on each link and information generated by the sink flows exactly once on each link. In order to create the flow graph, we first define a partial order between trunk nodes that are at the same distance from the sink (equal-distance nodes), on th shortest path tree. This way we will be able to properly assign uplinks and downlinks without creating loops in the directed graph.

But before introducing the ranking function used for the partial ordering we are going to define the following term: the *ancestor trunk* node and the *common ancestor trunk* nodes. An *ancestor trunk* node A of a trunk node T is a node that is closer to the sink than T and which can reach T without having to go through other nodes that are equal-distance to A or T. Note that a node cannot have more than two *ancestor trunk* nodes that are in equal-distance. The *common ancestor trunk* nodes $A1$ and $A1$ of two equal-distance trunk nodes $T1$ and $T2$ are two nodes so that: *(i)* $A1$ and $A1$ are also equal-distance nodes, *(ii)* $A1$ and $A1$ are connected with a path that goes only through equal-distance nodes, *(iii)* $A1$ is an *ancestor trunk* node of $T1$, *(iv)* $A1$ is an *ancestor trunk* node of $T2$, and *(v)* $A1$ is a different nod than $A2$. Note that two trunk nodes $T1$ and $T2$ can have at maximum four pairs of *common ancestor trunk* nodes.

Given the above terminology we now define a partial order between two trunk nodes that are in equal distance from the sink with the following ranking rule: A truck node $T1$ has a lower ranking from an equal-distance trunk node $T2$, denoted as $rank(T1) < rank(T2)$, if their common ancestor trunk nodes, $(A1,B1)$ and $(A2,B2)$ respectively, have the following ranking relation: $rank(A1) < rank(A2)$ and $rank(A1) < rank(B2)$ or $rank(A2) < rank(A2)$ and $rank(A2) < rank(B2)$. In the case that $T1$ and $T2$ are one hop away from the sink, the sink randomly assigns an order between them. Then a trunk node becomes an uplink if it has a lower ranking than the trunk node that is at the other end of the chain. Uplink indicates a direction from the lower distance node to the higher distance one (the uplink node), and downlink the reverse. Furthermore, the links that connect the equal-distance nodes have a direction from the uplink to the downlink node.

5.3 Robustness to Failures

The circulatory network topology has the following nice property. It requires minimum topology reconfiguration changes when a failure happens. In the case that the node is not a trunk node the required changes are the following. The two neighbors of the failed node will have to reconnect themselves with their neighbors that are one hop closer to the sink (on the shortest path tree). Note that both of them will have such a neighbor. Thus By connecting to their lower distance neighbor they become trunk nodes more specifically an uplink if the node had an incoming link from the failed node, or a downlink if the node had an outgoing link to the failed node. Note that if it happens that one of them is already a trunk node, then it is already connected to its lower distance neighbor, so no reconfiguration is required.

In the case that the failed node is a trunk node, the reconfiguration of it equal-distance node is exactly the same as before. The only difference is in the case that the failed node is connected also with with a higher distance node. In such a case the higher distance node requests from its neighbor to change the direction of their link. Consequently the neighbor tries to add a downlink if the new link direction points toward it, or an uplink otherwise. It can only succeed in case it has a neighbor that is at a shorter distance to the sink. In contrast, if it cannot create an uplink or downlink it will repeat the same procedure with its neighbor. Eventually, one of them will be able to create a downlink. If not they become an isolated island given that there is no link that it can connect them to the rest of the network.

The nice property of the above reconfiguration is that it tends to be very localized. Thus failures in one part of the network do not usually affect other parts. In contrast in the case of a shortest path tree topology the whole tree has to be recomputed every time that one node fails.

5.4   Energy Savings

In order to reduce the number of transmissions needed in the circulatory graph, the collection of sensor information by the processing center occurs in a polled manner. The processing center periodically sends out a data collection request. This request is forwarded along the circle, and at each stage the receiving sensor adds its measurement information to the information received and forwards it along the path. When a node has more than one inbound link in the circulatory graph, the node combines the information received from both of the inbound links and transmits them in a single cycle to the outbound node. When a node has more than one outbound link, it transmits the information on both the outbound links. In several configurations, it may be possible to combine the multiple outbound transmissions into a single multicast transmission, provided the receivers are awake and available to receive the message.

Another level of energy savings can be obtained by means of duty-cycling and synchronization of the message transmissions. If the frequency at which the polling cycle happens is known to all of the nodes, e.g. by sending a message with this information at the initiation of the transmission cycle, each node would know when to wake up to receive the next message from the transmitter. In this mode of operation, each of the sensor nodes is initialized to be up and receive a configuration message from the previous node in the distribution hierarchy. The central processor node initiates the transmission of the configuration message. The configuration message contains information about the delay expected for the next message transmission. The receiving node can sleep until some time before the next expected transmission. The amount of time to sleep would be dependent on the frequency of the polling by the central processor node.

When a node has more than one inbound degree, it will get the configuration message from more than one neighbor. It would need to wake up to receive both the messages. However, when transmitting further, it can combine the messages from both sources and send a single one message forward. This allows a larger time for the duty-cycling of the transmitters/receivers further along the chain.

If the processing center sends out a polling message with a period of τ, and the maximum difference in the clock rate drift of any pair of neighbors is δ, and a message transmission takes an average time of μ, then any node along the cycle need only be awake to receive message for the $(\mu + 2\delta/\tau)$ percentage of time. Messages in sensor systems are typically short, and δ is typically significantly smaller than τ, thus significant power savings can be obtained.

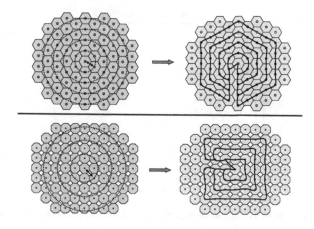

Fig. 4. Example topologies for with 6 neighbors (top) and 8 neighbors (bottom)

In the proposed scheme, the message length increases as the message travels around the loop. If the loop is of length λ, then the message in the beginning could be of size μ, and of size $\mu\lambda$ at the end of the message. If $\mu\lambda$ is smaller than τ, then the duty-cycling will still be beneficial in saving power. We would expect this assumption to be valid in most military wireless sensor deployments.

5.5   Analysis

In this section we analytically compare the circulatory approach of network formation against the shortest path tree, an approach that is usually adopted by current sensor networks. We compare the two topologies against the following metrics:

– *Node Degree*: We compute the average node degree. We consider only the in-degree given that out-degree is essentially always one due to the broadcast nature of the wireless radio.
– *Path Length*: We compute the average and maximum path lengths that a message has to follow in order to reach the sink.
– *Node Lifetime* We compute the time it takes for a node to exhaust its energy. We assume that energy is consumed during both the transmitting and the receiving phase.

For the analysis purposes we consider only regular topologies, such as the ones shown in Figure 4. We assume that all nodes have the same transmission power. In regular topologies a node is capable of communicating with C other nodes in its vicinity. For example in an arrangement of nodes where neighbors form hexagons C equals to 6, while in a grid topology C can be either 4 or 8, depending on the transmission range. If i indicates the number of hops that a node needs in order to reach the sink, then the number N_i of nodes that are in distance i from the sink is $N_i = Ci$. Thus in the case of a shortest path tree topology the average node degree D_i^t of the nodes that are i hops away from the sink is $D_i^t = N_{i+1}/N_i$, which leads to $D_i^t = 1 + 1/i$. In the

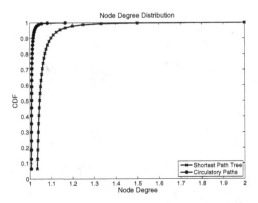

Fig. 5. Distribution of average node degree in each level in a network of 2790 with C=6

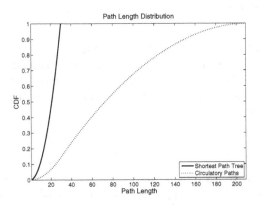

Fig. 6. Path length distribution in a network of 2790 with C=6

case of circulatory paths the average node degree D_i^c is $D_i^c = ((N_i - 1) + 2)/(N_i$, which leads to $D_i^c = 1 + 1/Ci$. Clearly, $D_i^c < D_i^t$ if $C > 1$, which is always true except from the degenerated case when the networks is a line. Figure 5 provides the cumulative distribution function (CDF) for the node degree of the shortest path tree and the circulation approach when $C = 6$ and the total number of nodes is 2790.

The path length distribution L_i^t for the shortest path tree is the same as the node degree distribution, that is $L_i^t = Ci$. For the case of circulatory paths, if a node is at distance i from the sink then the maximum path length is $L_{i,max}^c = (C+1)i - 1$, while the minimum distance is $L_{i,min}^c = i$. Thus, the average length of the paths for nodes that are in i distance form the sink is $L_{i,avg}^c = (L_{i,max}^c - L_{i,min}^c + 1)(L_{i,max}^c + L_{i,min}^c/2$ which leads to $L_{i,avg}^c = ((C + 2)i - 1)C/2$. Figure 6 gives the cumulative distribution function (CDF) for the path lengths, for the same network.

In order to compute the network lifetime we assume that the message size depends only on the number of reported events. That means that the total size of all messages propagated into the network is proportional to the total number of events that have been sensed. Also given that all nodes report events with the same frequency the energy

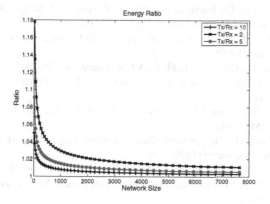

Fig. 7. Energy Consumption

consumed due to transmissions (and assuming no retransmissions) is the same independently of the network type (shortest path tree or circulatory paths). Thus the only difference is due to the energy consumed while receiving the messages, and we can safely assume that is proportional to the in-degree of a node. Thus the total energy consumed into the network during one duty cycle is: $E = \sum(E_{tx} + E_{rx}) = E_{tx}N + E_{rx}\sum_{i=1}^{l} D_i N_i$, where l is the maximum shortest path distance of a node from the source. In the case of shortest path tree network the total consumed energy is $E^t = E_{tx}N + E_{tr}C(l^2 + 3l)/2))$, while for the circulatory network it is $E^c = E_{tx}N + E_{tr}(l + C(l^2 + l)/2)$. Figure 7 shows the ration of energy consumed into the shortest path tree network over the energy consumed into the circulatory network during the same period of time, for the different network sizes. The three lines correspond to $a = 2, 5, 10$ with $E_{tx} = aE_{rx}$.

6   Conclusions and Future Work

This paper has attempted to provide a biologically inspired circulation model for solving a practical network design problem that is encountered frequently in real-life military wireless sensor networks. The proposed solution is efficient in maximizing the battery power and reducing the number of transmission needed at any node. The solution is an advancement into a field where the current state of the art heuristics are known to have specific deficiencies.

Future work we plan to conduct in this area includes a comparison of the performance of the heuristics against modifications of other design heuristics, and to attempt provable bounds on the performance of the heuristics as compared to an enumerative approach.

References

1. Brown, T., Bar-Noy, A.: Geometric Considerations for Optimally Placing Sensors in a Field. In: Conference on Unattended Ground, Sea and Air Sensor Technologies and Applications (2007)
2. Cahn, R.S.: Wide Area Network Design. Morgan Kaufmann Publishers, San Francisco (1998)

3. Esau, L., Williams, K.: On Teleprocessing System Design: A Method for Approximating the Optimal Network. IBM Systems Journal 5, 142–147 (1966)
4. Gabriel, K., Sokal, R.: A New Statistical Approach to Geographic Variation Analysis. In: Systematic Zoology (1969)
5. Gerla, M., Frank, H., Chou, W., Eckl, J.: A Cut-Saturation Algorithm for Topological Design of Packet-Switched Communication Networks. In: IEEE National Telecommunication Conference (1974)
6. Haas, Z., Tabrizi, S.: On some Challenges and Design Choices in Ad-Hoc Communications. In: IEEE MILCOM (1998)
7. Johnson, D., Maltz, D.: The Dynamic Source Routing Protocol for Mobile Ad-Hoc Networks. In: Mobile Computing (1999)
8. Joshi, A., Mishra, N., Batta, R., Nahi, R.: Ad-Hoc Sensor Network Topology Design for Distributed Fusion: A Mathematical Programming Approach. In: International Conference on Information Fusion (2004)
9. Kershenbaum, A., Chou, W.: A Unified Algorithm for Designing Multidrop Teleprocessing Networks. IEEE Transactions on Communications COM-22(11) (1974)
10. Megiddo, N.: On the Complexity of the One-Terminal Network design Problem. Operations Research Letters 1(3) (1982)
11. Ramanathan, R., Steenstrup, M.: Hierarchically Organized Multihop Mobile Wireless Networks for Quality-of-Service Support. In: ACM/Baltzer Mobile Networks and Applications (1998)
12. Verma, D., Nguyen, H.: The Constrained Overlay Network Generation Problem. In: International Symposium on Performance Evaluation of Computer and Telecommunication Systems (2001)
13. Warneke, B., Pister, K.: Exploring the Limits of System Integration with Smart Dust. In: International Mechanical Engineering Congress and Exhibition, Symposium on MEMS (2002)

Epcast: Controlled Dissemination in Human-Based Wireless Networks Using Epidemic Spreading Models

Salvatore Scellato[1], Cecilia Mascolo[2], Mirco Musolesi[3], and Vito Latora[4]

[1] Scuola Superiore di Catania
Via S. Nullo 5/i, 95123, Catania, Italy
sascellato@ssc.unict.it
[2] Computer Laboratory, University of Cambridge
15 JJ Thomson Avenue, Cambridge CB3 0FD, United Kingdom
cecilia.mascolo@cl.cam.ac.uk
[3] Dept. of Computer Science, Dartmouth College
6211 Sudikoff Laboratory, Hanover NH 03755 USA
musolesi@cs.dartmouth.edu
[4] Dipartimento di Fisica e Astronomia, Università di Catania
and INFN Sezione di Catania, Via S. Sofia 64, 95125, Catania, Italy
latora@ct.infn.it

Abstract. Epidemics-inspired techniques have received huge attention in recent years from the distributed systems and networking communities. These algorithms and protocols rely on probabilistic message replication and redundancy to ensure reliable communication. Moreover, they have been successfully exploited to support group communication in distributed systems, broadcasting, multicasting and information dissemination in fixed and mobile networks. However, in most of the existing work, the probability of infection is determined heuristically, without relying on any analytical model. This often leads to unnecessarily high transmission overheads.

In this paper we show that models of epidemic spreading in complex networks can be applied to the problem of tuning and controlling the dissemination of information in wireless ad hoc networks composed of devices carried by individuals, i.e., human-based networks. The novelty of our idea resides in the evaluation and exploitation of the structure of the underlying human network for the automatic tuning of the dissemination process in order to improve the protocol performance. We evaluate the results using synthetic mobility models and real human contacts traces.

Keywords: epidemic dissemination, human networks, mobile networks.

1 Introduction

Mobile human networks (i.e., ad hoc networks composed by devices carried by individuals) can be frequently and temporarily disconnected. Traditional routing protocol, including the basic flooding, fail to offer any sort of reliability when this happens. Epidemic-style protocols instead, being store and forward approaches

P. Liò et al. (Eds.): BIOWIRE 2007, LNCS 5151, pp. 295–306, 2008.
© Springer-Verlag Berlin Heidelberg 2008

and inherently delay tolerant [11], allow for communication in dynamic and mobile networks, also in presence of temporary disconnections or network partitions. A desired feature of the protocols is the ability to control the information spreading. For example, in emergency scenarios, when the network infrastructure has failed, it may be sufficient to send the messages only to a percentage of the rescue team members (e.g., 50% of the doctors). In other situations, there might be a need to reach all the deployed emergency personnel with the minimum overhead to avoid to collapse the network. Up to our knowledge, no solutions exploiting the minimal necessary and sufficient number of replicated messages, given the emergent network structure to guarantee a desired level of reliability exist.

The analogy between information dissemination in mobile systems and epidemics transmission in social systems is apparent. Information spreading can be modelled with a simple model for disease spreading, the so-called SIR (Susceptible-Infected-Recovered) model [2]: a host is initially *Susceptible* to new information, then it becomes *Infected* when he actually receives it, and finally it can stop the store-and-forward dissemination process becoming *Recovered* and, therefore, immune to further infections. Epidemics-inspired techniques have received huge attention in recent years from the distributed systems community [9]. These algorithms and protocols rely on probabilistic message replication and redundancy to ensure reliable communication. Epidemic techniques were firstly exploited to guarantee consistency in distributed databases [8]. More recently, these algorithms have been applied to support group communication in distributed systems. In particular, several protocols have been proposed for broadcasting, multicasting and information dissemination [10] in fixed networks.

A few attempts have been made to apply epidemic based techniques for information dissemination in mobile ad hoc networks [17,7,3]. However, existing epidemic algorithms do not permit to control the spreading of the information depending on the desired reliability and the network structure. This is partly due to the fact that these approaches are fundamentally based on empirical experiments and not on analytical models: the input parameters that control the dissemination process are selected by using experimental results and are not based on any mathematical model. This implies that the message replication process cannot be tuned with accuracy in a dynamic way: for instance, it is not possible to set the parameters of the dissemination process in order to reach only a certain desired percentage of the hosts in a prefixed amount of time. Moreover, these approaches do not exploit the information on the underlying network topology [1,4,5]. The use of epidemic spreading models based on the structure of the underlying network allows us to devise accurate mechanisms for controlling the message replication process. In other words, the number of the replicas in the network and their persistence can be tuned to achieve a desired delivery ratio.

In [15] we have presented initial results based on the so-called SIS (Susceptible-Infected-Susceptible), a model of disease spreading not considering the *recovered* state. In this paper, we propose a refined version of the algorithm based on a SIR model. The use of SIR, in coordination with the ability to decide to constrain the epidemy to a percentage of hosts, allows us to lower the message overhead considerably with respect to both our previous work and other approaches, as shown in our results section. We present an extended evaluation based on synthetic models and real traces of connectivity of the Dartmouth College [14] and National University of Singapore [16] campuses.

This paper is structured as follows. In Section 2 we describe the implementation of the middleware interface supporting the epidemic dissemination process. Section 3 presents briefly the models of epidemic spreading in complex networks that are at the basis of our dissemination algorithm. The implementation issues are discussed in Section 4. The proposed dissemination algorithm is evaluated analytically and by means of simulations in Section 5. Section 6 concludes the paper.

2 Primitives for Controlled Epidemic Dissemination

Our goal is to provide a set of primitives that allows developers to tune information dissemination in human networks according to their specific application requirements. Our aim is to ensure the spreading of information from a source A to a certain percentage Ψ of the mobile hosts of the system in a given interval time defined by a timeout t^*.

We introduce a primitive for *probabilistic anycast communication* as follows:

```
epcast(message,percentageOfHosts,time)
```

where `message` is the message that has to be sent to a certain percentage of hosts equal to the value defined in `percentageOfHosts` in a bounded time interval equal to `time`.

By using these basic primitives, more complex programming interfaces and communication infrastructures can be designed, such as publish/subscribe systems or service discovery protocols.

The infectivity of the epidemics (i.e., the probability of being infected by a host that is in the same radio range, like in human diseases spreading) can be used to control the anycast probabilistic communication mechanism. Given a percentage of hosts that has to be infected equal to Ψ, we are able to accurately calculate the value of the infectivity in order to obtain an infection rate equal to a proportion of the total number of the hosts in the network.

As we will discuss in the next section, these primitives rely on a probabilistic algorithm based on the transmission of a *minimal*, and, at the same time, sufficient, number of messages. Existing epidemic-style protocols usually achieve 100% delivery, but they waste resources by sending a large number of messages on the network, whereas our approach succeeds to send only the amount of messages necessary to inform the desired percentage of hosts in the given time.

3 Dissemination Techniques Based on Epidemic Models

In this section we introduce the mathematical models at the basis of the design of the communication API presented in Section 2. In order to model the message

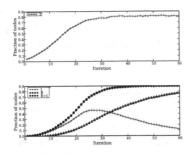

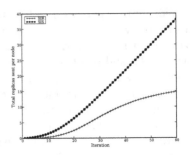

Fig. 1. Infection spreading comparation for SIS (top) and SIR (bottom) model with equal conditions ($\gamma = 0.05$, desired infection of 100%)

Fig. 2. Number of replicas per host per message for the SIS and the SIR model

replication mechanisms, we exploit mathematical models that have been devised to describe the dynamics of infections in human populations [2]. The study of mathematical models of biological phenomena has been pioneered by Kermack and McKendrick in the first half of the last century. Very recently, researchers in the area of complex networks theory have focused their attention on the problem of modeling epidemics spreading in networks characterised by well-defined structures [4,5].

According to the classic Kermack and McKendrick model, an individual can be in three states: *infected*, (i.e., an individual is infected with the disease) *susceptible* (i.e., an individual is prone to be infected) and *removed* (i.e., an individual is immune, as it recovered from the disease). This kind of model is usually referred to as the Susceptible-Infective-Removed (SIR) model [2]. Removing the possibility of permanently recovering from the disease a different version of the model is obtained, according to which individuals can exist in only two possible states, *infected* and *susceptible*. In the literature, this model is usually referred to as Susceptible-Infective-Susceptible (SIS) model [2].

The SIR model can guarantee the same delivery of the SIS model with a substantially lower number of messages as shown by the generic epidemic process depicted in Figures 1 and 2. This is due to the fact that the model introduces the possibility of having hosts that are recovered, i.e., hosts that will not participate in spreading the infection after having receiving a message M and deleted it from the buffer. In other words, in the SIR model the number of broadcasting nodes decreases after a given peak of infected nodes; instead in the SIS model, the number of broadcasting nodes at the end of the infection is (approximately) equal to the number of nodes to be infected (i.e., desired percentage of nodes in the +epcast primitive).

In the remainder of this paper we will substitute the term *individual*, used by epidemiologists, with the term *host*. A host is considered infected if it holds the message and susceptible if it does not. If the message is deleted from the host, the host becomes recovered and cannot be infected by the same message anymore. The information is spreaded among all infectives and recovered, while susceptibles are still unaware of that: it is now clear that the dissemination results depend on both infectives and recovered hosts, since these are the actual recipients of the messages that have been sent. It is useful to define a host as *reached* if it is either an infective or a recovered, since in both cases it has already received the message. Moreover it is worth noting that only infectives contribute to message replication and spreading, while recovered hosts do not.

The main assumptions of our model are the following:

- all susceptibles in the population are equally at risk of infection from any infected host (this hypothesis is usually defined by epidemiologists as *homogeneous mixing*);
- all infectives in the population have equal chances to recover;
- the infectivity of a single host, per message, is constant[1];
- the initial number of the nodes in the network is known *a priori* by each host[2];
- every host collaborates to the delivery process and no malicious nodes are present;
- each node has a buffer of the same size;
- the number of hosts is considered constant during the spreading of the infection[3];

Under the assumptions above, the system dynamics, in the case of a scenario composed of N active hosts, can be approximately[4] described by the following system of non-linear differential equations [2]:

$$\begin{cases} \frac{dS(t)}{dt} = -\beta S(t)I(t) \\ \frac{dI(t)}{dt} = \beta S(t)I(t) - \gamma I(t) \\ \frac{dR(t)}{dt} = \gamma I(t) \\ S(t) + I(t) + R(t) = N \end{cases} \tag{1}$$

where $S(t), I(t), R(t)$ are respectively the number of susceptible, infectives and removed hosts at time t, β is the average number of contacts with susceptible

---

[1] Note that the infectivity per single message (i.e., a disease) is constant, but not per single host. In other words, a host usually stores messages characterised by different infectivities in its buffer.

[2] The initial number of hosts can be usually estimated in occasion of sport events, rallies, etc. for example by evaluating the seating capacity of the venues or the size of the area when the event takes place. Statistical data are also usually available for many application scenarios, such as number of passengers that uses a station or an airport in a certain time of the day, etc. Alternatively, this number can be estimated using distributed algorithms for the calculation of the approximated network size such as [13].

[3] This is a realistic assumption, since users usually require that the information will be disseminated in a limited time.

[4] This is rigorously justifiable in a network only for complete graphs in large population limit. However, the model provides a good approximation also in scenarios composed of a limited number of hosts.

hosts that leads to a new infected host per unit of time per infective, and γ is the average rate of removal of infectives per unit of time per infectives in the population. The equations of the system state that the decaying rate of susceptibles and the growth rate of infectives are affected only by the infectivity β, the number of susceptibles $S(t)$ and the number of infectives $I(t)$; the decaying rate of infectives and the relative growth of recovered is proportional to the removal rate γ and the number of infectives $I(t)$. The last equation states that actually only two equation are needed to completely define the problem, since the sum of the three classes is constant. We furthermore set the initial conditions: $S(0) = S_0 = N - 1$, $I(0) = I_0 = 1$, and $R(0) = R_0 = 0$, with the condition $I_0 = 1$ representing the first copy of the message that is inserted in its buffer by the sender.

A numerical solution of the system (1) can be easily obtained by standard ODE solver routines. This allows to compute the number of infectives and recovered at instant t as a function of the infectivity β and of the removal rate γ. The value of γ is usually fixed by the local properties of the hosts [5]. Instead, the value of β, that is the fundamental parameter of the message replication algorithm, can be tuned in order to have, after a specific length of time t^*, a number of reached hosts (i.e., hosts that have received the message) equal to $I(t^*)+R(t^*)$ or, in other words, a fraction of reached hosts equal to $(I(t^*) + R(t^*))/N$.

In order to effectively exploit the model just described, the actual connectivity of each host should be kept into account. We will assume a mobile system with a homogeneous network structure, described by a connectivity distribution $P(k)$, strongly peaked at an average value $\langle k \rangle$. This is a realistic assumption in cases characterized by a high density of hosts, and where the movement is well described as an uncorrelated random process, such as in large outdoor spaces (i.e., squares, stations, airports or around sport venues) [12,15]. In this case, the degree k of each node can be approximated quite precisely with the average degree $\langle k \rangle$. In order to include the effect of the connectivity on the spreading, the system (1) can be rewritten by substituting β with $\lambda \frac{\langle k \rangle}{N}$ [4]:

$$\begin{cases} \frac{dS(t)}{dt} = -\lambda \frac{\langle k \rangle}{N} S(t) I(t) \\ \frac{dI(t)}{dt} = \lambda \frac{\langle k \rangle}{N} S(t) I(t) - \gamma I(t) \\ \frac{dR(t)}{dt} = \gamma I(t) \\ S(t) + I(t) + R(t) = N \end{cases} \quad (2)$$

where λ represents the probability of infecting a neighbouring host during a unit of time, and $\frac{\langle k \rangle}{N}$ gives the probability of being in contact with a certain host. In other words, in this model, by substituting β with $\lambda \frac{\langle k \rangle}{N}$, we have separated, in a sense, the event of being connected to a certain host and the infective process [4].

---

[5] If overflow phenomena do not occur (i.e., in the case of sufficiently large buffers), the model can be simplified with $\gamma = 0$ and, therefore, no host will never become recovered.

In conclusion, the main idea is to calculate the value of λ as a function of $I(t^*) + R(t^*)$ and $\langle k \rangle$. It is also interesting to note that in homogeneous networks, every host knows its value of k and, consequently, it has a good estimate of $\langle k \rangle$. We will exploit this property to tune the spreading of message replicas in the system.

4   Implementation

Every time the middleware primitive defined in Section 2 is invoked, the middleware calculates the value of the infectivity λ that is necessary and sufficient to spread the information to the desired fraction of hosts in the specified time interval (specified in the field `percentageOfHosts` of the `epcast` primitive), by evaluating the current average degree of connectivity and the current removal rate of messages from the buffer. The message identifiers, the value of the calculated infectivity, the timestamp containing the value specified in `time` expressing its temporal validity are inserted in the corresponding headers of the message in the *infectivity* field. Then, the message is inserted in the local buffer.

The epidemic spreading protocol is executed periodically with a period equal to τ. With respect to the calculation of the message infectivity, we assume τ as time unit in the formulae presented in Section 3. In other words, assuming, for example, $\tau = 10$, a timestamp equal to one minute corresponds to six time units. The value of τ can be set by the application developer during the deployment of the platform. Clearly, the choice of the values of τ influences the accuracy of the model, since it relies on a probabilistic process. For this reason, given a minimum value of timestamp equal to t_{MIN}, developers should ensure $\tau << t_{MIN}$. The number of rounds will be equal to t^*/τ. For the Law of the Large Numbers, we obtain a better accuracy of the estimation of the evolution of the epidemics as the number of rounds (i.e., from a probabilistic point of view, the number of trials) increases.

Every τ seconds each infected host broadcasts the message and its neighbours receive the message. If the message is not already present in their buffer they store it with a probability λ: moreover, they will not store it if the message has been already present in buffer in the past, although it is not present at current time. This behaviour maps quite well the SIR epidemics model, since a node receives a new message, actively spreads it for some time and then it deletes the message from the buffer (i.e. to make room for new messages), never accepting it again. Therefore, a node has to store the identifiers of all messages received in a defined time window, which is a reasonable given the limited occupation of the vector of the message identifiers.

5   Evaluation

5.1   Analytical Evaluation

An interesting quantitative parameter is the total number of messages needed to disseminate messages to a certain percentage of hosts. A message is broadcasted by an infective host in every round: as soon as the host deletes the message it does not accept the same message again.

Considering an infection process repeated for a number of times equal to r number of rounds, indicating with t_r the time length of the rth round, the total number of replicas per single type of message can be estimated as follows:

$$\text{Number Of Replicas} = \int_{t=0}^{t=t_r} I(t)dt \qquad (3)$$

From a graphical point of view, the number of copies is equal to the area under the curves in Figure 1 and 2. A comparison between SIR- and SIS-based protocols shows that while for both cases the formula 3 helds, in the former case the total number of replicas sent is much lower. This is the result of the recovering process, which enables hosts to stop message spreading when the epidemics is already growing but, at the same time, still assures that the final result will be guaranteed.

5.2   Experimental Evaluation

Description of the Simulation. In order to test the performance of these techniques, we defined a square simulation area with a side of 1 km and a transmission range equal to 200 m. The simulation was set to run several replicates for each mobile scenario in order to obtain a statistically meaningful set of results (with a maximum 5% error). All simulations are written in Python using NetworkX [6], a package for the creation, manipulation, and study of the structure, dynamics, and functions of complex networks. We analysed scenarios characterised by different number of hosts (more precisely 64, 128, 256, 512). These input parameters model typical deployment settings of mobile ad hoc networked systems. We do not model explicitly the failures in the system, since we assume that during the infection process, the number of hosts remains constant.

The movements of the hosts are generated using a Random Way-Point mobility model [6]; every host moves at a speed that is randomly generated by using a uniform distribution. The range of the possible speeds is $[1, 6]m/s$. We selected this mobility model, since as discussed in [12], its emergent topology has a Poisson degree distribution. Therefore, in this scenario, the properties of the network can be studied with a good approximation by assuming a homogeneous network model. The accuracy of the approximation increases as the density of population increases, since, considering the finite and limited simulated time, we obtain a scenario characterised by a time series of degree of connectivity values with lower variance. Moreover, the so-called border effects, due to the host that moves at the boundaries of the simulated scenarios, have less influence as the density of population increases.

Each node uses a buffer of 5 messages, managed as a FIFO queue, and 20 different messages are sent in the initial round by random chosen nodes.

---

[6] http://networkx.lanl.gov

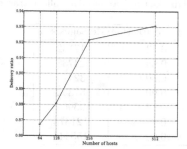

Fig. 3. Delivery ratio vs population density with desired reliability equal to 100

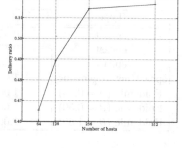

Fig. 4. Delivery ratio vs population density with desired reliability equal to 50

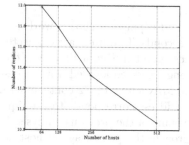

Fig. 5. Number of replicas per host per message vs population density with desired reliability equal to 100

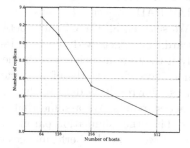

Fig. 6. Number of replicas per host per message vs population density with desired reliability equal to 50

Analysis of Simulation Results. In this subsection we will analyse the results of our simulations, discussing the performance of the proposed techniques. We will study the variations of some performance indicators, such as the delivery ratio and the number of messages sent as functions of the density of hosts (i.e., the number of the hosts in the simulation area).

Figures 3 and 4 show the delivery ratio (i.e., the desired percentage of hosts in the epcast primitive) in terms of population density, for the case of a desired percentage of hosts equal to 100 and 50, respectively, with $t^* = 10min$. The performance in terms of delivery ratio are close to the desired ones. Also in this case, the better approximation of the assumption of homogeneous network, obtained when the density of population increases, leads to better results (i.e., a more accurate estimation) for the case of 512 nodes.

The number of replicas per host per message are plotted in Figure 5 and 6. These diagrams illustrate the scalability of our approach, since the number of replicas is slightly decreasing when more nodes are added.

Table 1. Comparison of performances on the real dataset of Dartmouth College traces

Type	Desired fraction	Delivered fraction	Messages sent
epcast	0.50	0.43	17132
epcast	0.75	0.68	24738
epcast	1.00	0.90	32475
epcast(heterogeneous)	1.00	0.90	57342
Epidemic ($\beta = 0.25$)	1.00	0.64	95969
Epidemic ($\beta = 0.50$)	1.00	0.87	121873
Epidemic ($\beta = 1.00$)	1.00	0.92	155446

Evaluation with Dartmouth Traces. In order to evalute our approach on real data we run simulations using a source of data describing how real users move between different locations, i.e. wireless access points. A large amount of traces for Darmouth College's 802.11b campus network is available through the CRAWDAD project [14].

We selected all the contacts between 9 am and 6 pm in a chosen work day, discarding contacts with duration less than 60 seconds. Two users are connected only if they are associated with the same access point during a time slot: epidemics spreading is therefore performed among users co-located with access points. Our resulting data set had 2201 unique MACs and 11572 contacts with all access points. We assume that each MAC address corresponds to a unique user. The other simulation parameters are the same of the previous analysis. In Table 1 we show the performances of our approach: the percentage of host actually reached is slight less than the desired fraction of population and this can be explained by observing that these contacts are not always connected during all the simulation time and may be easily absent from the underlying network. In other words, the underpinning hypothesis of the epidemic spreading model that we are using are only approximately satisfied. We run a simulation with a standard epidemic approach where infectivity is not tuned using the SIR model but it is set to 0.25, 0.50 and 1.00 respectively. It is interesting to note that the number of messages is in all three cases higher; only the case with infectivity equal to 1.00, the standard epidemic protocol is able to reach all the hosts. This is also demonstrate how it is difficult to choose the right value of the infectivity in a purely heuristic way to reach all the hosts of the system.

We run also some simulations using a dataset from the National University of Singapore[16], which contains contact pattern of 22341 students inferred from the information on class schedules and class rosters for the Spring semester of 2006. Two students are connected if they attend the same class during a time slot. However, in this dataset a large fraction of students is not included in the instantaneous underlying network, since they are not attending any class. The result is that in this case the epidemics fails to start using our model based on the assumption of homogeneous mixing. Additional virtual point of aggregation can be included in the simulations, grouping a percentage of the students that are not attending lectures during a particular timeslot: this modification ensures homogeneous mixing, providing good results for our algorithm. However, this is only a conjecture given the nature of the traces.

Heterogeneous Networks. The results and the solutions discussed in this paper rely on the assumption of homogeneous networks, that are emerging from the random movements of the nodes. We now show that the proposed approach can be extended to the general case of heterogeneous networks. These structures are emerging in presence of small clusters of people or communities.

For heterogeneous networks the approximation $k \approx \langle k \rangle$ is not valid. However, the same probabilistic communication primitives introduced in Section 2 could be used, with a different semantics. This relies on the following observations: given k fluctuating in the range $[k_{MIN}, k_{MAX}]$, we observe that for a value of the infectivity corresponding to $k = k_{MIN}$, the obtained spreading of the infection $I(t^*, k_{MIN})$ will always be greater than the one obtained with another k. In other words, if k_{MIN} is selected in the calculation of the value of the infectivity, the value of `Reliability` can be considered approximately as a guaranteed lower bound of the reliability level.

The value of k_{MIN} can be dynamically retrieved and set by the middleware by monitoring the connectivity of the hosts composing the mobile system. We plan to investigate these adaptive mechanisms further in the future.

6   Concluding Remarks

In this paper we have shown how models of epidemic spreading in complex networks can be applied effectively to the problem of disseminating information to subset of hosts (or to all the hosts) in a wireless network, controlling at the same time the number of the copies in the system. We have presented an analytical and experimental evaluation of our approach using a synthetic random model and real traces, showing the effectiveness of our approach.

Acknowledgements. Cecilia Mascolo and Mirco Musolesi acknowledge the support of EPSRC through the CREAM Project. Salvo Scellato thanks UCL for the financial support as Visiting Student.

References

1. Albert, R., Barabasi, A.-L.: Statistical Mechanics of Complex Networks. Review of Modern Physics 74, 47–97 (2002)
2. Anderson, R.M., May, R.M.: Infectious Diseases of Humans: Dynamics and Control. Oxford University Press, Oxford (1992)
3. Baehni, S., Chabra, C., Guerraoui, R.: Frugal Event Dissemination in a Mobile Environment. In: Alonso, G. (ed.) Middleware 2005. LNCS, vol. 3790, pp. 205–224. Springer, Heidelberg (2005)
4. Barthélemy, M., Barrat, A., Pastor-Satorras, R., Vespignani, A.: Dynamic Patterns of Epidemic Outbreaks in Complex Heterogeneous Networks. Journal of Theoretical Biology (2005)

5. Boccaletti, S., Latora, V., Moreno, Y., Chavez, M., Hwang, D.-U.: Complex networks: Structure and dynamics. Phys. Rep. 424(4-5), 175–308 (2006)
6. Camp, T., Boleng, J., Davies, V.: A Survey of Mobility Models for Ad Hoc Network Research. Wireless Communication and Mobile Computing 2(5), 483–502 (2002)
7. Costa, P., Picco, G.P.: Semi-probabilistic Content-Based Publish-Subscribe. In: Proceedings of ICDCS 2005, pp. 575–585 (2005)
8. Demers, A., Greene, D., Hauser, C., Irish, W., Larson, J., Shenker, S., Sturgis, H., Swinehart, D., Terry, D.: Epidemic Algorithms for Replicated Database Maintenance. ACM SIGOPS Operating Systems Review 22(1) (January 1988)
9. Eugster, P.T., Guerraoui, R., Kermarrec, A.-M., Massouli, L.: Epidemic Information Dissemination in Distributed Systems. IEEE Computer (May 2004)
10. Eugster, P.T., Handurukande, S., Guerraoui, R., Kermarrec, A.-M., Kouznetsov, P.: Lightweight Probabilistic Broadcast. ACM Transactions on Computer Systems 21(4), 341–374 (2003)
11. Fall, K.: A delay-tolerant network architecture for challenged internets. In: Proceedings of the SIGCOMM 2003, pp. 27–34. ACM Press, New York (2003)
12. Glauche, I., Krause, W., Sollacher, R., Greiner, M.: Continuum Percolation of Wireless Ad Hoc Communication Networks. Physica A 325, 577–600 (2003)
13. Jelasity, M., Montresor, A.: Epidemic-style proactive aggregation in large overlay networks. In: Proceedings of the 24th International Conference on Distributed Computing Systems (ICDCS 2004), Tokyo, Japan, March 2004, pp. 102–109. IEEE Computer Society, Los Alamitos (2004)
14. Kotz, D., Henderson, T., Abyzov, I.: CRAWDAD data set dartmouth/campus (v. February 08, 2007) (February 2007), http://crawdad.cs.dartmouth.edu/dartmouth/campus
15. Musolesi, M., Mascolo, C.: Controlled Epidemic-style Dissemination Middleware for Mobile Ad Hoc Networks. In: Proceedings of MOBIQUITOUS 2006, ACM Press, New York (2006)
16. Srinivasan, V., Motani, M., Ooi, W.T.: CRAWDAD data set nus/contact (v. August 01, 2006) (August 2006), http://crawdad.cs.dartmouth.edu/nus/contact
17. Vahdat, A., Becker, D.: Epidemic Routing for Partially Connected Ad Hoc Networks. Technical Report CS-2000-06, Department of Computer Science, Duke University (2000)

Maintaining Spatial-Temporal Knowledge through Human Interaction

Halikul Lenando and Roger M. Whitaker*

School of Computer Science,
Cardiff University, 5 The Parade,
Roath, Cardiff, CF24 3AA
R.M.Whitaker@cs.cardiff.ac.uk

Abstract. Using wireless peer-to-peer interactions between portable devices, it is possible to locally share information and maintain spatial-temporal knowledge emanating from the surroundings. We consider the prospects for unleashing ambient data from the surrounding environment for information provision using two biological phenomena: human mobility and human social interaction. This leads to analogies with epidemiology and is highly relevant to future technology-rich environments. Here, embedded devices in the physical environment, such as sensors and wireless-enabled appliances, represent information sources that can provide extensive situated information. In this paper we address a candidate scenario where isolated sensors in the environment provide real-time data from fixed locations. Using simulation, we examine what happens when information is greedily acquired and shared by mobile participants through peer-to-peer interaction. This is assessed taking into account availability of source nodes and the effects of mobility with respect to temporal accuracy of information. The results reaffirm the need to consider a range of mobility models in testing and validating protocols.

Keywords: Opportunistic networking, mobile peer-to-peer networking, wireless.

1 Introduction

The introduction of nomandic wireless communication systems such as mobile ad-hoc networks (MANETs) [4] and sensor networks has followed an interesting pattern of development. Generally speaking, there has been an engineering assumption that the end-to-end networking connectivity, as seen in wired networking, should be facilitated. Establishing and maintaining connectivity is the goal of majority of protocols that have been developed in this area. This is a considerable challenge since unlike wired networks, uncertainty on node density and MANET node movement are present for the general deployment scenario.

Until relatively recently, network disconnectivity has been regarded as an inconvenience to be overcome using approaches such as delay tolerant networking

---

* Corresponding author.

P. Liò et al. (Eds.): BIOWIRE 2007, LNCS 5151, pp. 307–318, 2008.
© Springer-Verlag Berlin Heidelberg 2008

(e.g., [8]) rather than an inherent characteristic on which communication and services could be based. Approaches such as opportunistic networking [14] and mobile peer-to-peer (MP2P) networking [6] are embracing this alternative view and provide new opportunities to embed communication more closely with human activity based on the *assumption* of transient interactions between devices in the presence of natural disconnectivity.

In this paper we continue with this trend but consider providing services at a system level rather than facilitating communication. Specifically we look at the prospects for unleashing ambient data from the surrounding environment for information provision using two biological phenomena: *mobility* and *social interaction*. In this case the species is the human who carries a mobile wireless device. As compared to simple species such as ants, which are now frequently exploited for their emergent behaviour [5], humans exhibit much more complex social behaviour based on lots of factors including their environment. To model this as a first approximation, we consider a number of mobility scenarios that may be encountered.

Using mobile peer-to-peer interactions, classifiable as the most general case of opportunistic networking, it is possible to locally share information and maintain spatial-temporal knowledge emanating from the surroundings. This represents a paradigm shift from the conventional notion of a network where maintaining/facilitating end-to-end connectivity is paramout. Instead, flexibility for the user is increased since the information source and destination nodes need never be connected at the same point in time. We assume that peer devices interact on a strictly pairwise basis, without maintaining connectivity to any other node beyond the target peer. This provides a basic means by which peers may compare and share information.

We apply this assuming that the information for sharing has *temporal relevance*: that is, the quality of the information depletes with time such as is the case for real-time sensor data, transport information and information concerning the status/availability of environmental resources. This leads to analogies with epidemiology where a disease may spread due to the physical locality of participants and their interactions. However, unlike epidemiology, we aim to *maintain* the equivalent of infection through temporal updating from contact with peer devices.

We focus on scenarios where there are fixed devices, termed *information sources*, providing real-time information about a resource. Using simulation that has basic features of the IEEE 802.15.4 wireless PAN standard, we consider the characteristics of this paradigm and the temporal quality of information that can be maintained. This is not currently known and is valuable to establishing the types of scenarios for which services can be provided.

2   Related Literature

There are two general areas in which developments have been made. From the distributed data-base management perspective, the functionality for communication via mobile peer-to-peer interaction has been assumed and the acquisition and sharing of data has been considered for a range of scenarios (e.g., [24]) including sensors (e.g., [25]). The notion of a "data mule" has also been introduced (e.g., [17]) for data acquistion in sparse sensor deployments and economic models for the value of information have been introduced [26]. Resource discovery has also been explicitly considered [3], as has the problem of searching across mobile peers [23]. However issues of data dissemination [16,7,12,10], co-operation and resource utilisation [22,13] have currently reeived the most attention.

Beyond information provision, mobile peer-to-peer interactions have been considered has a general basis for communication applications such as messaging and data transfer. This has been conducted in the network/engineering community under the guise of opportunistic networking. The focus of most effort in this area has been on mechanisms to emulate functionality seen in connected networks, most notably routing. Focussing on opportunistic networks without any fixed infrastructure, developments can be largely categorised as being *dissemination-based* or *context-based*.

Dissemination-based routing seeks to diffuse a message all over the network. The heuristic behind this policy is that, since there is no knowledge of a possible path to the destination, nor an appropriate next-hop node, a message should be widely dispersed. High levels of dispersion can lead to congestion [18] and different attempts to control this have been made. These concern various ways of controlling the maximum number of relay hops, as considered by epidemic routing [20], "spray and wait" [19], the MV-protocol [1], the PROPHET protocol [9] and network coding [21] which generally seek to limit flooding by exploiting knowledge about contacts with destination nodes.

Context-based routing seeks to exploit more information about the context in which nodes are operating so as to identify suitable next hops towards eventual destinations. The underlying idea in this approach is that context can be used to control redundant message duplication, but potentially at the cost of higher delay. Additionally there is a higher computational cost in these approaches, both from processing and memory requirements. Developments using context-based strategies for routing include context-aware routing [11] and MobySpace routing [8].

3   The Model

We assume that all nodes are equipped with IEEE 802.15.4 type transmission capabilities and classify devices as being a fixed location *information source* or *nomadic* (i.e., possibly mobile, depending on the scenario). The *information sources* are assumed to provide real-time information about the status of a resource or an event.

A single isolated information source may represent a sensor or wireless-enabled device situated in the environment (e.g., a vending machine). Multiple local information sources represent the gateway points to a "backbone" network such as the internet, from which the information source nodes pull information and

Table 1. Global parameters

Parameter Description	Setting
Channel Bit rate	250 kbps
Discovery Success rate	0.95
Transmission Success rate	0.95
Channel Set-up time	0.5
meta-data size	0.1 kb
artefact size	3 kb
transmission range	30 metres
simulation time step	0.1 seconds
simulated duration	15 minutes
region size	$500m \times 500m$

make it available for distribution to mobile peers. In the case where multiple information sources occur, for the purposes of this paper we assume that they all provide copies of the same real-time updated information. We refer to the information provided relative to the time at which it was created, and call this an *artefact*. The description of an artefact (specifically its age) is defined by its *meta-data*, which may be exchanged between nodes independent of the artefact. We assume that each node has the storage sufficient for an artefact and at least two pieces of meta-data and the processing capability to compare meta-data from another node with its own.

We assume that the transmission technology applied is of the wireless personal networking variety, such as IEEE 802.15.4, which can be embedded in motes and other tiny wireless devices on a very cheap basis. We approximate protocol behaviour by modelling the time taken for discovery and channel set-up between peers (e.g., 0.5 seconds [15]). We assume that this occurs with 95% success. Random selection of the discovered peer is assumed, and at any one point in time, a node may maintain a link to at most one other node. A range of 30 metres is assumed, which is more conservative than specifications such as IEEE 802.15.4 to account for environmental impediments to transmission.

3.1 Information Exchange Protocol

To determine the potential quality of information from peer-to-peer interactions, we apply a fully opportunistic protocol for artefact exchange. The protocol is greedy in the sense that when a node is not engaged in a peer-to-peer interaction, it is engaged in peer discovery as described above. Note that this is not a resource efficient protocol and it is unlikely to be appropriate in practice. We are modelling it here merely to scope the *possible* performance in terms of information quality, that could be achieved.

Once a connection has been established and channel set-up is complete, the pair of nodes A, B exchange *meta-data* which describes the age of their current artefact or the absence of an artefact. The node initiating channel set-up, denoted A, initiates meta-data transmission and B replies. At this point each node can determine whether it is required to transmit or receive an update of the artefact. An artefact update occurs if either: (i) precisely one of the nodes has no artefact, in which case a copy of the artefact is transmitted; (ii) the nodes have artefacts of different ages, in which case the older artefact is updated. To maintain simplicity, transmissions are not acknowledged, and there is a 95% success rate of a successful transaction between A and B (assuming the nodes remain in transmission range).

3.2 Mobility Models

Mobility is important for MP2P networking because it governs the opportunities for data exchange. A common criticism of ad-hoc network research has been the lack of consideration of mobility when evaluating protocols (e.g., [2]) and the same is true of existing research on MP2P applications. Therefore we assess the effects of three different mobility models, namely random walk, random waypoint and the Gauss Markov model.

- **The Random Walk**
  Under this model, each mobile node chooses a direction and speed in which to travel. The distance travelled along this trajectory is also chosen, in this scenario as a fixed time value. When reaching the boundary of the simulation region, a node bounces off the simulation border with an angle determined by the incoming direction. This is applied in this paper adopting a constant mobile node speed of 5km/h, while travelling in a uniformly random direction for a duration of 30 seconds.
- **The Random Waypoint Model**
  Under this model, each node chooses a destination from the simulation region, and also a speed, which is uniformly distributed in the domain [*minspeed,maxspeed*]. When reaching the destination, the nodes remain static for a period of time prior to departing to a new destination. This is applied in this paper with *minspeed* of 4km/h and *maxspeed* of 8km/h, assuming a waiting time uniformly selected from the range [10 seconds, 30 seconds].
- **The Gauss Markov Model**
  As described in [2], this model overcomes the problem of linear movement patterns using a single tuning parameter α, where $0 < \alpha < 1$. Each node has a general direction and a general speed (called the mean direction and mean speed) that are updated over a fixed interval t. Between updating mean speed and mean direction, each node proceeds approximately in the mean direction and approximately at the mean speed with local variations, the magnitude of which is defined by α. The location of a node at iteration i, denoted (x_i, y_i), is defined as:

$$x_i = x_{i-1} + \bar{s} \cos \bar{d} \tag{1}$$
$$y_i = y_{i-1} + \bar{s} \sin \bar{d} \tag{2}$$

where $\bar{s}$ and $\bar{d}$ are the current mean speed and current mean direction. After each period of length t, new values are calculated for $\bar{s}$ and $\bar{d}$. Assuming this occurs at some iteration i then the update for $\bar{s}$ and $\bar{d}$ is defined as follows:

$$\bar{s} = \alpha s_{i-1} + (1 - \alpha)\bar{s} + \sqrt{(1 - \alpha^2)}s_{x_i} \qquad (3)$$

$$\bar{d} = \alpha d_{i-1} + (1 - \alpha)\bar{d} + \sqrt{(1 - \alpha^2)}d_{x_i} \qquad (4)$$

where s_{i-1} and d_{i-1} are the current speed and direction at iteration $i - 1$ and s_{x_i} and d_{x_i} are values taken from a Gaussian distribution. In this paper we use $\alpha = 0.5$ and allocate random initial starting values for $\bar{s}$ and $\bar{d}$, taking $\bar{s}$ in the range [1 km/h,10km/h]. If a node strays within a guard distance of the region edge, such a node is given a new random direction, away from the boundary. In this simulation a guard distance of 10 metres is applied.

3.3   Test Problem Scenarios

We adopt a $500m \times 500m$ region, which is sufficiently large to represent a large store, small shopping mall or city plaza. We specify the location(s) of up to three information sources having the same transmission range as mobile nodes (30 metres for this problem). For demonstration purposes, we assume that 100 mobile nodes are present, with starting positions selected on a uniform-random basis. For all experiments performed, we consider the behaviour of node movement and information acquisition based on 100 random trials. Each trial represents 15 minutes of system operation with an artefact only being held by the source node at the start of each trial.

3.4   Performance Metrics

In order to assessment the quality of information that peers can maintain, we define various metrics. Let S_j be the set of nodes at iteration j holding an artefact. Let a_{ij} denote the age of the artefact held by node i, $i \in S_j$. Then following is of interest:

- **Time to receiving an artefact**
  This metric measures how quickly information becomes available to node i, irrespective of the quality (i.e., age) of an artefact. For node i, this is defined as j such that $i \in S_j$ and $i \notin S_{j-1}$. Related to this, for n_s simulation steps with n_n nodes, the *occupancy* of nodes is defined as:

$$\frac{1}{n_s n_n} \sum_{j=1,\dots,n_s} |S_j|$$

- **Profile of artefact ages**
  This metric assesses artefact ages throughout all n_s steps in the simulation. The total instances of artefacts aged x is defined as:

$$\sum_{j=1,\dots,n_s} |\{i \in S_j : a_{ij} = x\}|$$

- **Frequency of artefact updating**
  This metric measures the periods that artefacts remain at nodes without updating. Assuming node i has an artefact, an update occurs at time step j if and only if $a_{i,j-1} > a_{ij}$. Then the *duration* before the *next* update for node i is d such that:

$$a_{i,j} < a_{i,j+1} < \cdots < a_{i,j+d} > a_{i,j+d+1} \tag{5}$$

4 Experimentation

We perform experiments to demonstrate the effect that different behaviour (i.e., mobility) has on the performance metrics. In this Section these are carried out using a single information source. This is carried out assuming a single information source with location $(0, 250)$, that is on the mid-point of a boundary. Figure 1 shows the difference on the spread of artefacts from the information source under the three mobility models. The error bars indicate a 90% confidence interval taken across a sample of 100 trials. It is notable that random way point model provides the slowest dissemination of artefacts which is due to the pause time invoked at destinations which reduces opportunities for artefact exchange. Under the random walk, approximately 50% of nodes receive an artefact within 2600 simulation steps (4.3 minutes) where as under the Gauss Markov assumptions on mobility, 3925 simulation steps are taken (6.5 minutes). Under assumptions of the random waypoint model, 8875 simulation steps are taken (14.8 minutes).

The profile of the set of artefact ages is displayed in Figure 2, with error bars indicating a 90% confidence interval across 100 trials. For each mobility model, the graph is characterised by a sharp initial peak indicating that the most frequent artefact age is in the 0 - 20 age range, which is most prominent

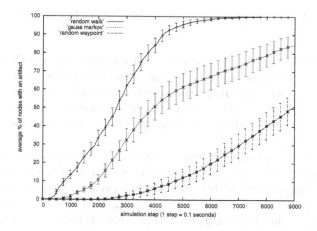

Fig. 1. Number of nodes with an artefact

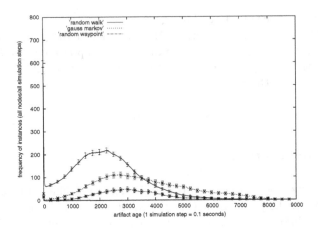

Fig. 2. Profile of artefact ages $(a_{i,j})$ over all time steps

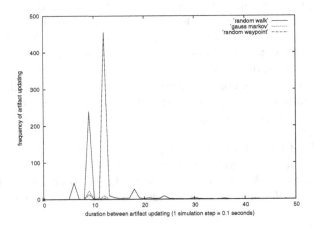

Fig. 3. Frequency of durations between artefact updates

for random walk, Gauss Markov and random waypoint respectively. This occurs from nodes local to the source acquiring information soon after it has been released. Beyond this, each distribution exhibits a secondary local maxima (2240, 2737, 3100 respectively). The profile of the distributions are given in Table 2. As the distributions are relatively flat, weak clustering around the average artefact age is evident.

In Figure 3 the average frequency of durations between artefact updates is presented across 100 trials. Note that this is shown using the range 0 - 50. Beyond this range, there are very low frequencies evident throughout (all average frequencies under 0.5 in the range 50 - 9000). Table 3 gives further statistics. In particular, it is evident that duration is high variable between mobility scenario as seen from the plots in Figure 3.

Table 2. Statistics on artefact age profile (100 trials)

mobility scenario	total artefact occupancy (%)	percentiles from 100 aggregated trials					Distribution of average artefact age (90% Confidence)
		1	5	50	95	99	
Random Walk	70.7	10	357	2169	4492	5785	2262.53 (±133.23)
Gauss Markov	46.7	516	1215	3305	3305	6612	3530.47 (±271.41)
Random Waypoint	14.9	633	1383	3081	5995	7382	3261.71 (±238.96)

Table 3. Statistics on duration between artefact updating (100 trials)

mobility scenario	percentiles from 100 aggregated trials					Distribution of average duration (90% Confidence)
	1	5	50	95	99	
Random Walk	6	9	12	2867	4255	576.90 (±51.92)
Gauss Markov	9	9	2172	7148	8349	2523.63 (±257.15)
Random Waypoint	9	9	4192	8492	8826	4074.44 (±302.70)

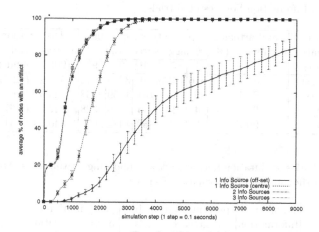

Fig. 4. Number of nodes with an artefact while varying number of information sources (100 trials)

4.1   Sensitivity to Information Sources

The number and location of information sources that are available in the region significantly impact on the maintenance of temporal quality. This is demonstrated by considering the the effects of four scenarios- a single information source in the centre; a single information source off-set at location $(0, 250)$; 2 information sources located at $(125, 125)$ and $(375, 375)$; 3 information sources located at $(125, 125)$, $(250, 250)$ and $(375, 375)$. All results are presented from 100 random trials with error bars showing a 90% confidence level. The gain in performance is shown in Figures 4 and 5 assuming the Gauss Markov mobility

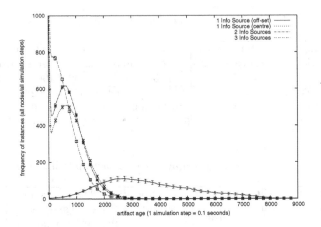

Fig. 5. Profile of artefact ages $(a_{i,j})$ over all time steps while varying number of information sources

Table 4. Statistics on artefact age profile for the Gauss Markov model while varying the number of Information Sources (100 trials)

Number of Information Sources	total artefact occupancy (%)	percentiles from 100 aggregated trials					Distribution of average artefact age (90% Confidence)
		1	5	50	95	99	
1 (off-set)	46.7	516	1215	3305	3305	6612	3530.47 ($\pm$271.41)
1 (centre)	81.1	6	40	751	1859	2438	823.65 ($\pm$63.33)
2	90.1	4	23	674	1736	2328	747.04 ($\pm$57.15)
3	90.7	2	12	450	1482	2075	540.68 ($\pm$41.06)

Table 5. Statistics on duration between artefact updating for the Gauss Markov model while varying the number of Information Sources

Number of Information Sources	percentiles from 100 aggregated trials					Distribution of average duration (90% Confidence)
	1	5	50	95	99	
1 (off-set)	9	9	2172	7148	8349	2523.63 ($\pm$257.15)
1 (centre)	6	6	18	1176	2077	217.80 ($\pm$18.77)
2	6	6	13	844	1520	152.94 ($\pm$10.71)
3	6	6	12	597	1240	103.40 ($\pm$8.46)

model. It is notable that an additional information sources beyond a single instance provide improvement in both immediacy of acquisition and maintenance of lower artefact ages. In the case of a single information source, there is substantial sensitivity to its location. Tables 4 and 5 further quantify the updating and age profiles.

5 Conclusions

While disconnectivity precludes applications dependent on a real-time quality of service, it is commensurate information sharing. This paper has identified some of the sensitivities that affect the *potential* temporal information quality that can be sustained from mobile peer-to-peer interactions. These sensitivities are significant and will influence operational protocols for peer-to-peer information exchange. Further related issues arising include the effect of information hetrogeneity and intelligent adaptibility for be-spoke deployment scenarios. These need to correlate with diverse patterns of behaviour and interactions that humans face in situated environments. Future research in this direction is supported by the EU FP7 SOCIALNETS project (social networking for adaptive behaviour).

References

1. Burns, B., Brock, O., Levine, B.N.: Mv routing and capacity building in distruption tolerant networks. In: Proceedings of IEEE INFOCOM (2005)
2. Camp, T., Boleng, J., Davies, V.: A survey of mobility models for ad hoc network research. Wireless communications and mobile computing 2, 483–502 (2002)
3. Cao, H., Wolfson, O., Xu, B., Yin, H.: Mobi-dic: Mobile discovery of local resources in peer-to-peer wireless network. IEEE Data Eng. Bull. 28(3), 11–18 (2005)
4. Chlamtac, I., Conti, M., Liu, J.J.N.: Ad-hoc networking: imperatives and challenges. Ad-hoc networks 1(1), 13–64 (2003)
5. Dorigo, M., Blum, C.: Ant colony optimization theory: A survey. Theoretical Computer Science 2/3, 243–278 (2005)
6. Kortuem, G., Schneider, J., Preuitt, D., Thompson, T.G.C., Fickas, S., Segall, Z.: When peer-to-peer comes face-to-face: Collaborative peer-to-peer computing in mobile ad hoc networks. In: Proceedings of the First International Conference on Peer-to-Peer Computing, P2P 2001 (2001)
7. Kurhinen, J., Vuori, J.: Information diffusion in a single-hop mobile peer-to-peer network. In: ISCC 2005: Proceedings of the 10th IEEE Symposium on Computers and Communications (ISCC 2005), pp. 137–142 (2005)
8. Leguay, J., Friedman, T., Conan, V.: DTN routing in a mobility pattern space. In: Proc. WDTN (2005)
9. Lindgren, A., Doria, A., Schelen, O.: Probabilistic routing in intermittently connected networks. In: Dini, P., Lorenz, P., Souza, J.N.d. (eds.) SAPIR 2004. LNCS, vol. 3126, pp. 239–254. Springer, Heidelberg (2004)
10. Luo, Y., Wolfson, O., Xu, B.: A spatio-temporal approach to selective data dissemination in mobile peer-to-peer networks. In: Proceedings of the 3rd International Conference on Wireless and Mobile Communications (ICWMC 2007), p. 50b (2007)
11. Musolesi, M., Hailes, S., Mascolo, C.: Adaptive Routing for Intermittently Connected Mobile Ad Hoc Networks. In: Proceedings of the IEEE 6th International Symposium on a World of Wireless, Mobile, and Multimedia Networks (WoWMoM 2005), Taormina, Italy (2005)
12. Nittel, S., Duckham, M., Kulik, L.: Information dissemination in mobile ad hoc geosensor networks. In: Egenhofer, M.J., Freksa, C., Miller, H.J. (eds.) GIScience 2004. LNCS, vol. 3234, pp. 206–222. Springer, Heidelberg (2005)

13. Papadopouli, M., Schulzrinne, H.: Effects of power conservation, wireless coverage and cooperation on data dissemination among mobile devices. In: Proceedings of the 2nd ACM international symposium on Mobile ad hoc networking & computing (MobiHoc 2001), pp. 117–127 (2001)
14. Pelusi, L., Passarella, A., Conti, M.: Opportunistic networking: Data forwarding in disconnected mobile ad hoc networks. IEEE Communications Magazine 44(11), 134–141 (2006)
15. Pering, T., Raghunathan, V.: Exploiting radio hierarchies for power-efficient wireless device discovery and connection setup. In: Proceedings of the 18th International Conference on VLSI design, pp. 774–779 (2005)
16. Repantis, T., Kalogeraki, V.: Data dissemination in mobile peer-to-peer networks. In: Proceedings of the IEEE 6th International on Mobile data management, pp. 211–219 (2005)
17. Shah, R.C., Roy, S., Jain, S., Brunette, W.: Data mules: modeling a three-tier architecture for sparse sensor networks. In: Proceedings of the First IEEE 2003 IEEE International Workshop on Sensor Network Protocols and Applications, pp. 30–41 (2003)
18. Spyropoulos, T., Psounis, K.: Spray and focus: Efficient mobility-assisted routing in heterogeneous and correlated mobility. In: Proceedings of IEEE Percom ICMAN Workshop, pp. 79–85 (2007)
19. Spyropoulos, T., Psounis, K., Raghavendra, C.: Spray and wait: An efficient routing scheme or intermittently connected mobile networks. In: Proceedings of ACM SIGCOMM workshop on delay tolerant networking, pp. 252–259 (2005)
20. Vahdat, A., Becker, D.: Epidemic routing for partially connected ad hoc networks (2000)
21. Widmer, J., Le Boudec, J.-Y.: Network coding for efficient communication in extreme networks. In: Proceeding of the ACM SIGCOMM workshop on Delay-tolerant networking (2005)
22. Wolfson, O., Xu, B., Prasad Sistla, A.: An economic model for resource exchange in mobile peer to peer networks. In: SSDBM, pp. 235–244 (2004)
23. Wolfson, O., Xu, B., Yin, H., Cao, H.: Search-and-discover in mobile p2p network databases. In: ICDCS, p. 65 (2006)
24. Xu, B., Wolfson, O.: Data management in mobile peer-to-peer networks. In: Ng, W.S., Ooi, B.-C., Ouksel, A.M., Sartori, C. (eds.) DBISP2P 2004. LNCS, vol. 3367, pp. 1–15. Springer, Heidelberg (2005)
25. Xu, B., Wolfson, O., Chamberlain, S.: Spatially distributed databases on sensors. In: ACM-GIS, pp. 153–160 (2000)
26. Xu, B., Wolfson, O., Rishe, N.: Benefit and pricing of spatio-temporal information in mobile peer-to-peer networks. In: Proceedings of the 39th Hawaii International Conference on Systems Science, p. 223b (2006)

Beta Random Projection

Yu-En Lu*, Pietro Liò, and Steven Hand

University of Cambridge Computer Laboratory
15 J J Thomson Avenue
Cambridge CB3 0FD, UK

Abstract. Random projection (RP) is a common technique for dimensionality reduction under L_2 norm for which many significant space embedding results have been demonstrated. In particular, random projection techniques can yield sharp results for R^d under the L_2 norm in time linear to the product of the number of data points and dimensionalities in question. Inspired by the use of symmetric probability distributions in previous work, we propose a RP algorithm based on the hyper-spherical symmetry and give its probabilistic analyses based on Beta and Gaussian distribution.

Keywords: Randomised algorithm, dimensionality reduction, multi-dimensional indexing.

1 Introduction

Dimensionality reduction is a common technique to simplify and accelerate large scale data processing, especially for applications such as information retrieval and data visualisation. In these applications, information retrieval in particular, documents are modelled as points in a high dimensional space in which each dimension captures a certain feature. For modern Web data, the number of dimensions could easily go beyond thousands[1]. This causes significant overhead in processing and storage which suffers from the so called "curse of dimensionality".

For instance, the nearest neighbour query, the standard routine in multimedia databases [12] and machine learning [4], often require either full scan over the database or significant amount of index proportional to the dimensionality for efficient evaluation [6]. Clearly, reduced dimensionality means simultaneous improvements in terms of computational complexity throughout storage, indexing and query processing. Moreover, from the systems point of view, an even more effective speedup comes from fitting more indices or points into main memory which is orders of magnitude faster than external storage, e.g., the hard disk.

In this paper, we present a family of randomised projection (RP) algorithms and analyse its properties for dimensionality reduction problems under the L_2 metric. One of the main advantages of randomised algorithms are simplicity. Standard statistical

---

* Corresponding author.

[1] For example, as we shall see later, LA Times documents from TREC 5 consist of 187K keywords as document features after stemming.

P. Liò et al. (Eds.): BIOWIRE 2007, LNCS 5151, pp. 319–331, 2008.
© Springer-Verlag Berlin Heidelberg 2008

methods such as singular value decomposition (SVD) or principle component analysis require repeated iteration throughout the data and thus suffers time complexity polynomial to the data dimensionality[2]. RP methods, on the other hand, require only seed values independent of the dataset and guarantee successful reduction in constant number of trials with high probability. This is especially beneficial for WWW and multimedia databases where the number of documents is so large that the cost to directly apply some super-linear algorithms becomes prohibitive.

A common problem for all dimensionality reduction schemes is that the measure of the document space varies from application to application. While the choice of such measure is beyond the scope of this paper, we provide two arguments below to motivate our use of L_2 norm. Firstly, the L_2 norm has been shown to be very expressive in practice and is pivotal for many applications such as web page indexing [2]. Secondly, and most importantly, algorithms constructed under L_2 are more generalised than one might expect initially. It can be shown that all norms are equivalent in the sense that they could simulate each other with up to some constant factor distortion, due to a special case of Weierstrass' theorem.

Owing to the theoretical advantage of L_2 norm, exact embedding algorithms from arbitrary norms have then became an active field in which many elegant results have been shown. For example, Bourgain showed that any n points in *metric* space can be embedded in $O(\log n)$ dimensions under L_2 norm [3]. For a comprehensive review of this field, see [5].

For the rest of this paper, we first present the high level ideas of RP in section 1.1 followed by a summary of contributions. In section 2, we formally present the Beta random projection and give derivations on its exact distributions and tail bounds.

1.1   Random Projection

In essence, a random projection is a function $H_A : R^d \rightarrow R^k$ where $A_{d,k}$ is a choice of d-dimensional *estimation vectors* in R^d. Its objective is, given a set of points, to produce a new image in R^k with estimation vectors such that the pair-wise distance simulates those in R^d. Typically, the new image is a projection onto the subspace spanned by the estimation vectors. One then may apply probabilistic arguments to estimate how much distortion this transform could introduce.

This possibility is first shown in the 80s when Johnson and Lindenstrauss showed that n points in R^d could be embedded into R^k with $1 \pm \epsilon$ factor distortion where $k = O(\log n/\epsilon^2)$ [8]. To see this bound is tight, consider the case for dimensionality reduction of 3 points each of which is of distance 1 to each other on a plane. It is not possible to find an embedding of these three points on a line in which their pair-wise distance could all be preserved. An optimal embedding could be to have three points with coordinate $0, 1, 2$ on a line, thus matching the bound with error equal to 1.

While the JL-embedding demonstrates the possibility of reduction, its proof is based on geometric approximations and thus makes it difficult to comprehend and design algorithms to discover such mappings. Indyk et al. [6] proposed that drawing the estimators

---

[2] For example, the standard SVD algorithm would require $O(d^3)$ time steps to decompose a d by d matrix due to its nature to find the optimal rank approximation.

from d-dimensional Gaussian distribution could lead to simpler proofs and constant factor improvements over the original JL-embedding bounds.

Later on, with an eye on the benefits of random projection, various work was proposed to give additional properties such as volume preservation and to accelerate random projection. Most notably, Achlioptas [1] demonstrated that one could yield as good an embedding as by Indyk's using *sparse* estimation vectors. Specifically, his estimators are drawn from a sparse distribution in which each coordinate of the estimation vector has equal probability of $1/6$ to be either $\sqrt{3}$ or $-\sqrt{3}$ and 0 otherwise. This significantly trims down the computational complexity, as $2/3$ of the coordinates are expected to be discarded. Surprisingly, this does not hurt the accuracy of the projection, at least in theory. In fact, the bound of k is pushed down to $\frac{6}{\epsilon^2/2-\epsilon^3/3}\log n$ by using moment methods in [1].

This sparse estimator approach is followed up by various further attempts to accelerate projection [10, 9] by trading off distortion with performance. Li et al. presented projections based on even more sparse estimators [10] and another based on marginal information [9]. Their results showed that further performance gain is possible by incurring some more distortion.

While sparse random projections are significantly faster in terms of computation, however, there are some drawbacks when faced with certain datasets. Consider two points $u, v \in R^{100}$ with $u = (1, 0, 0, \cdots, 0)$ and $v = (0, 0, \cdots, 1)$. Dropping attributes with probability 0.9 as in [10] would yield at least a 0.82^k probability that the two points have zero distance on a R^k projection. In fact, this problem would become more severe as the dimensionality increases due to the fact that the probability that each "valid" coordinate is dropped increases inversely proportioned to the square root of dimensionality. Whilst one might argue that bad datasets could be just rare, real world datasets are, in fact, much more sparse than one might expect. In TREC5, for example, only 0.084% of the La Times term document matrix is non-zero and that of FBIS is 0.077%. In FreeDB, an album database containing 21 million song names, 90% of the songs use less than 10 keywords out of a 13 thousand vocabulary and 98% use less than 20.

1.2 Contribution

In this paper, we present the hyper-spherical *Beta random projection* for vector spaces under the L_2 norm. Our estimation vectors are drawn from all unit vectors in R^d uniformly at random. Similar to the techniques used in Indyk and Achlioptas, the data points are projected onto the k dimensional space spanned by the estimation vectors. We show that k could in general be further improved from $\frac{12\log n}{\epsilon^2-2\epsilon^3/3}$ to $\frac{8\log n-2\log 4\pi}{\epsilon^2}$. Our key insight is to discover that, for an arbitrary vector, each such projection is in fact a random variable based on the beta distribution and its original norm, hence the name Beta random projection.

We later analyse the exact distributions of the distortion due to random projection. We first give tight bounds for the case in which k is large enough [3] to apply the central limit theorem (see section 2.1). For small k, we later present results based on beta

---

[3] In practice, it suffices to have k larger than 30. We shall use this value for distinguishing between the two approximations for the rest of the paper.

distribution approximation in section 2.2. We show that hyper-spherical random projection yields more accurate projections in both cases.

2   Beta Random Projection

Let A be a $k \times d$ matrix in which A_i, a uniformly random point on the unit d-dimensional sphere, is the i-th row of A. In practice, we generate each A_i by normalising each vector drawn from $N_d(0,1)$ where N_d is a d-dimensional Gaussian distribution [11]. We define the hyper-spherical random projection $H(v; d, j)$ for any point $v \in R^d$ as follows:

$$H(v; d, k) = \sqrt{\frac{d}{k}} \cdot A \cdot v \qquad (1)$$

We endeavour to show that the lower bound of k could be improved up to 40% as indicated by the theorem below.

Theorem 1. *For any $u, v \in S \subset R^d$ and $|S| = n$ there exists a mapping $H : R^d \to R^k$ such that*

$$(1 - \epsilon)|u - v| \le |f(u) - f(v)| \le (1 + \epsilon)|u - v|$$

with probability at least $1 - n^{-1}$, when

$k \ge \frac{8 \log n - 2 \log 4\pi}{\epsilon^2} \ge 30$ *such that $\sqrt{k}\epsilon \ge 2$.*

Proof. It suffices to show that for any vector v, $\Pr[\neg 1 - \epsilon \le |f(v)|^2/|v|^2 \le 1 + \epsilon] \le 2/n^2$, since we require this event to occur $C(n, 2)$ times which is the number of all distance pairs. Given Theorem 2 and Lemma 3, this amounts to solving for k such that $\frac{2}{\sqrt{\pi k}\epsilon}e^{-\frac{k}{4}\epsilon^2} = 2/n^2$.

In the following, we first show the distortion distribution in the case when k is large enough for central limit theorem in Theorem 2. For the case in which k is small, we present the approximation by the Beta distribution in Theorem 3. We would like to note that these two approximations is accurate in most cases, as we give the comparison between the actual and approximated distributions in Appendix 4. Following each approximation results, we derive the probability bounds for the event that the distortion is more than $1 \pm \epsilon$.

2.1   Normal Approximation

In this section, we present the results for BRP with k large enough for treatise using the Central Limit Theorem. The proof proceeds by first exercising the Lemma 7 that each $A_i \cdot v$ in (1) would yield a distortion obeying Beta distribution [4]. Then, we apply the theorem to obtain the aggregate distortion as the number of projection increases.

---

[4] We defer the proof of Lemma 7 until section 3 for smoother presentation.

Theorem 2. *There exists a random projection* $H : R^d \to R^k$ *under* L_2 *norm such that for any* $\epsilon > 0, k \geq 30$ *and* $v \in R^d$,

$$\Pr[1 - \epsilon \leq \frac{|H(v; d, k)|^2}{|v|^2} \leq 1 + \epsilon] \geq 2 \cdot \text{erf}(\frac{\sqrt{k}}{2}\epsilon) \tag{2}$$

where $\text{erf}(z) \equiv \frac{2}{\sqrt{\pi}} \int_0^z e^{-x^2} dx$ *is the error function.*

Proof. (2) First, we rewrite (1) into $H(v; d, k) = \sqrt{\frac{d}{k}} \cdot (b_1, b_2, b_3, \ldots, b_k)$. Without loss of generality, we assume an unit-vector v. From lemma 7, each b_i is a random variable such that $b_i^2 \sim \beta(\frac{1}{2}, \frac{d-1}{2})|v|^2$. Let $Y = \sum b_i^2$, we rewrite (2) as

$$\Pr[1 - \epsilon \leq \frac{|H(v; d, k)|}{|v|} \leq 1 + \epsilon] = \Pr[(1 - \epsilon)\frac{k}{d} \leq Y \leq (1 + \epsilon)\frac{k}{d}] \tag{3}$$

Since k is large enough, we invoke Lemma 8, hence

$$Y \sim N(\frac{k}{d}, \frac{2k}{d(d + 2)})$$

Thus (3) can be rewritten as standard normal distribution as in (5). Integrating through the standard normal distribution and taking into account the fact that the error function is monotonically decreasing, we arrive at the claim.

$$\Pr[(1 - \epsilon)\frac{k}{d} \leq Y \leq (1 + \epsilon)\frac{k}{d}] \tag{4}$$

$$= \Pr[-\frac{k/d}{\sqrt{2k/(d(d+2))}}\epsilon \leq Z \leq \frac{k/d}{\sqrt{2k/(d(d+2))}}\epsilon] \tag{5}$$

$$= \qquad 2 \cdot \text{erf}\left(\frac{k}{d}/2(\frac{k}{d(d+2)})^{1/2}\epsilon\right) \tag{6}$$

$$\geq \qquad 2 \cdot \text{erf}(\frac{\sqrt{k}}{2}\epsilon) \tag{7}$$

□

Tail Bounds. Here, we would like to bound the probability of the event $\mathcal{E}$ that a vector's projection has more than ϵ factor distortion to its original. For ease of comparison, we first present the sharp bounds given by Achlioptas in [1].

Lemma 1 (Achlioptas [1]). *Let* R *be a* $k \times d$ *matrix where* $\{r_{ij}\}$ *are i.i.d. random variables following the discrete density* $\Pr[r_{ij} = \sqrt{3}] = 1/6, \Pr[r_{ij} = 0] = 2/3$, *and* $\Pr[r_{ij} = -\sqrt{3}] = 1/6$. *Let* $f_A(v; k, d) = \frac{1}{k} \cdot R \cdot v$ *where* v *is an arbitrary vector in* R^d.

$$\Pr[\neg\mathcal{E}] \leq 2e^{-\frac{k}{4}\epsilon^2 + \frac{k}{6}\epsilon^3}$$

Below, we provide tail bounds for the case where the central limit theorem applies. Our approach is based on Taylor expansion on the error function which exhibits different convergence behaviour according to its argument $\frac{\sqrt{k}\epsilon}{2}$. Thus, we separate the two cases in Lemma 2 and 3.

Lemma 2. *Let $d > k > 30$, $\sqrt{k}\epsilon < 2$.*

$$\Pr[\neg\mathcal{E}] \leq exp(-\frac{2}{\sqrt{\pi}}\sqrt{k}\epsilon + \frac{1}{6}(\sqrt{k}\epsilon)^3)$$

Proof. Since Theorem 2 indicates $\Pr[\mathcal{E}] \geq 2erf(\frac{\sqrt{k}}{2}\epsilon)$, and let $z = \frac{\sqrt{k}}{2}\epsilon$, we have

$$\Pr[\neg\mathcal{E}] \leq \qquad\qquad 1 - 2erf(z) \tag{8}$$

$$= 1 - 2 \cdot \frac{2}{\sqrt{\pi}}\left(z - \frac{z^3}{3} + \frac{z^5}{10} + \dots\right) \tag{9}$$

$$\leq \qquad 1 - \frac{2}{\sqrt{\pi}}\sqrt{k}\epsilon + \frac{1}{6}(\sqrt{k}\epsilon)^3 \tag{10}$$

(9) is the standard Taylor expansion. Exercising the fact that $\sqrt{k}\epsilon < 2$ (hence $z < 1$), we arrive at the claim. For ease of comparison, observe that $1 - x \leq e^{-x}$, our claim is less than $exp(-\frac{2}{\sqrt{\pi}}\sqrt{k}\epsilon + \frac{1}{6}(\sqrt{k}\epsilon)^3)$. $\qquad\square$

Lemma 3. *Let $d > k > 30$, $\sqrt{k}\epsilon \geq 2$.*

$$\Pr[\neg\mathcal{E}] < \frac{2}{\sqrt{\pi k}\epsilon}e^{-\frac{k}{4}\epsilon^2}$$

Proof. Let $z = \frac{\sqrt{k}}{2}\epsilon$. Since $z \geq 1$, we can substitute the bound in Theorem 2 with asymptotic series

$$erf(z) = 1 - \frac{e^{-z^2}}{\sqrt{\pi}}\sum_{i=0}^{\infty}\frac{-1^i(2i-1)!}{2^i}z^{-2i-1}$$

Thus,

$$\Pr[!\mathcal{E}] = 1 - \left[1 - \frac{e^{-z^2}}{\sqrt{\pi}}(z^{-1} - \frac{1}{2}z^{-3} + \frac{3}{4}z^{-5})\right] \tag{11}$$

$$< \qquad\qquad \frac{2}{\sqrt{\pi k}\epsilon}e^{-\frac{k}{4}\epsilon^2} \tag{12}$$

$$\square$$

2.2 Beta Approximation

As the application does not always allow for $k \leq 30$, we give results based on Beta approximation in this section. Again, it can be shown that this approximation is precise as the error bound is given in [7]. We verify this in practice later in Appendix 4, Figure 4. Below, we establish the aggregate distortion distribution followed by bounds for the event that the distortion is larger than $1 \pm \epsilon$.

Theorem 3. *There exists a random projection $H : R^d \rightarrow R^k$ under L_2 norm such that for any $\epsilon > 0, 1 < k < 30, d \gg k$ and $v \in R^d$,*

$$\frac{|H(v; d, k)|^2}{|v|^2} \sim \beta(\frac{k}{2}, \frac{d}{2} + 1) \tag{13}$$

where $\beta(\alpha, \beta)$ is the beta distribution with parameter α, β.

Proof. (3) From Lemma 7, we know each $(A_i \cdot v)^2 \sim \beta(\frac{1}{2}, \frac{d-1}{2})$. Hence it remains to calculate $Y = \sum_{i=1}^{k}(A_i \cdot v)^2$.

To derive the distribution of Y, we invoke the beta-sum approximation below:.

Fact 4 (Johannesson and Giri [7]). *Let* $S = \sum_{i=1}^{k} X_i$ *where* X_i *are i.i.d. random variables of* $\beta(\alpha, \beta)$*. The distribution of* S *can be approximated by:*

$$\beta(e, f); e = Ff, f = \frac{F}{\sigma^2(1 + F)^3}$$

where $E = \sum EX_i$*,* $F = \frac{E}{1-E}$*, and* $\sigma^2 = \sum Var(X_i)$*.*

Since $d \gg k$, then $f \approx \frac{d+2}{2}$ Thus, we have $Y \sim \beta(\frac{k}{2}, \frac{d}{2} + 1)$.

Tail Bounds. Below, we show that the approximated probabilities for the case of small k where the resulting distribution is approximated by the Beta distribution. Due to the fact that beta distributions under our parameters exhibit significant skew, we bound the right and left tails separately in lemmas 4 and 5. In lemma 5, it is unexpected to see that it indicates that it is relatively unlikely to have a distortion *less* than the original. We present a case in simulation to demonstrate this in Figure 5.

In the following analyses, we shall separate $\mathcal{E}$ into two for easy discussion. Let $\mathcal{E}_R$ be the right tail, the event that $|H(v)|^2/|v|^2 \geq 1 + \epsilon$, and $\mathcal{E}_L$ be the left, the event that $|H(v)|^2/|v|^2 \leq 1 - \epsilon$.

Lemma 4. *Let* $d > 30 > k > 0$ *and* $\epsilon > 0$*.*

$$\Pr[\mathcal{E}_R] \leq \frac{2}{\sqrt{\pi k}}e^{-\frac{k}{2}(\epsilon - \ln(1+\epsilon))}$$

Proof. We denote the pdf of $\beta(\alpha, \beta)$ by $f(z; \alpha, \beta) = \frac{1}{B(\alpha,\beta)}z^{\alpha-1}(1 - z)^{\beta-1}$.

We can thus bound the right tail as an integral of f over $A = [(1 + \epsilon)\frac{k}{d}, 1]$:

$$\Pr[\mathcal{E}_R] = \int_A f(u)du \tag{14}$$

$$\leq \frac{\alpha}{\beta} \sum_{t=1+\epsilon,2+\epsilon,\dots}^{\infty} f(t\frac{\alpha}{\beta}) \tag{15}$$

$$\leq \frac{1}{B(\alpha,\beta)} \sum \left(t\frac{\alpha}{\beta}\right)^{\alpha-1}(1 - t\frac{\alpha}{\beta})^{\beta-1} \tag{16}$$

$$\leq d\frac{1}{B(\alpha,\beta)}\left(\frac{\alpha}{\beta}\right)^{\alpha}\sum \frac{t^{\alpha-1}}{e^{\alpha t}} \tag{17}$$

$$\leq \frac{\Gamma(\alpha+\beta)}{\Gamma(\alpha)\Gamma(\beta)}\left(\frac{\alpha}{\beta}\right)^{\alpha}\sum \frac{1}{e^{\alpha(t-\ln t)}} \tag{18}$$

$$\leq \frac{1}{\sqrt{2\pi\alpha}}e^{-\alpha(\epsilon-\ln(1+\epsilon))}\left(\frac{1}{1 - e^{-\alpha/2}}\right) \tag{19}$$

Differentiating $f(\cdot)$ show that it reaches maxima at $\max\{0, \frac{k-2}{d+k-4}\}$ and then drops exponentially, thus (15). Since d is large, $(1+\epsilon)\frac{k}{d}^{\frac{d}{2}-1} \to -(1+\epsilon)\frac{k}{2}$. Via a little calculus, we have $(1 - (1 + \epsilon)\frac{k}{d})^{d/2-1} \to e^{-(1+\epsilon)\frac{k}{2}}$ and hence (17). Notice that $t - \ln t > t/2$, therefore the rate at which this summation increases must be faster than $e^{-\alpha/2}$. Taking the first term of the original series, we have the the inequality (19). Observing that the term in parenthesis is strictly less than 2, we arrive at the claim.

For the special case in which $0 \leq \epsilon \leq 1$, one could obtain an improved bound over the classic result as in [1], via Taylor's expansion for $\ln(1 + \epsilon)$, $\ln(1 + \epsilon) \leq \epsilon - \epsilon^2/2 + \epsilon^3/3$. That is:

$$\frac{2}{\sqrt{2\pi\alpha}}e^{-\alpha(\epsilon^2/2-\epsilon^3/3))}$$

Lemma 5. *Let $d \gg 30 > k > 0$, $0 \leq \epsilon < 1$, and $\mathcal{E}_L$ be the event that $|H(v)|^2/|v|^2 \leq 1 - \epsilon$ where $0 \leq \epsilon \leq 1$ and $k(1 - \epsilon) \geq 1$.*

$$\Pr[\mathcal{E}_L] \leq \frac{1}{(k/2 - 1)!} \left(1 - e^{-k(1-\epsilon)/2}\right)$$

Proof. Let $\rho = 1 - \epsilon$. Notice that as $b \to \infty$ and $cb \to \lambda$, then $(1 - b)^a \to e^{-\lambda}$. Substituting x with $t = (\rho a/b)^{-1}x$, the basic beta-distribution pdf results in (21). Invoke Lemma 6, we have (22). Cleaning the equation yields our proposition.

$$\Pr[\mathcal{E}_L] = \frac{1}{B(a,b)} \int_0^{(1-\epsilon)a/b} x^{a-1}(1 - x)^{b-1}dx \tag{20}$$

$$\leq \frac{1}{B(a,b)}\rho^a \left(\frac{a}{b}\right)^a \int_0^1 t^{a-1}e^{-\rho at}dt \tag{21}$$

$$\leq \frac{(b+a-1)\ldots b}{(a-1)!}\rho^a \left(\frac{a}{b}\right)^a \frac{1}{a^a\rho^a}\left(1 - e^{-\rho a}\right) \tag{22}$$

$$\leq \frac{1}{(a-1)!}\left(1 - e^{-\rho a}\right) \tag{23}$$

Lemma 6. *Let $a > 0$, $\rho a \geq 1$, and $0 \leq \rho \leq 1$.*

$$\int_0^1 t^{a-1}e^{-\rho at}dt \leq \frac{1}{a^a\rho^a}\left(1 - e^{-\rho a}\right)$$

Proof. Since this integral does not have a closed form for arbitrary value of a, We prove this by mathematical induction. Firstly, for $a = 1$, LHS is induced to $\frac{1-e^{-\rho}}{\rho}$ by standard calculus – the desired inequality holds. For $a = 2$, LHS ends up with $\frac{1-e^{-2\rho}-2\rho e^{-2\rho}}{4\rho^2}$, the hypothesis holds.

Assuming the claim holds for $a = k$, the inequality can be rewritten as below by standard calculus.

$$\int_0^1 t^{k-1}e^{-\rho kt}dt = (\rho k)^{-1}\int_{\rho k}^{\infty} t_1^{k-1}e^{-t_1}dt_1 \leq \frac{1}{k^k\rho^k}(1 - e^{-\rho k})$$

Similarly, the case for $a = k + 1$ can be written as (24) which, with some calculus, becomes (25). Observe that $t^k e^{-\rho(k+1)t}$ is monotonically increasing. The additional area due to changing the starting point $\rho(k + 1)$ to ρk is at least $\rho(\rho k)^k e^{-\rho k}$. Expanding the integral again, we can establish the inequality of (26). The terms in parenthesis of (28) is strictly less than $1 - e^{-\rho(k+1)}$ after $\rho \geq 1/k$, as suggested by partial differentiatuion along ρ and checking the base case $\rho k = 1$. Finally, we arrive at the claim after cleaning up (28).

$$LHS = \int_0^1 t^k e^{-\rho(k+1)t} dt \tag{24}$$

$$= \frac{1}{\rho(k+1)} \int_{\rho(k+1)}^{\infty} t_2^k e^{-t_2} dt_2 \tag{25}$$

$$\leq \frac{1}{\rho(k+1)} \left\{ k \int_{\rho k}^{\infty} t_2^{k-1} e^{-t_2} dt_2 - (\rho k)^k e^{-\rho k} \rho \right\} \tag{26}$$

$$\leq \frac{1}{\rho^{k+1}(k+1)^{k+1}} \tag{27}$$

$$\left\{ \rho(1 + \frac{1}{k})^{k-1} k(k+1)(1 - e^{-\rho k}) - \rho^{2k+1}(k(k+1))^k e^{-\rho k} \right\} \tag{28}$$

$$\leq \frac{1}{\rho^{k+1}(k+1)^{k+1}} (1 - e^{-\rho(k+1)}) \tag{29}$$

3   Distribution of the Inner Product

In this section, we demonstrate our observation that each inner product of A_i and the points are, in effect, beta distributions with parameters $(1/2, (d-1)/2)$.

Lemma 7. *Let X be an uniformly random point on the surface of the unit d-dimension sphere and $v \in R^d$. Then, we have*

$$\frac{(X \cdot v)^2}{|v|^2} \sim \beta(\frac{1}{2}, \frac{d-1}{2})$$

where $\beta(\alpha, \beta)$ is the beta distribution.

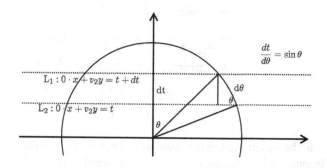

Fig. 1. An illustration of the inner product on a 2D sphere, i.e., a circle

Proof. (7) Without loss of generality, we consider the case in which v is an unit vector. Since the unit d-dimension sphere is symmetric and X spreads uniformly on its surface, it suffices to consider the case in which $X = (x_1, x_2, \ldots, x_{d-1}, t)$ and $v = (0, 0, \ldots, 0, 1)$. Notice that $x_1^2 + x_2^2 \cdots + x_{d-1}^2 = 1 - t^2$ which is a $(d-1)$-dimension hypersphere by definition.

We first give some intuitions on the proof. Consider each v is a hyperplane. Since we sample uniformly random from the unit hypersphere, the probability distribution of the inner product is essentially proportionate to the region of the surface that intersects with the hyperplane. We illustrate the 2D case in Figure 1.

That is:

$$\Pr[X \cdot v \le t] = \Pr[|x_d| \le t | \sum_{j=1}^{d} x_j^2 = 1]$$

We can denote x_i via hyper-spherical coordinates and choose $t = \cos \rho_1$, thus

$$x_1 = \quad \sin \rho_1 \cos \rho_2$$
$$x_2 = \sin \rho_1 \sin \rho_2 \cos \rho_3$$
$$\cdots$$
$$x_{d-1} = \quad \sin \rho_1 \ldots \cos \rho_d$$
$$t = x_d = \quad \cos \rho_1$$

Thus the cdf reduces to the following region of area:

$$\frac{\int_{-\rho}^{\rho} \int_0^{\pi} \cdots \int_0^{2\pi} \prod_{j=1}^{d-1} \sin^{d-1-j} \rho_j d\rho_j}{S_d(1)}$$

where $S_d(1)$ denotes the surface area of the unit d-sphere. The above equations thus reduce to

$$\Pr[|X \cdot v| \le t] = \quad \Pr[|X \cdot v| \le \cos \rho_1]$$
$$= \quad \frac{\Gamma(d/2)}{\sqrt{\pi}\Gamma(\frac{d-1}{2})} 2 \int_0^{\rho_1} \sin^{d-2}(\rho_1) d\rho_1$$
$$= \quad \frac{1}{\beta(\frac{1}{2}, \frac{d-1}{2})} \int (1 - \cos^2 \rho_1)^{\frac{d-3}{2}} d\cos \rho_1$$
$$= \quad \frac{1}{\beta(\frac{1}{2}, \frac{d-1}{2})} 2t \cdot {}_2F_1(\frac{1}{2}, -\frac{d-3}{2}; \frac{3}{2}; t^2)$$
$$= \quad \frac{1}{\beta(\frac{1}{2}, \frac{d-1}{2})} \beta_{t^2}(\frac{1}{2}, \frac{d-1}{2})$$
$$= \quad I_{t^2}(\frac{1}{2}, \frac{d-1}{2})$$

where ${}_2F_1(\cdot)$ is the generalised hyper-geometric function, $\beta(\cdot)$ is the beta function, $\beta_{t^2}(\cdot)$ is the incomplete beta function, and $I_{t^2}(\cdot)$ is the regularised beta function. By the definition of beta distribution, we obtain the claim.

4   Conclusion

In this paper, we present a novel random projection algorithm for dimensionality reduction under L_2. We show analytically that our algorithm further improves previous work by at least a constant factor. Also, our analyses demonstrates that the preferred behaviour still holds for very small distortions and very few dimensions.

In the future, we would like to further characterise the effect of random projection for highly correlated datasets and its applications in distributed indexing.

Acknowledgement

This research was sponsored by US Army Research laboratory and the UK Ministry of Defence and was accomplished under Agreement Number W912NF-06-3-0001. The views and conclusions contained in this document are those of the authors and should not be interpreted as representing the official policies, either expressed or implied, of the US Army Research Laboratory, the U.S. Government, the UK Ministry of Defense, or the UK Government.

References

1. Achlioptas, D.: Database-friendly random projections. In: PODS 2001: Proceedings of the twentieth ACM SIGMOD-SIGACT-SIGART symposium on Principles of database systems, pp. 274–281. ACM Press, New York (2001)
2. Bawa, M., Condie, T., Ganesan, P.: Lsh forest: self-tuning indexes for similarity search. In: WWW 2005: Proceedings of the 14th international conference on World Wide Web, pp. 651–660. ACM Press, New York (2005)
3. Bourgain, J.: On lipschitz embedding of finite metric spaces in hilbert space. Israel J. Math. 52, 46–52 (1985)
4. Cost, S., Salzberg, S.: A weighted nearest neighbor algorithm for learning with symbolic features. Mach. Learn. 10(1), 57–78 (1993)
5. Indyk, P., Matoušek, J.: Low-distortion embeddings of finite metric spaces. In: Handbook of Discrete and Computational Geometry, 2nd edn. (2004)
6. Indyk, P., Motwani, R.: Approximate nearest neighbors: towards removing the curse of dimensionality. In: STOC 1998: Proceedings of the thirtieth annual ACM symposium on Theory of computing, pp. 604–613. ACM Press, New York (1998)
7. Johannesson, B., Giri, N.: On approximations involving the beta distribution. Communications in statistics. Simulation and computation (Commun. stat., Simul. comput.) 24(2), 489–503 (1995)
8. Johnson, W.B., Lindenstrauss, J.: Extensions of Lipschitz mappings into a Hilbert space. In: Conference in modern analysis and probability, pp. 189–206 (1984)
9. Li, P., Hastie, T., Church, K.W.: Improving random projections using marginal information. In: Lugosi, G., Simon, H.U. (eds.) COLT 2006. LNCS (LNAI), vol. 4005, pp. 635–649. Springer, Heidelberg (2006)
10. Li, P., Hastie, T.J., Church, K.W.: Very sparse random projections. In: KDD 2006: Proceedings of the 12th ACM SIGKDD international conference on Knowledge discovery and data mining, pp. 287–296. ACM Press, New York (2006)

11. Muller, M.E.: A note on a method for generating points uniformly on n-dimensional spheres. Commun. ACM 2(4), 19–20 (1959)
12. Pentland, A., Picard, R., Sclaroff, S.: Photobook: Content-based manipulation of image databases. In: SPIE Storage and Retrieval for Image and Video Databases, vol. II(2185) (February 1994)

Appendix

Lemma 8. *Let $B_1, B_2, \ldots, B_k$ be i.i.d. random variables with distribution $\beta(\frac{1}{2}, \frac{d-1}{2})$. The sum of B_i, $S = \sum_{i=1}^{k} B_i$, has probability distribution*

$$\frac{S - k/d}{\sqrt{2k/d(d+2)}} \sim N(0,1)$$

where $k \geq 30$ and $N(0,1)$ is the standard normal distribution.

Proof. (8) Each B_i is a random variable of beta distribution with mean $1/d$ and variation $\frac{2}{d(d+2)}$. Since k is large enough, central limit theorem indicates that:

$$\frac{Y - k/d}{\sqrt{2k/d(d+2)}} \sim N(0,1)$$

where $N(0,1)$ is the standard normal distribution.

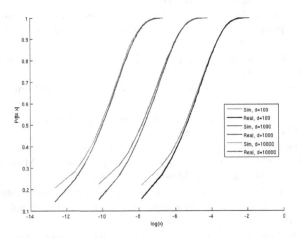

Fig. 2. An illustration of the theoretical bound and each projection $A_i \cdot v$ where A_i is an uniformly random unit vector and $v \in R^d$ obeys $\beta(\frac{1}{2}, \frac{d-1}{2})$. The gaps in each beginning x segments are due to loss of floating point precision at the extreme ends of the hypersphere with respect to v.

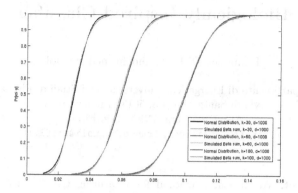

Fig. 3. The effects of normal approximation. We could observe that the approximation is accurate.

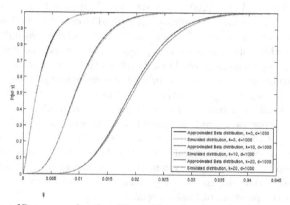

Fig. 4. The effects of Beta approximation. We could see that the approximation accuracy degrades a little as k approaches d. While the loss of precision is acceptable, it could be remedied by not exercising the assumption that $d \gg k$ in Theorem 3. This would only alter the value of k slightly in all our bounds used, which explains the minor under-estimation as k increases.

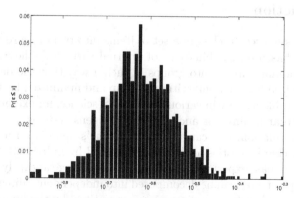

Fig. 5. Simulated results for the distortion distribution for d=1000 and k=5 where 99.5% of the data points are 0. Observe that there is an initial probability hike when x is very small. This is in accordance to the probability predicted in Lemma 5.

Biologically Inspired Classifier

Francesca Di Patti and Franco Bagnoli

Dipartimento di Energetica, Università degli Studi di Firenze,
via di Santa Marta 3, 50139 Firenze, Italy
Also CSDC and INFN, Sez. Firenze
f.dipatti@gmail.com, franco.bagnoli@unifi.it

Abstract. We present a method for measuring the distance among records based on the correlations of data stored in the corresponding database entries. The original method (F. Bagnoli, A. Berrones and F. Franci. Physica A 332 (2004) 509-518) was formulated in the context of opinion formation. The opinions expressed over a set of topic originate a "knowledge network" among individuals, where two individuals are nearer the more similar their expressed opinions are. Assuming that individuals' opinions are stored in a database, the authors show that it is possible to anticipate an opinion using the correlations in the database. This corresponds to approximating the overlap between the tastes of two individuals with the correlations of their expressed opinions.

In this paper we extend this model to nonlinear matching functions, inspired by biological problems such as microarray (probe-sample pairing). We investigate numerically the error between the correlation and the overlap matrix for eight sequences of reference with random probes. Results show that this method is particularly robust for detecting similarities in the presence of traslocations.

Keywords: knowledge network, microarray.

1 Introduction

Cluster analysis is used to classify a set of items into two or more mutually exclusive groups based on combinations of internal variables. The goal of cluster analysis is to organize items into groups in such a way that the degree of similarity is maximized for the items within a group and minimized between groups.

Clustering problems arise in various domains of science, for example in opinion formation, microarray analysis and antibody-antigens systems.

In opinion formation, one can assume that one's opinion on a certain item is given by the characteristics of the item, weighted by individual "tastes". The tastes result from past experiences, but they do not change abruptly from time to time. In principle, tastes can be decomposed into independent "dimensions". It is rather difficult to identify such dimensions, as testified by the limited success of market campaigns. However, it can be shown [1] that exploiting the correlations among the expressed opinions, it is possible to deduce the distance between the tastes of two individuals.

P. Liò et al. (Eds.): BIOWIRE 2007, LNCS 5151, pp. 332–339, 2008.

A DNA microarray is a collection of microscopic DNA spots of probes, commonly complementary to some region of a gene, arrayed on a solid surface by covalent attachment to a chemical matrix. DNA arrays are commonly used for expression profiling, namely monitoring expression levels of thousands of genes simultaneously, or for comparative genomic hybridization. Gene expression microarray experiments can generate data sets with multiple missing expression values. However, many algorithms for gene expression analysis require a complete matrix of gene array values as input, and may lose effectiveness even with a few missing values. Methods for imputing missing data are needed, therefore, to minimize the effect of incomplete data sets on analyses, and to increase the range of data sets to which these algorithms can be applied [2]. Moreover, comparison between a "forecasted" value based on correlations in the dataset, and the measured one, can be considered a consistency "check" of the dataset itself.

Antibodies are proteins that are used by the immune system to identify and neutralize foreign objects, such as bacteria and viruses. Classifying antibodies, based on the similarity of their binding to the antigens, is essential for progress in immunology and clinical medicine.

A striking feature of the natural immune system is its use of negative detection in which "self" is represented (approximately) by the set of circulating lymphocytes that fail to match self. This suggests the idea of a negative representation, in which a set of data elements is represented by its complement set. That is, all the elements not in the original set are represented (a potentially huge number), and the data itself are not explicitly stored. This representation has interesting information-hiding properties when privacy is a concern and it has implications for intrusion detection. One of the example where this idea has been concretised is the case of a negative database [3].

In a negative database, the negative image of a set of data records is represented rather than the records themselves. Negative databases have the potential to help prevent inappropriate queries and inferences. Under this scenario, it is desirable that the database supports only the allowable queries while protecting the privacy of individual records, say from inspection by an insider. A second goal involves distributed data, where one would like to determine privately the intersection of sets owned by different parties. For example, two or more entities might wish to determine which of a set of possible "items" (transactions) they have in common without reveling the totality of the contents of their database or its cardinality.

In this paper we use the microarray example to test the introduction of nonlinearities in the computation. Since in our model a datum is essentially stored as the set of matching items plus the set of nonmatching ones, our results can be applied both to positive and negative representation of data.

2 Matching Model

Let us first illustrate the problem summarizing the main results reported in [1].

Consider a population of M individuals experiencing a set of N products. Assume that each product is characterized by an L-dimensional array

$a = (a^{(1)}, a^{(2)}, \ldots, a^{(L)})$ of features, while each individual has the corresponding list of L personal tastes on the same features $b = (b^{(1)}, b^{(2)}, \ldots, b^{(L)})$. The opinion of individual m on product n, denoted by $s_{m,n}$, is defined proportional to the scalar product between b_m and a_n: $s_{m,n} = \lambda(L)\, b_m \cdot a_n$, where $\lambda(L)$ is a suitably chosen normalization factor. In general, $\lambda(L)$ should scale as L^{-1} and depend on the ranges of a and b.

In order to predict whether the person j will like or dislike a certain product a_n, *assuming to know* a_n, it is sufficient to obtain the individual tastes of that individual, i.e. the vector b_j. The similarity between tastes of two individuals i and j is defined by the overlap $\Omega_{ij} = b_i \cdot b_j$ between the preferences b_i and b_j.

One can build a knowledge network among people, using the vectors b_m as nodes and the overlaps Ω_{ij} as edges. Maslov and Zhang [4] (MZ) assume that a fraction p of these overlaps are known. They show that there are two important thresholds for p in order to be able to reconstruct the missing information. The first one is a percolation threshold, reached when the fraction of edges p is greater than $p_1 = 1/M - 1$ where M is the number of people. This means that there must be at least one path between two randomly chosen nodes, in order to be able to predict the second node starting from the first one.

Since vectors b_n lie in an L dimensional space, and a single link "kills" only one degree of freedom, a reliable prediction needs more than one path connecting two individuals. Maslov and Zhang show that there is a "rigidity" threshold p_2, of the order of $2L/M$, such that for $p > p_2$ the mutual orientation of vectors in the network is fixed, and the knowledge of the preferences of just one person is sufficient to reconstruct those of all the other individuals.

In general one does not have access to individuals' preferences, nor one knows the dimensionality L of this space. In order to address this problem, the authors define the correlation C_{ij} between the opinions of agents i and j by

$$C_{ij} = \frac{\sum_{n=1}^{N}(s_{in} - \bar{s}_i)(s_{jn} - \bar{s}_j)}{\sqrt{\sum_{n=1}^{N}(s_{in} - \bar{s}_i)^2 \sum_{n=1}^{N}(s_{jn} - \bar{s}_j)^2}}, \tag{1}$$

where $\bar{s}_i$ is the average of the opinion matrix S over column i. The elements C_{ij} can be conveniently stored in a $M \times M$ opinion correlation matrix C.

One can compute an accurate opinion anticipation $\tilde{s}_{mn}$ of a true value s_{mn} using this formula:

$$\tilde{s}_{mn} = \frac{k}{M}\sum_{i=1}^{M} C_{mi} s_{in} \tag{2}$$

where k is a factor that in general depends on L and on the statistical properties of the hidden components. However, if the components of a_n and b_m are independent random variables, k is independent of n and m, so it can be simply chosen in order to have $\tilde{s}_{mn}$ defined over the same interval as s_{mn}.

For large values of N and M, the factor k can be identified with the number of components L, and obtain an estimate for the average prediction error

$$\varepsilon = \sqrt{\frac{1}{MN} \sum_{mn} \left(\tilde{s}_{mn} - s_{mn}\right)^2} \simeq \gamma L^{3/2} \frac{\sqrt{M} + \sqrt{N}}{\sqrt{MN}}, \tag{3}$$

where

$$\gamma = \lambda(L)\sqrt{\langle a^2 \rangle \langle b^2 \rangle}. \tag{4}$$

Formula (3) implies that the predictive power of Eq. (2) grows with MN and diminishes with L. This fact is a consequence of the decay of the correlations among opinions with L, so that more amount of information is needed in order to perform a prediction as L grows. This condition can be compared with the "rigidity" threshold p_2 in the MZ analysis.

3   Test Case Microarray Inspired

In order to investigate the introduction of nonlinearities in the function used to model the process of opinion formation, we considered the case of a microarray.

As mentioned in section 1, microarray experiments can suffer from the missing values, and this fact represents a problem for many data analysis methods, which require a complete data matrix. Although existing missing value imputation algorithms have shown good performance to deal with missing values, they also have their limitations. For example, some algorithms have good performance only when strong local correlation exists in data, while some provide the best estimate when data is dominated by global structure [5].

Here we modified the model described in the previous section to investigate the relationship between the correlation and the overlap between sequences.

To do this we considered an alphabet of four symbols, namely A, T, G, C, corresponding to the four nucleotides that constitute the DNA. We used this alphabet to generate randomly M sequences of length L representing the probes of the microarray[1]. Then we generated N samples of length W representing the sequences to be hybridized on the microarray.

The correlation C_{ij} between sample i and sample j is defined by

$$C_{ij} = \frac{\sum_{k=1}^{M}(m_{ik} - \overline{m}_i)(m_{jk} - \overline{m}_j)}{\sqrt{\sum_{k=1}^{M}(m_{ik} - \overline{m}_i)^2 \sum_{k=1}^{M}(m_{jk} - \overline{m}_j)^2}} \qquad i,j = 1,\ldots,N, \tag{5}$$

where m_{ik} is the maximum complementary match between sample i and probe k without gaps.

The aim is to test the relationship between the correlation matrix C and the overlap matrix Ω constructed using the following idea of similarity. We hypothesized to infer the similarity between sequences based on the number of subsequences of length L in common. For this reason we defined the overlap Ω_{ij} between sequence i and sequence j as the number of subsequences of length L that appear in the both sequences, divided by $W - L + 1$ for normalization.

---

[1] The probes in real microarray are discriminated generally carefully chosen in order to genes of interest.

This matching function is nonlinear since the effect of a mismatch depends on its position in the subsequence.

To test our hypothesis, we considered eight referential sequences:

Seq. 0: This is the first reference sequence, completely random of length W.

Seq. 1: Equal to sequence 0, except for a mutation in the middle (this mimics the Affimetrix central mismatch mechanism for measuring the level of random pairing).

Seq. 2: Equal to sequence 0, but shifted of one basis.

Seq. 3: Equal to sequence 0, with shift and central mutation.

Seq. 4: First half of sequence 4 is equal to the second half of sequence 0, and vice versa.

Seq. 5: First half of sequence 0 is equal to the second half of sequence 0, the rest is random.

Seq. 6: Another reference sequence.

Seq. 7: Sequences 6 and 7 contains the same "gene", of length $W/3$, in different positions.

4   Results

To check the validity of the model described in the previous section, we measured the error ε_{ij} for the pair of sequences i and j defined as the absolute value of the

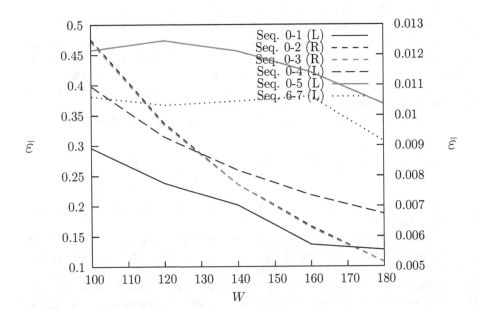

Fig. 1. The error $\bar{\varepsilon}$ as a function of the length W of the samples, averaged over 40 realizations, $N = 10$, $L = 30$, $M = 500$. The plots of sequences 0-2 and 0-3 refer to the right y-axis. One can observe that all errors diminish with W.

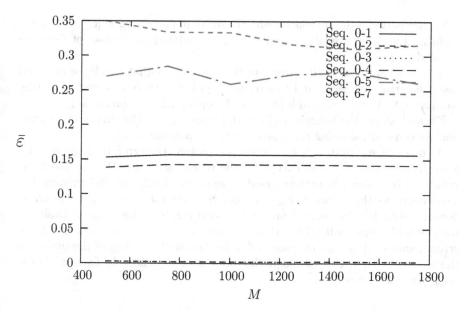

Fig. 2. The error $\bar{\varepsilon}$ as a function of the number of probes M, averaged over 40 realizations, $N = 10$, $L = 20$, $W = 150$. One can observe that errors do not vary with M.

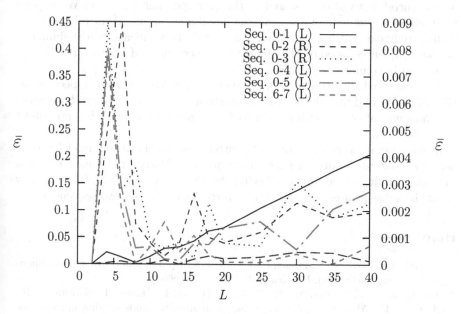

Fig. 3. The error $\bar{\varepsilon}$ as a function of the length L of the probes, averaged over 100 realizations, $N = 10$, $W = 200$, $M = 1000$. The plots of $\bar{\varepsilon}_{02}$, $\bar{\varepsilon}_{03}$ and $\bar{\varepsilon}_{04}$ refer to the right y-axis.

difference between the correlation and the overlap, namely $\varepsilon_{ij} = |C_{ij} - \Omega_{ij}|$. We performed various simulations and than we calculated the average of the error denoted by $\bar{\varepsilon}$.

In figure 1 we plotted the error $\bar{\varepsilon}$ vs W. One can see that for all the analysed cases the error decreases, and this result agrees with those reported in [1] (the parameter W here corresponds to M in the opinion formation model).

Figure 2 shows the behaviour of $\bar{\varepsilon}$ with respect to M. The curves are approximately constant, showing that the error is independent of M.

As one can see from figure 3, where we plotted the error vs L, $\bar{\varepsilon}$ does not follow a monotonous trend, except for the pair of sequences 0-4 for which the value of $\bar{\varepsilon}$ is almost constant and next to zero, and for $\bar{\varepsilon}_{01}$ which increases. For what concerns the values of $\bar{\varepsilon}_{02}$, $\bar{\varepsilon}_{03}$ and $\bar{\varepsilon}_{05}$, one can detect that the errors decrease until $L \simeq 10$, because probes too short can hybridize in many positions without a high specificity. Then they oscillate until $L \simeq 35$, and for larger L the errors increase. This last increase is due to the small coverage of the probes in the sequence space, since we kept the number of sequences M fixed while the sequence space grows as 4^L.

5   Comments

We have proposed a method for measuring the distance among records based on the correlations of data stored in the corresponding database. We applied the method to the case of a microarray modifying the model introduced in [1] with a nonlinear matching function. More precisely, we measured the similarity between sequences based on the number of subsequences of length L (the length of probes) in common.

We monitored the error for eight sequences of reference, with respect to M, W, and L. We find that the error is low in all cases, decreasing when W increase, and independent of M. With respect to L we find that the model is more robust for traslocation.

In conclusion we can say that the correlation matrix of our model can be used to estimate the distance between sequences. Moreover we point out that the same result can be found following the idea of negative database, namely using the subsequences of length L not in common between two sequences.

References

1. Bagnoli, F., Berrones, A., Franci, F.: De gustibus disputandum (forecasting opinions by knowledge networks). Physica A 332, 509–518 (2004)
2. Troyanskaya, O., Cantor, M., Sherlock, G., Brown, P., Hastie, T., Tibshirani, R., Botstein, D., Altman, R.: Missing value estimation methods for dna microarrays. Bioinformatics 17, 520–525 (2001)
3. Esponda, F., Ackley, E.S., Helman, P., Jia, H., Forrest, S.: Protecting data privacy through hard-to-reverse negative databases. In: Katsikas, S.K., López, J., Backes, M., Gritzalis, S., Preneel, B. (eds.) ISC 2006. LNCS, vol. 4176, pp. 72–84. Springer, Heidelberg (2006)

4. Maslov, S., Zhang, Y.C.: Extracting hidden information from knowledge networks. Physical Review Letters 87, 248701 (2001)
5. Gan, X., Liew, A.W.C., Yan, H.: Microarray missing data imputation based on a set theoretic framework and biological knowledge. Nucleic Acids Research 34, 1608–1619 (2006)

Human Heuristics for Autonomous Agents

Franco Bagnoli[1,*], Andrea Guazzini[1], and Pietro Liò[2]

[1] Department of Energy, University of Florence, Via S. Marta 3, 50139 Firenze, Italy
Also CSDC and INFN, sez. Firenze
[2] Computer Laboratory, University of Cambridge, 15 J.J. Thompson Avenue,
Cambridge, CB30FD, UK

Abstract. We investigate the problem of autonomous agents processing pieces of information that may be corrupted (tainted). Agents have the option of contacting a central database for a reliable check of the status of the message, but this procedure is costly and therefore should be used with parsimony. Agents have to evaluate the risk of being infected, and decide if and when communicating partners are affordable. Trustability is implemented as a personal (one-to-one) record of past contacts among agents, and as a mean-field monitoring of the level of message corruption. Moreover, this information is slowly forgotten in time, so that at the end everybody is checked against the database. We explore the behavior of a homogeneous system in the case of a fixed pool of spreaders of corrupted messages, and in the case of spontaneous appearance of corrupted messages.

1   Introduction

One of the most promising area in computer science is the design of algorithms and computer architectures closely based on our reasoning process and on how the brain works. Human neural circuits receive, encode and analyze the "available information" from the environment in a fast, reliable and economical way. The evolution of human cognition could be viewed as the result of a continuous improvement of neural structures which drive the decision making processes from the inputs to the final behaviors, cognitions and emotions. Heuristics are simple, efficient rules, hard-coded by evolutionary processes or learned, which have been proposed to explain how people make decisions, come to judgments, and solve problems, typically when facing complex problems or incomplete information. It is common experience that that much of human reasoning and decision making can be modeled by fast and frugal heuristics that make inferences with limited time and knowledge. For example, Darwin's deliberation over whether to marry provides an interesting example of such heuristic process [1,2].

Let us quickly review some widely accepted hypothesis about heuristics. In the early 1970s, Daniel Kahneman and Amos Tversky (K&T) produced a series of important papers about decisions under uncertainty [3,4,5,6,7]. Their basic claim

---

* To whom correspondence should be addressed.

P. Liò et al. (Eds.): BIOWIRE 2007, LNCS 5151, pp. 340–351, 2008.
© Springer-Verlag Berlin Heidelberg 2008

was that in assessing probabilities, *"people rely on a limited number of heuristic principles which reduce the complex tasks of assessing probabilities and predicting values to simpler judgmental operations"*. Although K&T claimed that, as a general rule, heuristics are quite valuable, in some cases, their use leads *"to severe and systematic errors"*. One of the most striking features of their argument was that the errors follow certain statistics and, therefore, they could be described and even predicted. The resulting arguments have proved highly influential in many fields, including computer science (and particularly in human-machine interaction area) where the influence has stemmed from the effort to connect algorithmic accuracy to speed of elaboration and, equally important, to the algorithmic understanding of the human logic [7]. If human beings use identifiable heuristics, and if they are prone to systematic errors, we might be able to design computer architectures and algorithms to improve human-computer interaction (and also to study human behavior).

K&T described three general-purpose heuristics: **representativeness**, **availability** and **anchoring**. People use the *availability* heuristic when they answer a question of probability by relying upon knowledge that is readily available rather than examine other alternatives or procedures. There are situations in which people assess the frequency of a class or the probability of an event by the ease with which instances or occurrences can be brought to mind. For example, one may assess the risk of heart attack among middle-aged people by recalling such occurrences among one's acquaintances. Availability is a useful clue for assessing frequency or probability, because instances of large classes are usually reached better and faster than instances of less frequent classes. However, availability is affected by factors other than frequency and probability. This is a point about how familiarity can affect the availability of instances. For people without statistical knowledge, it is far from irrational to use the availability heuristic; the problem is that this heuristic can lead to serious errors of fact, in the form of excessive fear of small risks and neglect of large ones.

The *representativeness* heuristic is involved when people make an assessment of the degree of correspondence between a sample and a population, an instance and a category, an act and an actor or, more generally, between an outcome and a model. This heuristic can be thought of as the reflexive tendency to assess the similarity of characteristics on relatively salient and even superficial features, and then to use these assessments of similarity as a basis of judgment. Representativeness is composed by categorization and generalization: in order to forecast the behavior of an (unknown) subject, we first identify the group to which it belongs (categorization) and them we associate the "typical" behavior of the group to the item. Suppose, for example, that the question is whether some person, Paul, is a computer scientists or a clerk employed in the public administration. If Paul is described as shy and withdrawn, and as having a passion for detail, most people will think that he is likely to be a computer scientist and ignore the "base-rate", that is, the fact that there far more clerk employed in public admin than computer scientists. It should be readily apparent that the representativeness heuristic will produce problems whenever people are ignoring base-rates, as they are prone to do.

K&T also suggested that estimates are often made from an initial value, or *anchoring*, which is then adjusted to produce a final answer. The initial value seems to have undue influence. In one study, K&T asked subjects to say whether the number that emerged from the wheel was higher or lower than the relevant percentage. It turned out that the starting point, though clearly random, greatly affected people's answers. If the starting point was 65, the median estimate was 45%; if the starting point was 10, the median estimate was 25%.

Several of recent contributions on heuristic have put the attention on the "dual-process" to human thinking [8,9,10,11,12]. According to these hypothesis, people have two systems for making decisions. One of them is rapid, intuitive, but sometimes error-prone; the other is slower, reflective, and more statistical. One of the pervasive themes in this collection is that heuristics and biases can be connected with the intuitive system and that the slower, more reflective system might be able to make corrections. The dual-process idea has some links with the experimental evidences of the presence of areas for emotions in the brain, for instance of fear-type. These "emotional" areas may be triggered before than the cognitive areas become involved.

We shall try to consider some of these concepts to model autonomous agents that have the task of processing messages from sources that are not always trustable. The agent is a direct abstraction of an human being, easily understandable by psychologists and biologist with the advantage of following a stochastic dynamics that can be combined with other approaches like ODE [14,15,16,13,17]. Here we make the analogy between the diffusion of hoaxes, gossips, etc., and that of computer viruses or worms.

The incoming information may be corrupted for many reasons: some agents may be infected by malware and particularly viruses, some of them may be programmed to provide false information or they may just be malfunctioning. Let us suppose that the processing of a corrupted information will infect the elaborated message, so that the corruption "percolates and propagates" into the connection network, unless stopped. We assume that an agent may contact a central database for inquiring about the reliability of a message, but this checkout is costly, at least in terms of the time required for processing the information. Therefore, an agent is confronted with two opportunities: either trust the sender, accept the message and the risk or passing false information and process it in a short time, or contact the central database, be sure of the correctness of the message but also waste more time (or other resources such as bandwidth) in elaborating it. This is analogous to the passport check when crossing a boundary: customers may either trust the identity card and let people pass quickly, or check them against a database, slowing down the queue.

This paper, which is motivated by the fact that human heuristics may be used to improve the efficiency of artificial systems of autonomous decision-makers agents, is structured as follows. In Section 2, we introduce a model where the above mentioned heuristics are implemented. Section 3 focuses on equilibrium and asymptotic conditions in the absence of infection. In Section 4, we describe the different scenarios which are considered (no infection, quenched infection and annealed infection); numerical results for different value of control parameters under infection are reported in Section 5. A discussion about the psychological implications of the model and conclusions are drawn in Section 6.

2 Model

Let us consider a scenario with N agents, identified by the index $i = 1, \ldots, N$. Each agent interacts with other K randomly chosen agents. The connections indicate messages transferred. In principle, one can have input connections with himself (meaning further processing of a given piece of information) and multiple connections with a given partner (more information transferred). An agent receives information from its connecting inputs, elaborates it and send the result to its output links. Let us assume for simplicity that this occurs in a synchronous way and at discrete time steps t. The information however can be tainted (corrupted), either maliciously (virus, sabotage, attack) or because it is based on incorrect data.

If an information is tainted, and it is accepted for processing, it contaminates the output. All agents have the possibility of checking the correctness of the incoming messages against a central database, but this operation is costly (say, in terms of time), and therefore heuristics are used to balance between cost and the risk of being infected.

An agent i has a dynamical memory for the reliability of its partners j, $-1 \leq \alpha_{ij} \leq 1$; this memory is used to decide if a message is acceptable or not. The greater $\alpha_{ij} > 0$, the more the partner is considered reliable, the reverse for $\alpha_{ij} < 0$. However, the trusting on an individual is not an absolute value, it has to be compared with the perception of the level of the infection. Let us denote by $0 \leq A_i \leq 1$ the perception of the risk *i.e.*, the perceived probability of message contamination, of individual i. A simple yet meaningful way of combining risk perception with uncertainty is to assume that each individual i decides according with its previous knowledge (α_{ij}) if $|\alpha_{ij}| > A_i$ and checks against the database (*i.e.*, get to know the truth) otherwise. If A_i is large, the agent i will check many messages against the database, the reverse for small values of A_i.

After checking the database, one knows the truth about his/her partner. This information can be used to increase or decrease α_{ij} and also to compute A_i. In particular, if the check is positive (negative), α_{ij} increases (decreases) of a given amount v_α. Finally A_i in increased by a quantity $v_A n_i / c_i$, where c_i is the cost (total number of checks for a given time step) and n_i the number of infected discovered. The idea is that A_i represents the perceived "average" level of infection, corresponding to the "risk perception" of being infected. We shall limit here to fixed and homogeneous responses, in an more realistic case, different classes of agents or individuals will react differently, according to their "programming" and their past experiences, to a given perception of the infection level.

Some of these quantities change smoothly in time. There is an oblivion mechanism on α_{ij} and A_i, implemented with the parameters r_α and r_A, respectively, such that the information stored τ time steps before the present time has weight $(1-r)^\tau$. New information is stored with weight r. This mechanism emulates a finite memory of the agent, without the need of managing a list.

The observable quantities are the total number of infected individuals, I, the cost of querying the database, C and the number of errors E, which are given by the number of tainted accepted messages and not-tainted refused messages.

In this model, we are only interested in the correctness of the message, not in its content. Actually, a real message should be considered as a set of 'atomic' parts, each of which can be analyzed, eventually with their relations, in order to judge the reliability of the message itself. For instance, the spam detection mechanism is often based on a score assigned to patterns (e.g., MONEY, SEX, LOTTERY) appearing in the message. Therefore, a more accurate model should represent messages as vectors or lists of items. We deal here with a simple scalar approximation.

We try to include the human heuristics in this simple model by means of A (representativeness) and α_{ij} (availability). The oblivion mechanism can moreover be considered the parameter corresponding to the "anchoring" experiences. In our present model, there is only one variable connected to affordability (from completely trustable to completely not trustable), and the categorization procedure consists essentially in trying to assess the placement of an individual on this axis. The trustability of an individual (α_{ij}) depends on the past interactions. Since A represents the average level of infectivity, the trustability of an individual is evaluated against it, in order to save the cost (or the time) of the check against the central database.

3 Relaxation to Equilibrium and Asymptotic State without Infection

In order to put into evidence the emerging features of our model, let us first study the case without infection. Without "stimulation", the threshold A_i is fixed, and takes the value v_A for all individuals. The only dynamical variables are the α_{ij}.

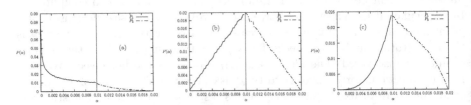

Fig. 1. The asymptotic distribution $P(\alpha)$ for $a < 2r$ ($a = 0.006$ and $r = 0.01$) (a); $a = 2r$ ($a = 0.01$ and $r = 0.005$) (b); $a > 2r$ ($a = 0.02$ and $r = 0.005$) (c)

Starting from a peaked (single-valued) distribution of α_{ij}, the model exhibits oscillatory patterns and long transients towards an equilibrium distribution (Fig. 1). We found that by increasing the connectivity K, the peaks become thinner and higher, following a linear relationship. The affinities α_{ij} in the asymptotic state have a non trivial distribution, ranging from 0 to $2v$. Let us call $P(\alpha)$ the probability distribution of α. From numerical simulation (see Fig. 1), one can see that $P(\alpha)$ can be divided into two branches, $P_1(\alpha)$ for $0 \le \alpha \le v$ and $P_2(\alpha)$ for $v \le \alpha \le 2v$. The evolution of $P(\alpha)$ is given by the combination of two phases: control against the database, that in the mean field approach occurs with probability $a = K/N$ for all $\alpha \le v$ (and therefore for P_1), and the oblivion mechanism, that multiplies all α by $(1 - r_\alpha)$. Combining the two effects, one finds for the asymptotic state

$$P_1(\alpha) = \frac{1-a}{1-r_\alpha} P_1\left(\frac{\alpha}{1-r_\alpha}\right), \tag{1}$$

$$P_2(\alpha) = \frac{a}{1-r_\alpha} P_1\left(\frac{\alpha}{1-r_\alpha} - v\right) + \frac{1}{1-r_\alpha} P_2\left(\frac{\alpha}{1-r_\alpha}\right). \tag{2}$$

From Eq. (1), one gets easily that $P_1(\alpha) \propto \alpha^x$, with

$$x = \frac{\ln(1-a)}{\ln(1-r_\alpha)} - 1 \simeq \frac{a}{r_\alpha} - 1.$$

In particular, the value $x = 1$ (Fig. 1-b) corresponds to $a = 2r_\alpha$. We were not able to express the asymptotic distribution $P_2(\alpha)$ in terms of known functions.

The process of relaxation to the equilibrium is in general given by oscillations, whose period is related to r_α. A rough estimation can be obtained by considering that a pulse of agents with the same value of $\alpha = 2v$ will experience the oblivion at an exponential rate $(1-r_\alpha)^T$, until $\alpha = v$, after which a fraction a of the pulse is re-injected again to the value $\alpha = 2v$. The condition for the pseudo-periodicity (for the fraction a of agents) is

$$2v(1 - r_\alpha)^T = v,$$

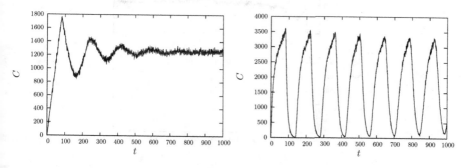

Fig. 2. Relaxation to equilibrium for the cost C for $N = 500$, $r_\alpha = 0.005$ and $K = 5$ ($a = 0.01$) (left); $K = 50$ ($a = 0.1$) (right)

from which the period T can be estimated

$$T \simeq -\frac{\ln(2)}{\ln(1 - r_\alpha)} \simeq \frac{\ln(2)}{r_\alpha}$$

in the limit of small r_α.

Since the re-injected fraction is given by a, the larger is its value, the larger the oscillations and the slower is the relaxation to the asymptotic distribution, as is shown in Fig. 2. One can notice that the period is roughly the same (same value of r_α), but the amplitude of oscillations is much larger in the plot to the right (larger a).

Since $a = K/N$, these large oscillations make difficult to perform measurements on the asymptotic state on small populations, but large values of N require longer simulations. One may say that the model is intrinsically complex.

The asymptotic cost is given by $C_\infty = a \int_0^{v_\alpha} P_1(\alpha) \propto a v_\alpha^{a/r_\alpha}$. As one can see from Fig. 1, there is a cost even in the absence of infection, since the agents have to monitor the level of infection against the database. The lower values of the cost are associated to values of r_α smaller than a.

4   Infectivity Scenarios

The source of infection may be quenched, *i.e.*, a fraction p of the population always emits tainted messages, or annealed, in which case the fraction p of the spreaders is changed at each time step. Let us first study the case of a pulse of infection (with $p = 1$) in the asymptotic state and a duration $\Delta t = 20$. For large values of the asymptotic cost, The infection is removed in just a few time steps, as shown in Fig. 3.

For smaller values of the cost, the fate of the infection is related to the scenario (quenched or annealed infectors). If the infection level is small, and the infectors are quenched, the rising of the corresponding α_{ij} efficiently isolate the contagion.

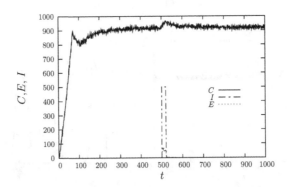

Fig. 3. Temporal behavior of the cost C, infection I and error level E for $K = 5$, $n = 500$ ($a = 0.01$), $r = 0.005$. The pulse is at the time 500.

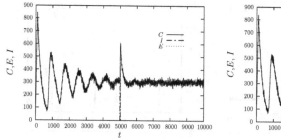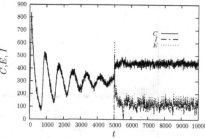

Fig. 4. Temporal behavior of the cost C, infection i and error level E for $K = 2$, $n = 500$ ($a = 0.004$), $r = 0.001$. Left: $v_A = 10^{-3}$ (eradication). Right: $v_A = 10^{-4}$ (endemic infection). The pulse occurs at time 5000.

In the case of a "pulse" of infection, or for annealed infectors, the fate of the contagion is mainly ruled by the quantity A_i. If A_i grows rapidly (v_A sufficiently large), a temporary increase of the cost is enough to eradicate the epidemics, see Fig. 4. In the opposite case, the infection becomes endemic even for non-persistent infectors: it is maintained by the spreading mechanism. The increment used in the following investigations is small enough so that we can observe the persistence of the infection.

If the infectors are persistently renewed, the contagion cannot get eradicated but only kept under control. The role of the two heuristics is different in the two cases.

The representativeness heuristic (α_{ij}) is the optimal strategy to detect agents which are constantly less reliable than the others (quenched case), but it is completely useless in the annealed case. The availability heuristic (A_i), considering at each time step the average infection of the system, is able to control the spread of infection in the annealed case.

The oblivion mechanism, related to the anchoring heuristic, is a the key parameter governing the speed of adaptation to variable external conditions. It controls the oscillations of the cost (Fig. 2) and it is fundamental to minimize the computational load of the control process. The oblivion of α_{ij} (representativeness parameter) controls the computational cost at the equilibrium in both cases. High values of r_α correspond to a conservative behavior of the system, in this case a large computational cost and a corresponding low number of infected and errors characterized the equilibrium. Low values of r_α correspond to the dissipative behavior for which the system minimizes the computational cost but allows large fluctuations of infected and high values of errors.

5   Dynamical Behavior

We run extensive numerical simulations and recorded the asymptotic cost C, number of infected people I, and errors E as function of the oblivion parameters (r_α and r_A), the probability and the pulse of infection (p), and the density of

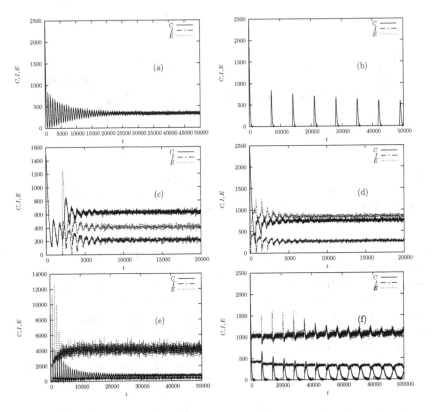

Fig. 5. Cost C, Infection level I and Errors E vs. time t for two different values of the parameter r_α, $r_\alpha = 10^{-3}$ (a,c,d,e); $r_\alpha = 10^{-4}$ (b,f), different values of p, $p = 0$, (a,b); $p = 10^{-2}$ (d,e,f); $p = 10^{-6}$, (c) and for some value of the connectivity ($K = 5$ for plots (a,b,c,d,f), $K = 30$ for plot (e)) and population $N = 500$.

contacts (K). In these simulation we kept $v_A = 10^{-6}$ in order to stay in the endemic phase, and therefore r_A did not play any role.

Fig. 5 shows the effect of infection with different values of r_α and contact density for for $K = 5$ and $N = 500$. Plots (a) and (b) show the oscillatory patterns without infection ($p = 0$) for $r_\alpha = 10^{-3}$ (a), $r_\alpha = 10^{-4}$ (b). The oblivion r_α (in the presence of infection) changes both the oscillatory frequency (as studies in the previous section) and the oscillatory delay before convergence to a basic fluctuation pattern. Note that increasing r_α the frequency of the oscillations increases. When $r_\alpha = 0.0001$, (a), the period T is $T = \ln 2 10^4 \approx 7000$; for $r_\alpha = 0.001$, (b), $T \approx 700$). By adding infection (annealed version), we obtain a quicker convergence the basal fluctuation equilibrium (c). We found that the time to reach the basic fluctuation equilibrium does not depend on the infection probability and the level of the fluctuation remain unchanged even for long runs (d). Plots (e) and (f) show that with the same value of the infection probability, increasing the density of contacts produces larger fluctuations, a quicker convergence of the cost ($K = 30$ for plot (e) with respect to $K = 5$ for all others). The two scenarios have different oblivion ($r_\alpha = 10^{-3}$ (e), $r_\alpha = 10^{-4}$ (f)). Then, increasing p, the frequency of the oscillations remains the same but the peaks broaden.

6   Discussion and Conclusions

In this paper we have been modeled the cognitive mechanisms known as availability and representativeness heuristics. The role of the first one in the human decision making process seems to be to produce a probability estimation of an event based on the relative observed (registered) frequency distribution. The second heuristic, representativeness, acts inferring certain attributes from others easier-to-detect. Both heuristics are liable or "noise affected", but surely they represent a very fast way to analyze environmental data using little quantity of memory and time. But the very interesting aspect, and not underlined enough in literature, is the role of the cooperation between heuristics. The co-occurrence of their activities could be coordinate also in the human cognition, but of course it is very interesting from a computational point of view.

We supposed that the availability heuristic corresponds to a mean field estimation of the "risk", while representativeness partially maintains the memory of the previous interactions with the others. In the quenched and annealed scenarios we can capture the effect of the heuristics coordination. The quenched scenario considers the case of "systematic spreaders" where same agents emits at each time step a tainted message. In this case the availability heuristic would fails to minimize cost and infection if representativeness was absent.

On the contrary in the annealed scenario the spreaders are completely chosen at random at each time step. In this extreme case there is no information contained in the previous history of the system, and representativeness heuristic became completely useless. In this situation the only available information is the rate of infection, and availability heuristic is the most efficient way to minimize both cost and risk of infection.

The oblivion mechanism associated to the two heuristics determines both the cost of the control process and a sort of its reactivity. In average the cost, which represents the number of operations/computation to cope the task, is proportional to the oblivion value, the number of infected and errors are inversely proportional to the oblivion. If the cost as so as it happens in the biological domain, is considered as a quantity which the system has to minimized, it means that will exist an optimal value of both oblivion parameters for each possible condition. The reactivity of the control process could be defined as the time needed from the system to reduce to zero a new infection. In our model the oblivion of both the two heuristics appears to control also the size of "cost oscillations". We found that the larger the oblivion level, the lower the oscillations and the time needed to reach the asymptotic equilibrium.

Our simulations show that under the infection, the cost reaches its asymptotic value much earlier than without infection. This suggests that a low value of infection level may even provide some advantages for the quick dumping of the oscillatory behavior resulting in an improved cost predictability.

The investigation of heuristics exploits a major overlap between artificial intelligence (AI), cognitive science and psychology. The interest in heuristics is based on the assumption that humans process information in ways that computers can emulate and heuristics may provide the basic bricks for bridging from brains to computers . Our model framework approach is quite general and offers some points of reflections on how the study of complex systems may become help developing new areas of AI. In the past years the AI community has debated as to whether the mind is best viewed as a network of neurons (connectionism), or as a collection of higher-level structures such as symbols, schemata, heuristics, and rules, i.e., emphasizing the role of symbolic computation. Nowadays the symbolic representations to produce general intelligence is in slightly decline but the "neuron ensemble" paradigm has also shifted towards more complex models particularly taking into account and combining findings from both fNMR and cognitive psychology fields ([19,20]).

Here we show that the incorporation of simple heuristics in a small network of agents leads to a rich and complex dynamics.Our model does not take into account mutation and natural selection which is of key importance for the emergence of complex behavior in animal societies and in the brain development (see for example Pinker and the follow up debate [21]).

A multi-agent model, where each agent represent a message/modifying person/neuron, can serve as a very natural abstraction of communication networks, and hence be easily used by psychologists as well as computer scientists. Such a model also allows the tracking of single agent fates so that communities with low member numbers are easily dealt with and these models also provide for much more detailed analysis compared to average population approaches like continuous differential equations.

Heuristics may have even greater value in case of environmental challenges, i.e. organisms need to adapt quickly to environmental fluctuations, for example starvation and high competition, they must be able to make inferences that are fast, frugal, and accurate. These real-world requirements lead to a new conception of what proper reasoning is: ecological rationality. Fast and frugal heuristics that are matched to particular environmental structures allow organisms to be ecologically rational. The study of ecological rationality thus involves analyzing the structure of environments, the structure of heuristics, and the match between them.

References

1. Healey, E.: Emma Darwin: The Inspirational Wife of a Genius. Headline Publishing Group, London (2001)
2. Litchfield, H.: Emma Darwin, a century of family letters, pp. 1792–1896. D. Appleton And Company, London (1940)
3. Tversky, A., Kahneman, D.: Availability: A heuristic for judging frequency and probability. Cognitive Psychology 5, 207 (1973)

4. Tversky, A., Kahneman, D.: Judgment under uncertainty: Heuristics and biases. Science 185, 1124 (1974)
5. Kahneman, D., Tversky, A.: Prospect theory: An analysis of decision under risk. Econometrica 47, 263 (1979)
6. Tversky, A., Kahneman, D.: The framing of decisions and the psychology of choice. Science 211, 453 (1981)
7. Kahneman, D., Slovic, P., Tversky, A. (eds.): Judgment under uncertainty: heuristics and biases. Cambridge University Press, New York (1982)
8. McNeil, B.J., Pauker, S.G., Soxand, H.C., Tversky, A.: On the elicitation of preferences for alternative therapies. New England Journal of Medicine 306, 1259 (1982)
9. Shelley Taylor, E.: The Availability Bias in Social Perception and Interaction. In: Kahneman, D., Slovic, P., Tversky, A. (eds.) Judgment Under Uncertainty: Heuristics and Biases. Cambridge University Press, New York (1982)
10. Slovic, P., Fischoff, B., Lichtenstein, S.: Facts versus Fears: Understanding Perceived Risk. In: Kahneman, D., Slovic, P., Tversky, A. (eds.) Judgment Under Uncertainty: Heuristics and Biases. Cambridge University Press, New York (1982)
11. Gigerenzer, G., Todd, P.M., ABC Research Group: Simple Heuristics That Make Us Smart. Oxford University Press, Oxford (1999)
12. Christensen, C., Abbott, A.S.: Team Medical Decision Making. In: Chapman, G.B., Sonnenberg, F.A. (eds.) Decision Making in Health Care: Theory, Psychology, and Applications. Cambridge University Press, New York (2000)
13. Sun, R. (ed.): Cognition and multi agent interaction: from cognitive modeling to social simulation. Cambridge University Press, New York (2006)
14. Weiss, G.: Multiagent Systems: A Modern Approach to Distributed Artificial Intelligence. The MIT Press, Cambridge (2000)
15. Wooldridge, M.: Introduction to MultiAgent Systems. Wiley, New York (2002)
16. d'Inverno, M., Luck, M.: Understanding Agent Systems. Springer Series on Agent Technology. Springer, Berlin (2003)
17. Merelli, E., et al.: Agents in bioinformatics, computational and systems biology. Brief Bioinform. 8, 45 (2007)
18. Dreyfus, H.: What Computers Still Can't Do. MIT Press, Cambridge (1992)
19. Gazzaniga, M.S.: Cognitive Neuroscience. MIT Press, Cambridge (1995)
20. Rosenzweig, M.R., Breedlove, S.M., Watson, N.V.: Biological Psychology: An Introduction to Behavioral and Cognitive Neuroscience. Sinauer Associates, Sunderland (2004)
21. Pinker, S.: How the Mind Works, Norton, New York (1997)

Designing Biological Computers: Systemic Computation and Sensor Networks

Peter J. Bentley

Department of Computer Science, University College London, Malet Place, London
p.bentley@cs.ucl.ac.uk

Abstract. Biological computation may or may not be Turing Complete, but it is clearly organized differently from traditional von Neumann architectures. Computation (whether in a brain or an ant colony) is distributed, self-organising, autonomous and embodied in its environments. Systemic computation is a model of computation designed to follow the "biological way" of computation: it relies on the notion that systems are transformed through interaction in some context, with all computation equivalent to controlled transformations. This model implies a distributed, stochastic architecture, and in this work it is proposed that a physical implementation of this architecture could be achieved through the use of wireless devices produced for sensor networks. A useful, fault-tolerant and autonomous computer could exploit all the features of sensor networks, providing benefits for our understanding of "natural computing" and robust wireless networking.

Keywords: biological computing, systemic computation, sensor networks, natural computation.

1 Introduction

Nature is full of examples of biological computation that adapt to their environments and show remarkable tolerance to damage or faults. Whether an ant colony, immune system or brain, these complex natural systems evolved by learning to survive their hazardous environments, resulting in highly adaptive and robust structures that rely on large numbers of simpler elements interacting in order to achieve a higher purpose.

Computer science and engineering has long sought to create devices with similar properties. While several levels of redundancy, or provably correct implementations, or careful dependability risk assessments may provide useful levels of protection, no human-designed method approaches the efficiency or fault tolerance of nature. Yet with computers increasingly taking on critical roles in our lives, whether in autopilots, stock markets, power or food distribution, the creation of fault tolerant computation is seen as increasingly important.

In an attempt to exploit desirable natural properties within a computer (e.g., see table 1), in 2005 a novel model of computation called *systemic computation* was developed [2]. The result of considerable research into bio-inspired computation and biological modelling, the model has been developed into a working computer architecture [2,3]. A

P. Liò et al. (Eds.): BIOWIRE 2007, LNCS 5151, pp. 352–363, 2008.
© Springer-Verlag Berlin Heidelberg 2008

parallel systemic computer based on this architecture is designed to run on hundreds or preferably thousands of processors, with all computation emerging through interactions, just as the overall result of biological processes emerges through the interaction of thousands of simpler elements. A systemic computer should share some of the same capabilities of biological systems and provide fault-tolerant computation through self-organising, parallel and distributed processing.

Table 1. Features of conventional vs natural computation. Increasingly researchers are aiming to create computers with more of the properties of natural systems.

Conventional	Natural
Deterministic	Stochastic
Synchronous	Asynchronous
Serial	Parallel
Heterostatic	Homoestatic
Batch	Continuous
Brittle	Robust
Fault intolerent	Fault tolerant
Human-reliant	Autonomous
Limited	Open-ended
Centralised	Distributed
Precise	Approximate
Isolated	Embodied
Linear causality	Circular causality
Simple	Complex

To date, two simulations of this architecture have been developed, with corresponding machine and programming languages, compilers and graphical visualiser [2,3]. Research is ongoing in the improvement of the PC-based simulator, refining the systemic computation language and visualiser. However, the systemic computation model defines a highly parallel, distributed form of computer. While simulations on conventional computers enable the improvement of the model and associated programming tools [3], a demonstration of, for example, fault tolerance would be heavily restricted by the conventional underlying architecture and operating system. The same drawbacks are true for larger scale simulations using networked PCs, Beowulf or GRID computing – the brittleness of underlying operating systems and network protocols could not be circumvented using higher-level code [14].

Consequently, this work proposes the use of clusters of numerous, and potentially smaller, simpler processors, organised at the lowest level possible according to the bio-inspired architecture and communication protocol of systemic computation. Systemic computing views interactions between the environment and the device as part of the computation and essential for self-adaptation, so the ideal platform should also enable a large variety and diversity of senses from environment to computer. Thus the use of wireless sensor network technology is ideal for such a computer.

Wireless sensor networks and ubiquitous computing are becoming one the most popular areas of computer science research today, with over 20 conferences each year dedicated to the field. There are regular predictions that pinhead-sized wireless computers will one day be in every product, that they will cost just a few pence to make, and that the future of the Internet will be ubiquitous computing. Whether the predictions come true or not, it seems highly likely that such ubiquitous devices will become cheaper, faster and more prevalent in the near future. With this in mind, a computer architecture able to exploit large clusters of very simple wireless devices is desirable, for the fields of ubiquitous computing and bio-inspired computing.

2 Systemic Computation

"Systemics" is a world-view where traditional reductionist approaches are supplemented by holistic, system-level analysis of the interplay between components and their environment at many different levels of abstractions [16]. *Systemic Computation* takes this approach [2,3], aiming to provide a more "natural" or bio-inspired model of computation compared to conventional von Neumann architectures. It uses the following assertions:

- Everything is a *system*
- Systems can be transformed but never destroyed.
- Systems may comprise or share other nested systems (see figure 1).
- Systems *interact*, and interaction between systems may cause transformation of those systems, where the nature of that transformation is determined by a contextual system.
- All systems can potentially act as context and affect the interactions of other systems, and all systems can potentially interact in some context.
- The transformation of systems is constrained by the scope of systems and systems may have partial membership within the scope of a system.
- Computation is transformation.

Systemic computation has been shown to be Turing Complete [2] and thus is directly equivalent to other models of computation. For example, it can also be regarded as a form of bigraph model [2,18]. Although the origins of systemic computation come from studies of natural systems, each system may be viewed as equivalent to an asynchronous bigraph node, and schemata (see later) may be viewed as dynamic links between nodes. Because of this, systemic computation may be rewritten in bigraph form (and also may benefit from expression in π-calculus [17,18]) for future theoretical investigations. In addition, some useful systemic computation transformation functions resemble those widely used in membrane computing and brane calculus [15], while in approach, it is comparable to the ideas of cellular automata [11] (but influenced by the views of Varela [22]).

Instead of the traditional centralised view of computation, here all computation is distributed. There is no separation of data and code, or functionality into memory, ALU, and I/O. Everything in systemic computation is composed of *systems*, which may not be destroyed, but may transform each other through their interactions, akin to collision-based computing [12]. Two systems interact in the context of a third system, which defines the result of their interaction. This is intended to mirror all conceivable natural processes, e.g:

- molecular interactions (two molecules interact according to their shape, within a specific molecular and physical environment)
- cellular interactions (intercellular communication and physical forces imposed between two cells occurs in the context of a specific cellular environment)
- individual interactions (evolution relies on two individuals interacting at the right time and context, both to create offspring and to cause selection pressure through death)

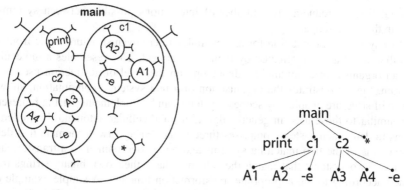

Fig. 1. Systemic computation calculation: PRINT((A1-A2)*(A3-A4)). Left: graph-based notation. Right: the tree of scope memberships for this calculation. (Systemic computation also permits networks of memberships that may be difficult to draw, e.g. "print" could be placed in the scope of "A1" in addition to "main", without being in the scope of "c1").

Systems have some form of "shape" (i.e., distinguishing properties and attributes and may encompass anything from morphology to spatial position) that determines which other systems they can interact with, and the nature of that interaction. The "shape" of a contextual system affects the result of the interaction between systems in its context. This encompasses the general concept that a resultant transformation of two interacting systems is dependent on the context in which that interaction takes place. A different context will produce a different transformation. Since everything in systemic computation is a system, context must be defined by a system.

In order to represent these notions computationally, the notions of schemata and transformation functions are used. The "shape" of a system in this model is the combination of schemata and function, so specific regions of that "shape" determine the meaning and effect of the system when behaving as a context or interacting. Thus, each system comprises three elements: two schemata that define the possible systems that may interact in the context of the current system, and the transformation function, which defines how the two interacting systems will be transformed.

Systemic computation also exploits the concept of *scope*. In all interacting systems in the natural world, interactions have a limited range or scope, beyond which two systems can no longer interact (for example, binding forces of atoms, chemical gradients of proteins, physical distance between physically interacting individuals). In cellular automata this is defined by a fixed number of neighbours for each cell. Here, the idea is made more flexible and realistic by enabling the scope of interactions to be defined and altered by another system. Interactions between two systems may result in one system being placed within the scope of another (akin to the *pino* membrane computing operation [15]), or being removed from the scope of another (akin to the *exo* membrane computing operation [15]). So just as two systems interact according to (in the context of) a third system, so their ability to interact is defined by the scope they are all in (defined by a fourth system). Scope is designed to be infinitely recursive so systems may contain systems containing systems and so on. Scopes may overlap or have fuzzy boundaries; any systems can be wholly or partially contained within the scopes of any other systems. Scope also makes this form of computation tractable in simulation by reducing the number of interactions possible between systems to those in the same scope.

Most systemic computation forms a complex interwoven structure made from systems that affect and are affected by their environment. This resembles a molecule, a cell, an organism or a population, depending on the level of abstraction used in the program. Fig. 1 illustrates the organisation of a real systemic computation, showing the use of structure enabled by scopes. Systems can be implemented using representations similar to those used in genetic algorithms and cellular automata. In implementations to date, each system comprises three binary strings: two schemata that define sub-patterns of the two matching systems and one coded pointer to a transformation function. Two systems that match the schemata have their own binary strings transformed according to the appropriate transformation function. A simple example of a (partially interpreted) system string (where S1 is the first schema and S2 is the second schema of a system) might be:

"0??????? [S1$_1$=SUM(S1$_1$,S1$_2$); S2$_1$=SUM(S2$_1$,S2$_2$); S2$_1$=0; S2$_2$=0] 0???????"

meaning: for every two systems with most significant bit of "0" that interact in the context of this system, add their two S1 values, storing the result in S1 of the first system and add the two S2 values, storing the result in S2 of the first system, then set S1 and S2 of the second system to zero. Given a pool of inert data systems, able to interact but with no ability to act as context, for example (where NOP means "no operation"):

"00010111 NOP 01101011" and "00001111 NOP 00010111"

after a sufficient period of interaction, the result will be a single system with its S1 and S2 values equal to the sum of all S1 and S2 values of all data systems, with all other data systems having S1 and S2 values of zero. (The program performing exactly this operation was described in [2].)

Systems within the same scope are currently presented to each other randomly just as most interactions in the natural world have a stochastic element. The computation proceeds asynchronously and in parallel, distributed amongst all the separate systems, structurally coupled to its environment, with parallelism and embodiment providing the same kind of speedup seen in biological systems. Computation is continuous (and open-ended) with homeostasis of different systems maintaining the program.

Work to date has resulted in the creation of two systemic computation compilers (one for PCs, one for Macs), a language and visualiser. Several systemic programs have been written, showing that simulations of this parallel computer can perform tasks from investigations of neurogenesis to a self-adaptive genetic algorithm solving a travelling salesman problem. Work on the language and refinements to systemic computation and its use for modelling are underway.

3 Illustration of Systemic Computation

In [3], a genetic algorithm was implemented in order to demonstrate how a systemic computer can be programmed to evolve solutions to the travelling salesman problem in a very simple way. It was further demonstrated that a very simple extension to the systemic program converted the standard genetic algorithm into a self-adaptive version able to modify its own genetic operators.

Here we provide a summary of the systemic program as illustration. A genetic algorithm uses a population of solutions (in this case, for the TSP) which are evaluated and then the better solutions are used as parents to generate a new population of solutions that inherit features from both parents. Thus in a GA, solutions interact in two ways: they compete for selection as parents, and once chosen as parents, pairs produce new offspring. The use of contextual systems (which determine the effects of solution interaction) for the genetic operations is thus highly appropriate, as shown in Fig 2. These perform selection and reproduction, enabling a pair of solutions to reproduce, replacing the worst parent with a better offspring. If alternate methods of reproduction are to be investigated, more genetic operator systems are simply added into the computation space with the evolving solutions. To enable self-adapting operators, new 'genetic operator adapter' systems are added (fig 3), which act as context for interactions between operators, enabling the program to adjust and adapt the type and number of genetic operators being applied to a population, based on the success rate of the operators.

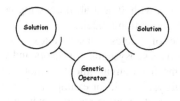

Fig. 2. An operator acts as a context for two interacting solutions

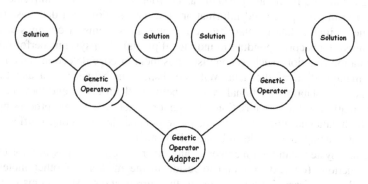

Fig. 3. A self-adaptive approach to the TSP using a GA on a SC architecture: a genetic operator adapter is added to the current interaction scheme to adapt genetic operators during computation

4 Wireless Systemic Computation

From the illustration in the previous section it should be evident that all functional and data parts of a systemic computer can be adapted (evolved), with the built-in redundancy of multiple systems spread over multiple processors. If a sufficiently flexible adaptive systemic computer is trained in an environment where a rich amount of information is available via sensors and where the different resources of the computer are likely to be temporarily or permanently disabled, then the computer will adapt its own organisation and processing to maintain functionality despite the damage. In other words, given an entire computer that can evolve itself like a genetic algorithm, and given a fitness function (feedback providing a measure of how well the computer is currently performing), the computer will adjust itself, exploiting any useful sensor data that may correlate with forthcoming damage, to maximise its performance. (Although a program or systemic structure could conceivably be evolved from scratch using this approach, here we focus on the creation of predefined computation capable of learning to exploit all useful information and resources available in order to maintain its functionality.) Fault tolerant evolved programs have been demonstrated in previous work using a similar evolutionary method [14], but these were severely limited by the brittle nature of the underlying operating system and single processor (only 0.05% damage to the binary executable was survivable 10% of the time by the best evolved method, compared to total failure for any damage to ordinary code) [14]. In the computer architecture introduced here, there would be a minimal conventional operating system with almost all processing performed over many processors using the systemic computation model.

A typical systemic model of wireless sensor motes in an environment would treat each mote as a system that uses subsystems of sensors and wireless communicators to transfer data sub-subsystems from environment to mote and from mote to mote. A single mote would have subsystems of, for example, camera, wireless communicator, internal processing systems, and adaptive systems, which adapt the internal processing systems. For example, consider the architecture for a systemic computer collating visual data gathered from multiple cameras in order to perform basic movement tracking and transmit the processed information to a base station. If a data system holds one frame of image data (containing overlapping subsystems of columns and rows, containing systems corresponding to individual pixels), then systems performing data transformation functions such as contrast adjustment or edge detection can enable the parallel manipulation of the data. With no distinction between data and function, motes may pass data or functional systems between themselves, and they may transform or rearrange the organisation of data or functions during processing. This enables both data and functionality to be moved to and shared amongst different processors, reorganising in a simple and 'natural' manner.

Since any system can be a member of any other system, it is possible for transformation functions (context systems) to reside in the memory of other motes while affecting the transformation of systems in the current mote. All systems can potentially be moved between motes, and memberships of systems within systems can be altered at any time. Linking these changes to systems corresponding to sensor inputs (such as low battery warning, high temperature, high vibration, excessive noise), and evolving the self-adaptive systems, the configuration of the program and multi-processor architecture could automatically adjust itself to maximise fitness (by minimising disruption to the desired end result).

4.1 Hardware Implementation Overview

The underlying motivations of this work are to produce a practical architecture suitable for use as a real computer (not a theoretical analysis akin to bigraphs or brane calculus, despite the surface similarities). Four implementation-specific features enable systemic computers to be tailored to a given application:

(i)   the word-length / coding method,
(ii)   the transformation function set / schemata matching method
(iii) the order of system interactions and
(iv)   the scope definition method.

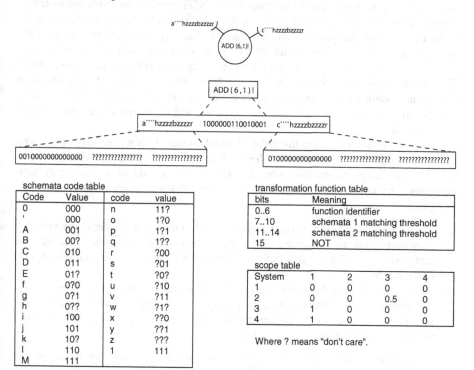

schemata code table

Code	Value	code	value
0	000	n	11?
'	000	o	1?0
A	001	p	1?1
B	00?	q	1??
C	010	r	?00
D	011	s	?01
E	01?	t	?0?
f	0?0	u	?10
g	0?1	v	?11
h	0??	w	?1?
i	100	x	??0
j	101	y	??1
k	10?	z	???
l	110	1	111
M	111		

transformation function table

bits	Meaning
0..6	function identifier
7..10	schemata 1 matching threshold
11..14	schemata 2 matching threshold
15	NOT

scope table

System	1	2	3	4
1	0	0	0	0
2	0	0	0.5	0
3	1	0	0	0
4	1	0	0	0

Where ? means "don't care".

Fig. 4. Graphical representation of one system with function "ADD" (top). Method of coding used in systems to enable two 16-character schemata to define 48-character systems to be matched, and to enable one 16-character transformation function to define function, matching thresholds and NOT operator (middle). Schemata code table (bottom left) defines the codes used in schemata, the transformation function table (middle right) gives the meaning of the 16-bit number. A scope table is given (bottom right), indicating non-fuzzy scopes where systems 3 and 4 are completely within system 1, and fuzzy scope where system 2 is half within system 3.

In the implementation described here, (i) schemata and transformation functions are defined by strings of 16 characters of alphabet 29, resulting in each system being 48 characters long;[1] (ii) some thirty transformation functions have been implemented, using partial matching against thresholds; (iii) system interactions occur randomly except where a system is changed, in which case changed systems are chosen for subsequent interaction first; (iv) scopes are held globally in a system scope table.

---

[1] The word-length is user-definable; here 16 characters was chosen.

In more detail: integers are coded using pure binary coding. A wildcard (defined by any character not a 1 or 0) enables the schemata of systems to define partial matches. Triplets of system string characters are coded using ASCII values 96 to 122 (characters: ` to z), fig 4. In this way a 16-character schemata string is used to define a 48-character system, enabling each system to define the two (types of) systems it could transform with complete precision. Two matching threshold values enable the accuracy of the required match to be defined, with a NOT operator inverting the matching requirements (i.e., systems have to match the schemata correctly for the function not to be carried out). Scopes may implemented globally (per device) in a scope table with column entries defining which of the systems are contained within the systems listed in the rows, fig 4 (other scope representations using pointers are also possible). The numeric value of each entry defines the degree of membership of each system in the parent, enabling partial or fuzzy membership. Hamming distance is used to calculate the difference between schemata and systems, and compared against the function threshold values.

The procedure determining the order in which system interactions takes place in a systemic computer is equivalent to the fetch-execute cycle of a conventional computer. In this design, interaction order is random (context and two interacting systems determined by pseudo-random number generator seeded by current time), with two exceptions:

(i)   when picking the parent system which contains the currently interacting systems, only scope systems containing 3 or more systems are chosen; when picking context systems, only those that define a function not equivalent to NOP (no operation) are chosen.

(ii)   any system changed by a function is added to a queue; if the queue is not empty, the system at the front is picked with a higher probability as one of the systems which may interact in the current cycle, with context and second system being picked from within the same parent system.

The content-sensitive selection improves efficiency, ensuring that most context systems will perform some transformation to the two interacting systems. The use of a queue provides speedup when the same system is used more than once in a computation, ensuring that changed systems are more likely to engage in subsequent interactions, enabling a cascade or diffusion effect through all other computations that may be required to the same system. Systems are held as a fixed numbered array in memory; as systems are never lost, this array structure and order is never modified. The use of a scope table enables systems to be moved between scopes using negligible computation time and zero change to system ordering in memory. (The implementation would be modified for a hardware-level implementation: systems could be held as non-addressable shifting memory chains passing through multiple schema matching stations; systems could also hold their scope information locally in much the same way that schemata are stored.)

4.2 Hardware Choices

While wireless microsensors of just a few millimetres in size have now been demonstrated, today some of the smallest commercially available "motes" are the

MICA2DOTs, at 25mm in diameter. These devices have highly limited memory (64K Flash memory), slow 8 bit processors (16Mhz) and slow communications. Systemic computing is designed to operate with such limited hardware, but the lack of SDRAM severely limits the lifetime and speed of repeated memory storage. So in order to achieve a more practical computing performance, in this work we suggest that hardware such as the slightly larger Imote2 would be a viable choice. The Imote2 is a high-performance wireless node, with an onboard Intel PXA271 XScale processor running at up to 416Mhz, 32Mb SDRAM and a USB interface. It is anticipated that 10 or more Imote2 motes working together would compare favourably with modern 3Ghz dual-core PCs in terms of processor speed, while future motes are likely to retain this performance, but at millimetre scales. (Alternative solutions include the "Atific Helicopter" Multi-Radio Wireless Sensor Network Development Kit, which provides the ability to implement a hardware parallel architecture. This device has four parallel independent digital radio modules, Altera Cyclone EP1C20F324C6 FPGA and preinstalled 8051 MCU reference design for a jump start.) Mounted in a single desktop-sized housing with outer shielding to minimise impact on wireless networks around it, the small size of the devices would allow a compact, fully connected, multi-processor computer to be achievable today.

As described above, systemic computation relies on the notion of systems (stored as binary strings) interacting with each other. The current model uses two schemata per system, each 16 characters in length, with alphabet 29. Each of the 29 possible character values is a compressed definition of one triplet of characters, each with an alphabet of 3 (0,1, or wildcard). In this way 16 characters defines a schema of 48 characters in length, and this schema (of alphabet size 3) can match a 48-character system (that has an alphabet size of 29). In hardware this means that four bits are needed for each character of the schemata. To enable a reasonably rich instruction set and enable the internal storage of several parameter values, four bits can also be used for each character of the kernel (middle functional part). So 192 bits are needed for each system. The memory of each mote will be filled with these 192-bit systems. However, since everything in systemic computation is a system, so the very motes themselves are regarded as systems. Thus each mote will contain many interacting systems, motes themselves may interact as systems, and motes may behave as scopes or contexts for other motes or systems.

TinyOS, the freely available operating system developed by Berkeley, enables a low-overhead wireless network to be maintained on the Imote2 devices (for example using Direct Diffusion, a common data-centric transmission protocol). To enable the operation of the systemic computation architecture, it must be implemented by an extension to the TinyOS operating system. This involves the following operations: (i) Maintenance of scope tables (which system is within the scope of which other system); (ii) interaction protocol (which systems interact with each other within the current mote); (iii) system transmission (which systems are expelled from the scope of the current system, or mote, and which are absorbed into another system, or mote); (iv) transformation function set (how two systems are transformed in a given context, e.g. a low-level transformation such as binary addition, or a higher-level transformation such as a mathematical operation over several parameters).

It is anticipated that the proposed asynchronous and parallel communications may cause interference and packet loss making transmission of systems unreliable.

Protocols to ensure that a transmitted system has been successfully received and stored by one and only one mote will be necessary. For example, the mote receiving a system and "absorbing" it must do so in one "system exchange" with the sender of the system, making that system unavailable to all other motes. Following the principles of systemic computation, this equates to two (wireless mote) systems interacting and modifying each other through the exchange of an internal system (data). Since the entire available memory of motes will always be full of systems (whether they are actively used for processing or not), the gain of a new system must involve the loss of an existing one and vice versa, so all transmissions between motes must be mutual exchanges of systems.

5 Summary

Natural computation points the way to many attractive properties in computing. Fault tolerance, self-organisation, embodied and distributed computers, that operate according to natural principles may well be the future of reliable information processing. In this work the notion of systemic computation was introduced, and a summary of the architecture for a "biological computer" using wireless sensor network hardware was provided. While still at the design stage, this form of "natural" distributed wireless computation already shows promise in simulation, with adaptability, fault tolerance and autonomous behaviour emerging naturally from the novel architecture [2][3].

References

1. Bentley, P.J.: Digital Biology. How nature is transforming our technology and our lives. Simon & Schuster, USA Hardback (2002) ISBN: 074320
2. Bentley, P.J.: Systemic Computation: A Model of Interacting Systems with Natural Characteristics. In: Adamatzky, A., Teuscher, C., Asai, T. (eds.) Special issue on Emergent Computation in Int. J. Parallel, Emergent and Distributed Systems (IJPEDS), vol. 22(2), pp. 103–121. Taylor & Francis pub., Oxon (2007)
3. Le Martelot, E., Bentley, P.J., Lotto, R.B.: A Systemic Computation Platform for the Modelling and Analysis of Processes with Natural Characteristics. In: Proc. of GECCO 2007 Workshop: Evolution of Natural and Artificial Systems - Metaphors and Analogies in Single and Multi-Objective Problems (2007)
4. Haroun Mahdavi, S., Bentley, P.J.: Innately adaptive robotics through embodied evolution (extended version). Journal of Adaptive Robotics (2005)
5. Kim, J., Bentley, P.J., Wallenta, C., Ahmed, M., Hailes, S.: Danger is Ubiquitous: Detecting Misbehaving Nodes in Sensor Networks using the Dendritic Cell Algorithm. In: Bersini, H., Carneiro, J. (eds.) ICARIS 2006. LNCS, vol. 4163, pp. 390–403. Springer, Heidelberg (2006)
6. Musolesi, M., Mascolo, C., Hailes, S.: EMMA: Epidemic Messaging Middleware for Ad Hoc Networks. Personal and Ubiquitous Computing 10(1), 28–36 (2006)
7. Greenhalgh, A., Hailes, S.: A summary of complex behaviour from simple power conserving protocols. In: 16th Annual IEEE Symposium on Personal Indoor and Mobile Radio Communications. IEEE, Germany (2005)

8. Hailes, S., Pagtzis, T., Chakravorty, R., Crowcroft, J., Kirstein, P.: Proactive Mobile IPv6 for Context-aware all-IP Wireless Access Networks. In: IEEE International Conference on Wireless Networks, Communications, and Mobile Computing, WirelessCom (2005)

9. Musolesi, M., Hailes, S., Mascolo, C.: Adaptive routing for intermittently connected mobile ad hoc networks. In: Proceedings of the IEEE International Syposium on a World of Wireless, Mobile and Multimedia Networks, pp. 183–189. IEEE Computer Society Press, Los Alamitos (2005)

10. Ahmed, M., Hailes, S.: Modelling Interactions in Ubiquitous Environments. In: Proc. 2nd UK-Ubinet Workshop (2004)

11. Adamatzky, A.: Computing in Nonlinear Media and Automata Collectives, p. 410. IoP Publishing, Bristol (2001)

12. Adamatzky, A. (ed.): Collision-Based Computing, vol. XXVII, p. 556. Springer, Heidelberg (2002)

13. Arvind, D.K., Wong, K.J.: Speckled Computing: Disruptive Technology for Networked Information Appliances. In: Proceedings of the IEEE International Symposium on Consumer Electronics (ISCE 2004), UK, pp. 219–223 (September 2004)

14. Bentley, P.J.: Investigations into Graceful Degradation of Evolutionary Developmental Software. Journal of Natural Computing 4, 417–437 (2005)

15. Cardelli, L.: Brane calculi. interactions of biological membranes. In: Danos, V., Schachter, V. (eds.) CMSB 2004. LNCS (LNBI), vol. 3082, pp. 257–280. Springer, Heidelberg (2005)

16. Eriksson, D.: A Principal Exposition of Jean-Louis Le Moigne's Systemic Theory. Review Cybernetics and Human Knowing 4, 2–3 (1997)

17. Milner, R.: The Polyadic pi-Calculus: a Tutorial. In: Hamer, F.L., Brauer, W., Schwichtenberg, H. (eds.) Logic and Algebra of Specification. Springer, Heidelberg (1993)

18. Milner, R.: Axioms for biographical structure. Journal of Mathematical Structures in Computer Science 15, 1005–1032 (2005)

19. Stepney, S., Braunstein, S., Clark, J., Tyrrell, A., Adamatzky, A., Smith, R., Addis, T., Johnson, C., Timmis, J., Welch, P., Milner, R., Partridge, D.: Journeys in Non-Classical Computation I: A Grand Challange. Int. Jnl. of Parallel, Emergent and Distributed Systems 20(1), 5–19 (2005a)

20. Stepney, S., Braunstein, S., Clark, J., Tyrrell, A., Adamatzky, A., Smith, R., Addis, T., Johnson, C., Timmis, J., Welch, P., Milner, R., Partridge, D.: Journeys in Non-Classical Computation II: Initial Journeys and Waypoints. Int. Journal of Parallel, Emergent and Distributed Systems (2005b)

21. Tyrrell, A.M., Sanchez, E., Floreano, D., Tempesti, G., Mange, D., Moreno, J.-M., Rosenberg, J., Villa, A.E.P.: POEtic Tissue: An Integrated Architecture for Bio-inspired Hardware. In: Tyrrell, A.M., Haddow, P.C., Torresen, J. (eds.) ICES 2003. LNCS, vol. 2606, pp. 129–140. Springer, Heidelberg (2003)

22. Varela, F.J., Maturana, H.R., Uribe, R.: Autopoiesis: The organization of living systems, its characterization and a model. BioSystems 5, 187–196 (1974)

A Rule System for Network-Centric Operation in Massively Distributed Systems

Falko Dressler and Reinhard German

Computer Networks and Communication Systems,
University of Erlangen-Nuremberg, Germany
{dressler,german}@informatik.uni-erlangen.de

Abstract. Sensor and Actor Networks (SANETs) represent a specific class of massively distributed systems in which classical communication protocols often fail due to scalability problems. New control paradigms are needed in this place. This paper outlines biological communication techniques as known cellular biology, which are known as cellular signaling pathways. We show the adaptation of these principles to the world of SANETs by discussing a rule-based control system for network-centric communication and data processing. This system is able to perform data pre-processing such as data aggregation or fusion as well as data-centric communication based on rules that are distributed throughout the entire network. First simulation results demonstrate that this system is able to outperform classical routing approaches in specific SANET scenarios.

Keywords: sensor and actor networks, self-organization, bio-inspired networking, rules-based sensor network, cellular signaling.

1 Introduction

Recent advances in microelectronics enabled the development of even smaller and cheaper devices that are primarily used in the domain of Wireless Sensor Networks (WSNs). At the beginning of this research, the envisioned scenario has been smart dust [1], i.e. the deployment of millions of tiny sensor nodes that cooperate on monitoring a given area. Whereas this scenario has not yet become reality, a multitude of algorithms for operation of massively distributed sensor systems have been developed. Based on the research the need for network-centric data preprocessing has been identified as a key challenge due to the observation that communication is much more expensive in terms of energy requirements compared to local processing. Similarly, Sensor and Actor Networks (SANETs) introduced further challenges and requirements. SANETs represent a specific class of sensor networks enriched with network-inherent actuation facilities [2]. In addition to the requirements known from sensor networks, actuation devices, usually named actors [3], are included into the scenario. This requires real-time operation in massively distributed systems and coordination capabilities on a higher abstraction layer.

In this paper, we present a system for network-centric operation in WSNs and SANETs that we named Rule-based Sensor Network (RSN). This system

P. Liò et al. (Eds.): BIOWIRE 2007, LNCS 5151, pp. 364–375, 2008.
© Springer-Verlag Berlin Heidelberg 2008

is the result from studies in the context of *bio-inspired networking* – precisely, in the context of cellular signaling cascades. In the following, we summarize the requirements in the domain of massively distributed systems and introduce the biological background. Then, we present RSN in detail and conclude with two first application scenarios.

1.1 Requirements in Massively Distributed Systems

Whereas other application domains exist, we concentrate on WSNs and SANETs. Starting with the first domain, we can identify *scalability* and *energy efficiency* as the most challenging characteristics. Self-organizing algorithms have been developed relying for example on clustering and aggregation techniques to improve scalability and network lifetime [4]. In SANETs, *coordination* aspects need to be solved for sensor-actor coordination as well as for actor-actor coordination [3]. This includes additional communication constraints for network-wide coordination or, at least, local decision taking strategies that lead to an emergent behavior on a higher abstraction layer. Additionally, *real-time constraints* need to be considered as demanded by feedback control in sensor-actor coordination. Some of these challenging requirements are addressed by RSN. This approach basically provides the building blocks for developing network-centric operation and control techniques needed in massively distributed systems such as SANETs.

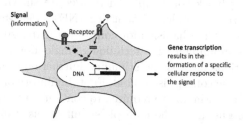

Fig. 1. Cellular signaling refers to the specific reaction according to received signaling molecules; shown is the multilevel transcription of a received protein

1.2 Biologically Inspired Operation

In the last few years, bio-inspired networking has become a new trend for addressing yet unsolved problems by adapting solutions known in nature [5]. Whereas a broad range of techniques and methods have been studied (e.g., the artificial immune system, swarm intelligence, and evolutionary algorithms), we focus in this paper on a rather new domain, the adaptation of communication and coordination techniques from cellular signaling. Figure 1 sketches the principles of cellular information exchange. Information particles, e.g. proteins, are received by a cell according to the specific binding to a locally expressed receptor (or even a set of receptors) [6]. The activation of the receptor initiates a signaling cascade in which new proteins are created or activated and, finally, a cellular response can be observed, which represents the *specific reaction* of the cell according to the received information. Thus, cellular processes are regulated by interactions between various types of molecules, e.g. proteins. A key challenge for biology in the 21st century is to understand the structure and the dynamics of the complex intercellular web of interactions that contribute to the structure and function of a living cell [7]. To uncover these structural design principles, *network motifs* have been defined as patterns of interconnections occurring in complex networks at numbers that are significantly higher than those in randomized networks [8].

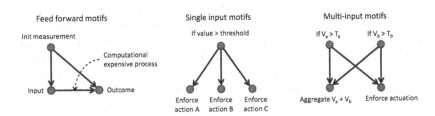

Fig. 2. Typical network motifs in integrated cellular networks

The concept of network motifs is depicted in Figure 2. Please note that this is only a small sample of network motifs in integrated cellular networks [8]. The three basic building blocks of complex networks are shown in this figure together with application examples relevant in SANETs. Feed-forward motifs represent network-inherent mechanisms for controlling (expensive) processes. This can also be seen as an amplification technique. Single-input motifs allow to initiate multiple reactions on a single stimulus. Furthermore, multi-input motifs are depicted. The basic concept is twofold. First, inhibitory or controlling effects can be achieved as two stimuli are required to continue in the signaling cascade. Secondly, if the threshold, i.e. the multiple simultaneous stimuli, is exceeded, a number of parallel actions can be initiated at once.

2   RSN – Rule-Based Sensor Network

Inspired by the capabilities of cellular signaling, i.e. the specific reaction to received information and the possibility to build signaling networks defining complex reaction pattern, we developed a rule-based programming system for application in SANETs. The primary design goals were a small footprint to enable the application of RSN on small embedded systems, easily transferable code, flexibility, and scalability for network-wide operations (basically, RSN provides the tools and concepts but the specific application needs to be designed properly as well). The rule-system greatly helps in designing distributed algorithms for use in self-organizing massively distributed systems. Additionally, RSN was inspired by early rule-based systems that have been developed in the context of active networking solutions [9]. Examples are the mobile object system [10] and communicating rules [11].

2.1 Basic Concept

The key objectives motivating the development of RSN were improved scalability and real-time support for operation in SANETs. RSN is based on the following three design objectives:

- *Data-centric communication* – Each message carries all necessary information to allow data specific handling and processing without further knowledge, e.g. about the network topology.
- *Specific reaction on received data* – A rule-based programming scheme is used to describe specific actions to be taken after the reception of particular information fragments.
- *Simple local behavior control* – We do not intend to control the overall system but focus on the operation of the individual node instead. Simple state machines have been designed, which control each node (being either sensor or actor).

These goals are achieved by using a simple rule system that enables the node to process received messages and to initiate adequate state and message specific operations. Thus, all received messages are stored in a buffer (source set). Periodically, after a configurable timeout Δt, all messages in the source set are processed by the instructions defined by the rules. Every rule has the form `if CONDITION then { ACTION }` as depicted in Figure 3. Each rule specifically selects messages from the source set to apply the corresponding action. Details about the actions and further RSN parameters are described in the following.

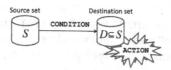

Fig. 3. Each rule selects a number of messages form the source set (`CONDITION`) and applies a (set of) actions to the selected messages (`ACTION`)

2.2 Available Actions

The following actions have been implemented in the current version of RSN. Basically, the following categories of actions can be distinguished: *rule execution,* i.e. operations on the received messages; *node control,* i.e. control of the local node behavior (e.g., addition of sensors); and *simulation control,* i.e. actions needed for experiment control without influence on the node behavior.

Rule Execution. The following actions are meant to be used for network-centric processing of messages. All these actions work on the source message set that has been created by the condition element, i.e. by selecting messages according to a well-defined specific pattern. Examples for the application of the described actions are provided in the next section.

- !stop – Early termination of the rule execution. Depending on the current state (i.e., the number and kind of received messages), it may be necessary to stop the current processing of the rule set. The next iteration will start with the first available rule.
- !drop – Erases all messages in the current set. Needs to be called if messages have been successfully processed.
- !dropDuplicates – All duplicates are discarded according to a unique identifier in each message. This command is needed to emulate for example standard gossiping algorithms.
- !return – A new message is created and appended to the source message set.
- !returnAll – Copies of all messages in the current set are created and stored in the source message set.
- !send – A new message is created and submitted to the lower layer protocol for transmission to neighboring nodes.
- !sendAll – Copies of all messages in the current set are created and submitted to the lower layer protocol for transmission to neighboring nodes.
- !actuate – A message is sent to locally connected actuators.

Node Control. Besides the actions for message processing, actions have been integrated to control the local node behavior. Such node control actions allow to enable/disable locally attached sensors and actuators as well as to modify the current rule set, i.e. the local programming of a node.

- !controlSensor – A control message is sent to all attached sensors. According to the submitted attributes in $control, the behavior of the sensors can be controlled: rsnSensorEnable and rsnSensorDisable enable or disable the sensor, rsnSensorSetType updates the type field of the sensor, and rsnSensorSetMeasuringInterval changes the sampling frequency.
- !controlActuator – Similarly, this command controls locally attached actuators. The attribute $control defines the action: the actuator is enabled or disabled by rsnActuatorEnable and rsnActuatorDisable, respectively, and rsnActuatorSetType updates the type field of the actuator.
- !controlManagement – The management plane defines the rule set itself. Again, the $control attribute is used to specify the intended action: the rule interpretation can be started or stopped by rsnManagementEnable and rsnManagementDisable, respectively, the rule set can be replaced in order to modify the behavior of this node using rsnManagementFromRsnString or rsnManagementFromRsnFile, and the evaluation interval can be configured by rsnManagementSetEvaluationInterval.

Simulation Control. The following actions have been integrated for simplified control of simulation experiments. These actions are not working on a given set of messages. Nevertheless, it is possible to initiate these actions based on the current state of the node, e.g. after the reception of a specific message.

- !recordAll – Statistics are recorded for all messages in the current working set. In particular, the following information are stored: ID of the current node, ID of the node that generated the message, node specific ID of the message, globally unique ID of a message, hop count, current time, and delay (elapsed time since message creation).
- !endSimulation – This action terminates an experiment. In our OMNeT++ based implementation, the simulation core is notified accordingly.

2.3   Variables and Variable Handling

All the described conditions and actions work on a set of message parameters or local variables describing the state of the node. In the following, some of the most important variables are introduced. Additionally, selected statistical preprocessing techniques for data aggregation have been integrated into the current version of RSN in order to enable selected application examples. In the following section, we describe and analyze two application examples that inherently benefit from the network-centric preprocessing features provided by RSN.

Message Attributes. Each message is specifically encoded to allow receiving nodes to determine the meaning of the message and the necessary behavior. This encoding can be changed according to the application scenario. Possible parameters (currently used in the RSN implementation) are listed in Table 1.

Table 1. Currently implemented message attributes

Attribute	Description
$name	Descriptive name of the message
$type	Type of the message; describes the content
$position	Position of the source node
$hopCount	Number of traversed nodes
$priority	Importance factor of this message
$length	Length of the message
$creationTime	Timestamp describing the creation of the message
$value	Message type specific value
$text	Further informative text, e.g. to qualify the value

Node Attributes. Each node can store and update state information locally. In the context of self-organization, this refers to the local state of an autonomous system. Such information can be updated according to received messages or by other local observations. Table 2 lists the currently implemented node attributes.

Preprocessing Features. Data aggregation is an important issue in massively distributed systems. Usually, statistical measures are used to describe results received from several nearby nodes. RSN supports such data aggregation techniques by providing a set of preprocessing techniques as summarized in Table 3. All the listed operations process the messages in the current working set.

Table 2. Currently implemented node attributes

Attribute	Description
:count	Number of messages in the current working set
:totalMessageCount	Number of all messages received by the node
:hostName	ID of the current host
:position	Position of the node
:random	Random value for probabilistic decisions

Table 3. Implemented preprocessing features

Command	Description
@minimum	Minimum of the selected value
@maximum	Maximum of the selected value
@sum	Sum of the selected value
@average	Average of the selected value
@median	Median of the selected value
@count	Number of the selected value

2.4   Implementation

We implemented RSN in form of a C++ library. This library contains all functionality that is necessary to process RSN statements. RSN statements are formulated in a flexible script language. We integrated the RSN library into the OMNeT++ simulation framework in order to execute intensive tests and experiments with different algorithms for data aggregation, probabilistic data communication, and distributed actuation control. OMNeT++ 3.3 is a discrete event simulation environment free for non-commercial use. We also used the INET Framework 20060330, a set of simulation modules released under the GPL. Scenarios in OMNeT++ are represented by a hierarchy of reusable modules written in C++. Their relationships and communication links are stored as Network Description (NED) files. Simulations are either run interactively in a graphical environment or executed as command-line applications.

The developed simulation model is depicted in Figure 4 (left). A single node is depicted consisting of a number of modules. Bottom-up, a wireless communication module is included (WLAN) as well as the rsnRouting module that is currently represented by a simple broadcast module (routing issues can be handled by RSN). The rsnManagement contains all core functions of RSN, i.e. message handling and rule processing. Finally, the rsnDispatcher module interconnects attached rsnSensor and rsnActuator modules with the rsnManagement.

3   Applicability of RSN

We evaluated the applicability of RSN in two scenarios. First, we explored network-centric data aggregation as an option to improve the efficiency of probabilistic data communication (gossiping). Secondly, we investigated the capabilities of network-centric actuation control in SANETs in terms of scalability and real-time behavior.

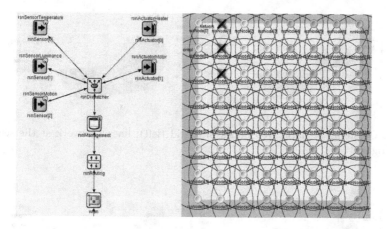

Fig. 4. Simulation model of RSN: integration in OMNeT++ and example setup

3.1   Data Aggregation Scenario

Aggregation as a major building block for efficient and scalable data communication and preprocessing in WSNs and SANETs because communication is much more expensive (in terms of energy consumption) compared to processing. In particular, we investigated a typical probabilistic communication approach using RSN: gossiping [12]. The principles are shown in the following RSN program. If a message travels further than DIAMETER, it will be discarded. In the first 4 hops, the message is flooded. In all other cases, the message is forwarded according to a random experiment.

```
if $hopCount >= DIAMETER then {
    !drop;
}
if ANY ($hopCount < 4 || :random > GOSSIP-PROB) then {
    !sendAll;
}
!drop;
```

We prepared two scenarios as discussed in [12]. In the first scenario, 100 nodes are distributed on a grid – one corner node is generating 300 messages to be transmitted by the network. Thus, the probability to duplicate messages is very high (according to the initial flooding for the first four hops). In the second scenario, three nodes are turned off in order to build a small linear network at the sending corner node as depicted in Figure 4 (right). Figure 5 shows the number of messages forwarded by a host. Obviously, the nodes close to the source are getting overloaded while distant nodes receive the messages with a pretty low probability.

Then, we installed a simple aggregation rule that controls the transmission of a single message if multiple messages have been received. The results are depicted in Figure 6. This time, much less duplicates are transmitted and, thus, the load

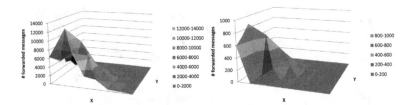

Fig. 5. Gossiping scenario: 100 nodes in a grid (left); linear network at the sending node (right)

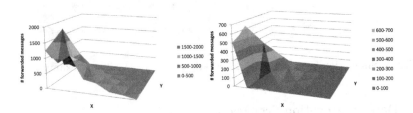

Fig. 6. Gossiping scenario with data aggregation: 100 nodes in a grid (left); linear network at the sending node (right)

of the network is reduced. At the same time, the dispersion of the messages in the network is increased. Unfortunately, the gossiping algorithm drops all messages (including aggregated ones) with the same probability. Thus, additional rules need to be implemented that correct this behavior, e.g. by defining priorities for aggregated and non-aggregated messages.

```
if :count > 1 then {
    !send($hopCount := @minimum of $hopCount,
        $value := @average of $value);
    !drop;
}
```

3.2   Network-Centric Actuation

In the following, we present an excerpt from extensive simulations to study network-centric actuation using RSN published in [13]. We compared network-centric actuation control with a classical base station scenario. For the latter one, we used the Dynamic MANET on Demand (DYMO) routing protocol to transmit messages from sensor nodes to a base station and the results back to one out of four available actors. In the RSN scenario, the sensor nodes have been configured with the following program – it represents a simple version of gossiping.

```
if $hopCount >= DIAMETER then {
  !drop;
}
if :random <= GOSSIP-PROB then {
  !sendAll;
}
!drop;
```

The actors have an even simpler programming. For each received message, they check whether the THRESHOLD (set to 50, 70, and 90, respectively) has been exceeded and, if necessary, local actuation is initiated.

```
!recordAll;
if $value > THRESHOLD then {
  !actuate($type:=rsnActuatorLightSource,
           $value:=@average of $value);
}
!drop;
```

A number of simulations have been executed with the primary objective to analyze the following characteristics of both evaluated communication and control approaches:

- Real-time support, i.e. the overall latency between measuring a value higher than the particular threshold and the time the message successfully arrived at the actuators. In this context, also the path length is of interest, which is directly proportional to the end-to-end latency and to the message loss probability.
- Overhead, i.e. the number of messages that need to be processed by all the nodes to transmit the necessary data messages. This includes protocol overhead from routing protocols as well as overhead due to duplicated messages for gossiping approaches.

First, the latency of the application messages has been analyzed. We measured the time from creating a sensor message until it was successfully received by the actor. Because only messages exceeding a given threshold are of interest for the actors, we just analyzed the latency after identifying the message as matching this criterion. In Figure 7 (left and middle), results for the RSN scenario are shown in form of boxplots. The graphs differentiate between the deployment scenarios (we evaluated grid and random deployment) and the gossiping probability (set to 0.2, 0.5, and 0.8, respectively). If only the reception of the first copy of the message is considered, the end-to-end delay slightly oscillates around 1.4 ms. The measured maximum is at about 16 ms. The results are nevertheless only meaningful, if all sensor messages can be differentiated, e.g. by a unique id. If this is not possible, the reception of further copies cannot be distinguished form the first one. The measurement results taking this effect into account slightly oscillate around 2.2 ms with a maximum peak at 33 ms.

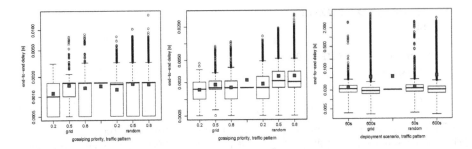

Fig. 7. End-to-end latency. Left (RSN): time until the first copy of a message arrives; middle (RSN) time until any copy arrives; right (base station): end-to-end latency as observed from the application.

If we compare these results to the base station scenario as shown in Figure 7 (right), we obviously see that the delays in this scenario are significantly higher (median: 20 ms, mean: 55 ms, and max: 5.700 ms). There are two reasons for this behavior. First, the mean path length is essentially longer as discussed below and, secondly, the on-demand routing protocol takes some time for setting up the routing path before being able to transmit a message. This effect is shown by the comparison between the 60 s and 600 s message generation setups. The route timeout of DYMO has been configured to 120 s. Thus, in the 600 s scenario, almost always the route towards the base and towards the actor nodes will timeout and needs to be reestablished. Further results and more details are available in [13].

4   Conclusion

In this paper, we investigated techniques for network-centric data processing in WSNs and SANETs. Based on a sketched overview to cellular information processing, we developed RSN, a rule-based system for sensor network programming. The application range of this approach is manifold; we outlined the advantages based on two examples: data aggregation for optimized probabilistic communication and network-centric actuation control.

The main advantages of RSN are the small footprint of rules and the simple local programming of nodes – making self-organization possible even in large scale sensor and actor networks. In particular, this system allows the quick and heterogeneous reprogramming of (individual) nodes. Therefore, network-centric optimization of the placement of computational intensive rules becomes possible – some concepts can be adapted from the database community: the data stream query optimization problem. Our future work in the context of RSN includes further evaluation of aggregation techniques, the implementation on sensor nodes for "real world" experiments, and intensified investigations of reprogramming techniques.

References

1. Kahn, J.M., Katz, R., Pister, K.: Emerging Challenges: Mobile Networking for "Smart Dust". Journal of Communications and Networking 2(3) (September 2000)
2. Akyildiz, I.F., Kasimoglu, I.H.: Wireless Sensor and Actor Networks: Research Challenges. Elsevier Ad-Hoc Networks 2, 351–367 (2004)
3. Melodia, T., Pompili, D., Gungor, V.C., Akyildiz, I.F.: A Distributed Coordination Framework for Wireless Sensor and Actor Networks. In: 6th ACM International Symposium on Mobile Ad-Hoc Networking and Computing (ACM Mobihoc 2005), Urbana-Champaign, Il, USA, pp. 99–110 (May 2005)
4. Dressler, F.: A Study of Self-Organization Mechanisms in Ad Hoc and Sensor Networks. Elsevier Computer Communications 31(13), 3018–3029 (2008)
5. Dressler, F., Carreras, I.: Advances in Biologically Inspired Information Systems - Models, Methods, and Tools. Studies in Computational Intelligence (SCI), vol. 69. Springer, Heidelberg (2007)
6. Alberts, B., Bray, D., Lewis, J., Raff, M., Roberts, K., Watson, J.D.: Molecular Biology of the Cell, 3rd edn. Garland Publishing, Inc. (1994)
7. Barabási, A.L., Oltvai, Z.N.: Network biology: understanding the cell's functional organization. Nature Reviews Genetics 5(2), 101–113 (2004)
8. Milo, R., Shen-Orr, S., Itzkovitz, S., Kashtan, N., Chklovskii, D., Alon, U.: Network Motifs: Simple Building Blocks of Complex Networks. Nature 298, 824–827 (2002)
9. Calvert, K.L., Bhattacharjee, S., Zegura, E.W., Sterbenz, J.: Directions in Active Networks. IEEE Communications Magazine 36(10), 72–78 (1998)
10. Vitek, J., Tschudin, C.: MOS 1996. LNCS, vol. 1222. Springer, Heidelberg (1997)
11. Mackert, L.F., Neumeier-Mackert, I.B.: Communicating Rule Systems. In: 7th IFIP International Conference on Protocol Specification, Testing and Verification, pp. 77–88 (1987)
12. Haas, Z.J., Halpern, J.Y., Li, L.: Gossip-Based Ad-Hoc Routing. In: 21st IEEE Conference on Computer Communications (IEEE INFOCOM 2002), pp. 1707–1716 (June 2002)
13. Dressler, F., Dietrich, I., German, R., Krüger, B.: Efficient Operation in Sensor and Actor Networks Inspired by Cellular Signaling Cascades. In: 1st ICST/ACM International Conference on Autonomic Computing and Communication Systems (Autonomics 2007), Rome, Italy (October 2007)

Field-Based Coordination for Pervasive Computing Applications

Marco Mamei and Franco Zambonelli

Dipartimento di Scienze e Metodi dell'Ingegneria,
University of Modena and Reggio Emilia
Via Allegri 13, 42100 Reggio Emilia, Italy
{mamei.marco,franco.zambonelli}@unimo.it

Abstract. Emerging pervasive computing technologies such as sensor networks and RFID tags can be embedded in our everyday environment to digitally store and elaborate a variety of information. By having application agents access in a dynamic and wireless way such distributed information, it is possible to enforce a notable degree of context-awareness in applications, and increase the capabilities of interacting with the physical world. In particular, biologically inspired field-based data structures such as gradients and pheromones are suitable to represent information in a variety of pervasive computing applications. This paper discusses how both sensor networks and RFID tags can be used to that purpose, outlining the respective advantages and drawbacks of these technologies.

Keywords: Field-based coordination, Ad-hoc networks, RFID infrastructures, Bio-inspired computing.

1 Introduction

Environment-mediated interaction (aka stigmergic interaction [1]) plays an important role in nature. Indeed, the spreading and sensing of pheromones in an environment to organize the activities of ant colonies, the process of morphogenesis as enforced by diffusion of chemicals in the embryo, the movement of masses induced by gravitational fields, are all examples of stigmergic interactions [2]. In the last few years, however, stigmergic models of interactions have been recognized as very powerful to facilitate interactions in dynamic distributed systems. Indeed, stigmergic models of interactions, whether relying on synthetic pheromones, on diffusion of digital chemicals, or on spreading of virtual computational fields, are being proposed to facilitate the enforcement of adaptive interaction patterns in dynamic distributed systems and to promote self-organization and self-adaptation of activities [1,3,4].

In the case of agents situated in a computational environment (e.g., the Web, a P2P network, or the Grid), supporting the interaction of agents with such an environment is a rather natural process. Simply, multiagent systems are computational entities the same as the environment, and once proper data formats

P. Liò et al. (Eds.): BIOWIRE 2007, LNCS 5151, pp. 376–386, 2008.

and interaction protocols are established, the access to the computational environment (and possibly the exploitation of such environment as an infrastructure in which to store the units of stigmergic interactions) becomes rather easy: the "sensors" and the "effectors" that the agents may use to interact reduce to a set of APIs or programming constructs.

The problem is totally different in the case of a physical environment. In this case, to access the physical environment, agents must be somehow be capable of perceiving and affecting physical properties. To this extent, an agent (whether in the form of an autonomous robot, or of an embedded controller, or of some software running on a mobile devices) must be necessarily supported by some hardware sensors and effectors to properly interact with the world.

Traditionally, most approaches for physically situated agents, assume that agents are augmented with the necessary capabilities for sensing and effecting the physical world. For instance, in the case of autonomous robots, traditional approaches assume that the robot itself is equipped with video-cameras, temperature sensors, location sensors (e.g., GPS), and robotic hands. Such approach tends to notably increase the internal complexity of agents. In fact, agents not only have to perform the computational activities associated to deciding how to accomplish a goal, but have also to take care of properly internalizing and interpreting the data coming form the associated sensors, and of properly controlling their effectors to actualize their actions.

Another drawback of the above approach is that the physical environment can hardly be used to support stigmergic models of interactions, unless one adopt rather tricky solutions. If the environment is purely physical, in fact, stigmergic interactions should occur by physically affecting the properties of the environment. For example, to mimic the behavior of ants, robots would be forced to actually pollute the environment with some kind of marker, and would have to be equipped with sensor to perceive such marks [5].

The advent of pervasive computing technologies dramatically changes this scenarios. The availability of small-scale and low-cost devices that can be distributed in physical environment in a non intrusive way, that can be devoted to sense (or affect) specific properties in the environment, and that enable to interact with them in a wireless way (a capability to be easily provided to agents), enables agents to externalize all the activities devoted to interpret and control their physical activities. Simply, sensing and effecting the environment reduces in properly accessing some digital services. The result is in a notable reduction of complexity in agents, both at the hardware and at the software level.

In addition, the presence in an environment of embedded computational resources, as those that can be provided by the embedded computing devices, can be fruitfully exploited as an infrastructure to support stigmergic models of interactions. In fact, stigmergy can take place without actually affecting the physical environment, but simply by exploiting the distributed embedded resources as stores for those data structures that are at the basis of stigmergy, e.g., pheromones, fields, etc.

Clearly, depending on the specific technologies and devices adopted, the interactions with the environment and the support of stigmergic coordination models can be more or less facilitated. In the following of this paper, we analyze in detail two different classes of devices (sensor networks (in Sect. 2) and RFID tags (int Sect. 3), discuss how they can be exploited, and outline their respective advantages and drawbacks.

2   Ad-Hoc and Sensor Network

Future pervasive computing scenarios comprise a huge number of heterogeneous devices interacting with each other to achieve complex distributed applications. Sensor networks and networks of handheld computers could be employed in a variety of applications including environmental monitoring [6,7], navigation [8], and human interaction support [9].

In general terms, sensor networks are an ideal platform to augment the physical environment with digital information.

- Sensors can store data to represent some kind of contextual information. Moreover, they can deliver such data to agents (e.g., users with PDA) passing nearby.
- Sensors can perform computations to support and facilitate the agents' fruition to that data. For example, sensors can propagate and diffuse data across the network. They can automatically delete old and possibly corrupted information. They can combine and transform data to let it become more expressive and easy to use.

Other than providing contextual information coming from the "outside" world, sensor network can also be used to store and convey information produced by the agent themselves. Moreover, relying on the sensor networking capabilities it is possible to spread distributed data structures across the environment.

In particular, we can imagine that each component of the system (software agent, wireless device, embedded sensor, etc.) is capable of generating and propagating field-like data structures that convey some information about their context. Agents can perceive these fields and react accordingly. The idea is that components are simply driven by these force fields as if they were particles under the action of a gravitational field.

Field-based data structures are distributed data structures encoding specific aspects of the application components' operational environment. These fields are propagated across a network by a component in order to represent and "communicate" its own activities. Field data structures are easily accessible by the components and provide easy-to-use context information (i.e., the overlays are specifically conceived to support their access and fruition).

The strength of these overlay data structures is that they can be accessed piecewise as the application components visit different places of the distributed environment. This lets the components to access the right information at the right location. In addition, overlay data structures decouple components' activities from the underlying network dynamism. Components interacting and perceiving their operational environment by means of these overlay data structure can disregard the underlying physical network and its dynamics.

To clarify these concepts let us focus on the problem of coordinating the movements of some autonomous components (i.e., agents) in a distributed environment [10]. Hereafter we will use the term agent to refer to any autonomous real-world or software entity with computing and networking capability (e.g., a user carrying on a Wi-Fi PDA, a robot, or a modern car). In particular, we focus on the simple application of having two persons, provided with a PDA, moving across an environment instrumented with an ad-hoc network infrastructure. The goal of the application is to allow one person to be guided by the PDA, to follow the other person. A simple solution based on overlay data structures is the let the person to-be-followed to spread in the environment (i.e., ad-hoc network) a data structure that increases an integer value by one at every hop as it gets farther from the source. This creates a sort of gradient that can be followed downhill by the other person to complete the application [10] (see Fig. 1(a)). If the person to-be-followed moves, it is important that the overlay data structure adjust its shape accordingly, so that the gradient leads to that person anyway (see Fig. 1(b)). The power of this approach is that the overlay data structure provides expressive contextual information tailored for that specific task. The agent running on the PDA does not need to know any map of the environment, nor it has to execute complex algorithms to decide where to go. It just blindly follows the overlay data structure.

Beside this exemplary application, overlay data structures are general purpose and can be applied in a wide range of application scenarios, ranging from robotics to network routing [10,11].

2.1 Pros and Cons

The power of this approach is that the distributed data structure provides expressive contextual information tailored for that specific task. The agent running on the PDA does not need to know any map of the environment, nor it has to execute complex algorithms to decide where to go. It just blindly follows the field data structure. All the complexity of the application is moved away form from the agents and diverted into the environment-infrastructure.

Sensor networks are a powerful technology to support environment abstractions in multi-agent systems. In the long run, once current technological problems will be properly addressed, it will be the leading infrastructure of environment applications. Its main strength is that it is an *active* infrastructure: sensor nodes can run (distributed) algorithms to process data as required. For example, sensor nodes can proactively delete old information or run algorithm to aggregate data on needs. At present, however, this is also sensor network main weakness. Nodes suffer, in fact, from battery-exhaustion problems, they are costly and failure prone.

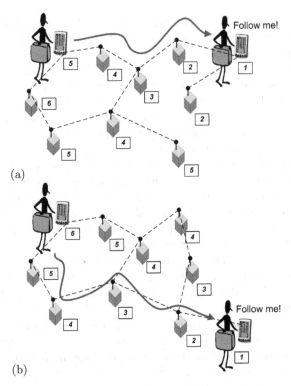

Fig. 1. (a) A gradient data structure enables an agent to follow another one. (b) The data structure is updated to reflect the new agent position.

2.2 Related Work

A number of recent proposals address the problem of defining supporting environments for the development of adaptive, dynamic, context-aware distributed applications, suitable for pervasive computing.

The TinyLime middleware [12] proposes accessing the environmental data collected by a sensor network via an associative tuple-based mechanisms. When a user with a mobile device "walks-through" a network of distributed sensors, all the data collected by the in-range sensors automatically feeds a local tuple space of the mobile device, which thus can perceive sensorial data collected by sensors simply by reading in the local tuple space.

ObjectPlaces [13] is an interesting middleware infrastructure that offers support to exchange and share information among nodes in mobile and ad-hoc networks. Specifically, in ObjectPlaces, agents communicate indirectly through the exchange of objects that can be temporarily stored across suitable object-places (that are virtual containers stored in the ad-hoc network itself). Agents invoke operations to add and remove objects, or to observe the content of a specific object-place (via a pattern-matching process). Agents can also create object-places dynamically, and link them together to form a graph-like environment connecting related object-places.

TOTA [10] and Smart Messages [14] are two architectures for computation and communication in large networks of embedded systems. Communication is realized by sending "smart tuples" in the network, i.e., tuples which include code to be executed at each hop in the network path. These models comply with the general idea of putting intelligence in the network by letting tuples and messages execute hop-by-hop small chunk of code to determine their propagation.

Lime [15] and XMIDDLE [16] exploits transiently tuple spaces as the basis for interaction in dynamic network scenario. Each mobile device, as well as each network nodes, owns a private tuple space. Upon connection with other devices or with network nodes, the privately owned tuple spaces can merge in a federated tuple space, to be used as a common data space to exchange information.

3   RFID Technology

Advances in miniaturization and manufacturing have yielded postage-stamp sized radio transceivers called Radio Frequency Identification (RFID) tags that can be attached unobtrusively to objects as small as a toothbrush. The tags are wireless and battery free. Each tag is marked with an unique identifier and provided with a tiny memory, up to some KB for advanced models, allowing to store data. Tags can be purchased off the shelf, cost roughly 0.20 Euro each and can withstand day-to-day use for years (being battery-free, they do not have power-exhaustion problems). Suitable devices, called RFID readers, can access RFID tags by radio, either for read and write operations. The tags respond or store data accordingly using power scavenged from the signal coming from the RFID reader. RFID readers divide into short- and long-range depending on the distance within which they can access RFID tags. Such distance may vary from few centimeters up to some meters. Deploying RFID technology requires that a number of places in the environment (e.g. doors, corridors, etc.) or objects (e.g. beds, washing machines, etc.) are tagged with RFID tags. Tagging a place or an object involves sticking an RFID tag on it, and making a database entry mapping the tag ID to a name. It is worth emphasizing that current trends indicate that within a few years, many household objects and furniture may be RFID-tagged before purchase, thus eliminating the overhead of tagging [17]. Moreover, some handheld devices start to be provided with RFID read and write capabilities (the Nokia 5140 phone can be already equipped with a RFID reader [18]).

The set of RFID tags deployed across the environment can be regarded as an infrastructure to store and deliver digital information.

From a general perspective, accessing the RFID tags nearby is a powerful source of context information. For example, RFID tags can reveal the location of agents in that tags can be associated to uniquely identified places. So reading the tag associated with "Prof. Smith desk" can let an agent infer its location as "Prof. Smith office". More in general, the knowledge of RFID tags (and thus objects) nearby can possibly identify a specific application context (e.g. reading a LCD-projector tag, and a microphone tag can let the agent infer of being in a meeting room).

In addition, given the fact that RFID tags can be written on-the-fly, agents can use the tags as a distributed shared memory with which to exchange information. For example, RFID tags can be accessed as if they were distributed tuple spaces [19,20]. A particulary significant development of this idea is related to spreading pheromone-inspired distributed data structures across the tags in the environment. The basic scenario consists of human users and robots carrying handheld computing devices, provided with a RFID reader, and running an agent-based application. The agent, unobtrusively from the user, continuously detects in range tags as the user roams across the environment. Moreover, the agent controls the RFID reader to write pheromone data structures (consisting at least in a pheromone ID) in all the tags encountered. This process creates a digital pheromone trail distributed across the tags. More formally, let us call $L(t)$ the set of tags being sensed at time t. It is easy to see that the agent can infer that the user is moving if $L(t) \neq L(t-1)$ (see Fig. 2).

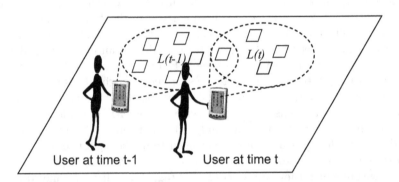

Fig. 2. When the user moves, its agent gets in range with a different set of location-tags (here represented as white rectangles), and recognizes the motion

If instructed to spread pheromone O, the agent will write O in all the $L(t)$-$L(t-1)$ tags as it moves across the environment. For the majority of applications a pheromone trail, consisting of only an ID, is not very useful. Like in ant foraging, most applications involve agents to follow each other pheromone trails to reach the location where the agents that originally laid down the trail were directed (or, on the contrary, to reach the location where they came from). Unfortunately, an agent crossing an-only-ID-trail would not be able to choose in which direction the agent that laid down that trail was directed. From the agent point of view, this situation is like crossing a road without knowing whether to turn left or right.

To overcome this problem, the agent stores in the tags also an ever increasing hop-counter associated with O - we will call this counter $C(O)$. In particular, if an agent decides to spread pheromone O at time t, the agent reads also the counter C(O) in the L(t) set (if C(O) is not present, the agent sets C(O) to a fixed value zero). Upon a movement, the agent will store O and C(O)+1 in the tags belonging to L(t+1) that do not have O or have a lower C(O). In addition, the basic pheromone idea requires a pheromone evaporation mechanism to discard old - possibly corrupted - trails. To this end we store in the tag also a value T(O) representing the time where the pheromone O has been stored. (see Fig. 3)

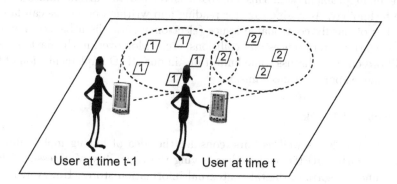

Fig. 3. Writing of pheromone information in RFID tags

To read pheromones, an agent trivially accesses neighbor RFID tags reading their memories. Since RFID read operations are quite unreliable, the agent actually performs a reading cycle merging the results obtained at each iteration. Given the result, the agent will decide how to act on the basis of the perceived pheromone configuration. To realize pheromone evaporation, after reading a tag, an agent checks, for each pheromone, whether the associated timestamp is, accordingly to the agent local time, older than a certain threshold T. If it is so, the agent deletes that pheromone from the tag. This kind of pheromone evaporation leads to two key advantages:

1. Since the data space in RFID tags is severely limited, it would be most useful to store only those pheromone trails that are important for the application at a given time; old, unused pheromones can be removed.
2. If an agent does not carry its personal digital assistant or if it has been switched off, it is possible that some actions will be undertaken without leaving the corresponding pheromone trails. This cause old-pheromone trails to be possibly out-of-date, and eventually corrupted. In this context, it is of course fundamental to design a mechanism to reinforce relevant pheromones not to let them evaporate.

With this regard, an agent spreading pheromone O, will overwrite O-pheromones having an older T(O). From these considerations, it should be clear that the threshold T has to be tuned for each application, to represent the time-frame after which the pheromone is considered useless or possibly corrupted.

3.1 Pros and Cons

The main point in favor of this approach is its extremely low cost since it uses technologies (RFID) that are likely to be soon embedded in the scenario independently of this application. Relying on such an implementation, a wide range of application scenarios based on pheromone interaction can be realized ranging from multi-robot coordination [21], to monitoring of human activities [22].

The main problem with this approach is related to current limitations of RFID technology. Accessing tags for reading and writing operations can fail for a number of hardly controllable issues (electromagnetic disturbances, metallic objects nearby, interferences and collisions, etc.). Moreover, in the next section, we will present and discuss some limitations in our RFID implementation of the pheromone evaporation mechanism.

3.2 Related Work

Several proposals, as well as ours, consider the idea of having mobile devices integrated with a RFID reader, thus having the capability of accessing RFID tags around, as sorts of digital contextual information stores. However, rather than considering the possibility of storing new information in RFID tags and enforcing coordination through them, most approaches exploit RFID tags only for reading pre-existent environmental/contextual information. For instance, the system described in [23] proposes associating location information with tags (e.g., "I am the tag of the living room") that can be read by mobile robots carrying on a RFID reader to roughly localize themselves.

The system described in [22] exploits RFID tags for inferring information about contextual activity in an environment. Users are assumed to wear an RFID reader connected with a Wi-Fi portable device so that, when the user moves and acts in the environment, the type and the sequence of tags read by the reader can suggest what the user is doing. For example, reading the tag associated to the user boss and of a video projector can let infer that the user is in a sort of important meeting with his/her boss

Pheromones spread in the environment can enable a group of users (both humans and robotics) to coordinate their respective movements. An exemplary application would be distributed environment exploration. Users could decide to explore a specific area if there are not pheromones pointing in that direction (the area is truly unexplored). In this context, it is important to remark that this approach clearly requires the presence of RFID tags before pheromones can be spread. If the environment does not contain tags at all, this approach could not be used. However, on the one hand, RFID tags are likely to be soon densely present in everywhere (embedded in tiles, bricks, furniture, etc.). On the other hand, it is possible to conceive solutions where agents physically deploy RFID tags while exploring the environment to be used for subsequent coordination. For instance, future development in plastic (and printable) RFID technology [24] let us envision the possibility of enriching an agent with a simple RFID printer to dynamically print in pavements, walls, or any type of surface, RFID tags.

4   Conclusion and Future Work

This paper presented the role of sensor network and RFID-based infrastructures to support environment abstraction and field-based coordination in pervasive computing scenarios. These infrastructures not only allow agents to acquire context information, but also can serve as suitable media to support their coordination activities.

Our future work in this direction is twofold. On the one hand, we will try to solve technological problems related to current hardware limitations. On the other hand, we will try to apply such mechanisms and abstractions to several pervasive computing scenarios.

Acknowledgments. Work supported by the project CASCADAS (IST-027807) funded by the FET Program of the European Commission.

References

1. Parunak, V.: Go to the ant: Engineering principles from natural agent systems. Annals of Operations Research 75, 69–101 (1997)
2. Bonabeau, E., Dorigo, M., Theraulaz, G.: Swarm Intelligence. From Natural to Artificial Systems. Oxford University Press, Oxford (1999)
3. Babaoglu, O., Meling, H., Montresor, A.: A framework for the development of agent-based peer-to-peer systems. In: 22nd International Conference on Distributed Computing Systems, pp. 15–22. IEEE CS Press, Vienna (2002)
4. Mamei, M., Zambonelli, F.: Physical deployment of digital pheromones through rfid technology. In: IEEE Swarm Symposium, IEEE CS Press, Pasadena (2005)
5. Svennebring, J., Koenig, S.: Building terrain covering ant robots: a feasibility study. Autonomous Robots 16(3), 313–332 (2004)
6. Paskin, M., Guestrin, C., McFadden, J.: A robust architecture for inference in sensor networks. In: International Symposium on Information Processing in Sensor Networks. ACM Press, Los Angeles (2005)
7. Werner-Allen, G., Lorincz, K., Ruiz, M., Marcillo, O., Johnson, J., Lees, J., Welsh, M.: Deploying a wireless sensor network on an active volcano. IEEE Internet Computing 10, 18–25 (2004)
8. Patterson, D., Liao, L., Fox, D., Kautz, H.: Inferring high-level behavior from low-level sensors. In: International Conference on Ubiquitous Computing. ACM Press, Seattle (2003)
9. Choudhury, T., Pentland, A.: Sensing and modeling human networks using the sociometer. In: International Symposium on Wearable Computers. IEEE CS Press, White Plains (2003)
10. Mamei, M., Zambonelli, F.: Programming pervasive and mobile computing applications with the tota middleware. In: Proceedings of the International Conference On Pervasive Computing (Percom). IEEE CS Press, Orlando (2004)
11. Stoy, K., Nagpal, R.: Self-reconfiguration using directed growth. In: 7th International Symposium on Distributed Autonomous Robotic Systems. Springer, Heidelberg (2004)
12. Curino, C., Giani, M., Giorgetta, M., Giusti, A., Murphy, A., Picco, G.: Tinylime: Bridging mobile and sensor networks through middleware. IEEE CS Press, Los Alamitos (2005)

13. Weyns, D., Schelfthout, K., Holvoet, T.: Exploiting a virtual environment in a real-world application. In: Weyns, D., Van Dyke Parunak, H., Michel, F. (eds.) E4MAS 2005. LNCS, vol. 3830, pp. 218–234. Springer, Heidelberg (2006)
14. Riva, O., Nadeem, T., Borcea, C., Iftode, L.: Context-aware migratory services in ad hoc networks. IEEE Transaction on Mobile Computing (to appear, 2007)
15. Picco, G., Murphy, A., Roman, G.: Lime: a coordination model and middleware supporting mobility of hosts and agents. ACM Transactions on Software Engineering and Methodology 15, 279–328 (2006)
16. Mascolo, C., Capra, L., Zachariadis, Z., Emmerich, W.: Xmiddle: A data-sharing middleware for mobile computing. Wireless Personal Communications 21, 77–103 (2002)
17. Smart-Mobs, http://www.smartmobs.com
18. Nokia-Mobile-RFID-Kit, http://www.nokia.com/nokia/055738,00.html
19. Mamei, M., Zambonelli, F.: Pervasive pheromone-based interaction with rfid tags. ACM Transactions on Autonomous and Adaptive Systems 2, 1–28 (2007)
20. Mamei, M., Quaglieri, R., Zambonelli, F.: Making tuple spaces physical with rfid tags. In: Proceedings of the Symposium on Applied Computing (SAC). ACM Press, Dijon (2006)
21. Payton, D., Daily, M., Estowski, R., Howard, M., Lee, C.: Pheromone robotics. Autonoumous Robots 11, 319–324 (2001)
22. Philipose, M., Fishkin, K., Perkowitz, M., Patterson, D., Fox, D., Kautz, H., Hahnel, D.: Inferring activities from interactions with objects
23. Kulyukin, V., Gharpure, C., Nicholson, J., Pavithran, S.: Rfid in robot-assisted indoor navigation for visually impaired. In: Proceedings of the International Conference on Intelligent Robots and Systems. IEEE CS Press, Los Alamitos (2004)
24. Collins, G.: Next stretch for plastic electronics. Scientific American (August 2004)

Coalition Games and Resource Allocation in Ad-Hoc Networks

R.J. Gibbens[1] and P.B. Key[2]

[1] Computer Laboratory, University of Cambridge, William Gates Building,
15 JJ Thomson Avenue, Cambridge, UK, CB3 0FD
Richard.Gibbens@cl.cam.ac.uk
[2] Microsoft Research Cambridge, Roger Needham Building,
7 JJ Thomson Avenue, Cambridge, UK, CB3 0FD
Peter.Key@microsoft.com

Abstract. In this paper we explore some of the connections between cooperative game theory and the utility maximization framework for routing and flow control in networks. Central to both approaches are the allocation of scarce resources between the various users of a network and the importance of discovering distributed mechanisms that work well. The specific setting of our study is ad-hoc networks where a game-theoretic approach is particularly appealing. We discuss the underlying motivation for the primal and dual algorithms that assign routes and flows within the network and coordinate resource usage between the users. Important features of this study are the stochastic nature of the traffic pattern offered to the network and the use of a dynamic scheme to vary a user's ability to send traffic. We briefly review coalition games defined by a characteristic function and the crucial notion of the Shapley value to allocate resources between players. We present a series of experiments with several test networks that illustrate how a distributed scheme of flow control and routing can in practice be aligned with the Shapley values which capture the influence or market power of individual users within the network.

1 Introduction

In this paper we explore some of the connections between cooperative game theory and the utility maximization framework for routing and flow control in networks. Central to both approaches are the allocation of scarce resources between the various users of a network and the importance of discovering distributed mechanisms that work well.

The specific setting of our study is ad-hoc networks and we examine the scheme proposed in [5].

The paper is organized as follows. In Sect. 2 we explain the basic model and quantites of interest largely following the notation used in [5]. We discuss the underlying motivation for the primal and dual algorithms that assign routes and flows within the network and coordinate resource usage between the users.

P. Liò et al. (Eds.): BIOWIRE 2007, LNCS 5151, pp. 387–398, 2008.

Important features of this study are the stochastic nature of the traffic pattern offered to the network and the use of a dynamic scheme to vary a user's ability to send traffic. We also review coalition games defined by a characteristic function and the crucial notion of the Shapley value to allocate resources between participants.

Section 3 presents a series of experiments with several test networks that illustrate how a distributed scheme of flow control and routing can in practice be aligned with the Shapley values which capture the influence or market power of individual users within the network.

2 Models

In this section we outline the basic models and quantities of interest. The essential features follow those given in [5].

2.1 Basic Models

Let N be the set of nodes and let $\mathcal{R}$ be a set of routes. For each $j \in N$ write $\mathcal{R}^S(j) \subset \mathcal{R}$ for the set of routes which start at j and $\mathcal{R}^D(j) \subset \mathcal{R}$ for the set of routes which end at j.

For each source s we will denote by x_s the flow starting at s which flows at rate y_r on route $r \in \mathcal{R}^S(s)$ with

$$x_s = \sum_{r \in \mathcal{R}^S(s)} y_r . \tag{1}$$

Then the amount of flow, c_j, through a node $j \in N$ is given by

$$c_j = \sum_{r : j \in r \wedge r \in \mathcal{R}^S(j) \cup \mathcal{R}^D(j)} y_r + \sum_{r : j \in r \wedge r \notin \mathcal{R}^S(j) \cup \mathcal{R}^D(j)} 2 y_r \tag{2}$$

where the first term aggregates all flows either starting or ending at j and where the second term aggregates all flows transiting both in and out of node j (and thus contribute twice to the quantity c_j). A node is constrained by some capacity, C_j, for aggregate flow so that

$$c_j \leq C_j \qquad \forall j \in N . \tag{3}$$

Similarly, we suppose that receiving and transmitting flows by a node consumes electrical power and we write γ_j for the power consumed at node j which we express in terms of the flows as

$$\gamma_j = \sum_{r \in \mathcal{R}^S(j)} y_r e_{jr}^T + \sum_{r \in \mathcal{R}^D(j)} y_r e^R + \sum_{r : j \in r \wedge r \notin \mathcal{R}^S(j) \cup \mathcal{R}^D(j)} y_r \left(e^R + e_{jr}^T \right) . \tag{4}$$

Here we suppose that the power consumed at node j by a flow of rate y_r is $y_r e^R$ for receiving and $y_r e_{jr}^T$ for forwarding from j to the next node along the route r

after node j. We further suppose that receiving power consumption does not depend on the identity (and hence location) of the transmitter whereas transmitting does depend on the identity of the receiver. The total power consumed at node j is constrained by a quantity Γ_j, that is

$$\gamma_j \leq \Gamma_j \qquad \forall j \in N. \tag{5}$$

2.2 Optimization Framework

Here we shall describe how flows x_s are determined given the constraints on bandwidth C_j and power consumption Γ_j.

We shall suppose the existence of prices μ_{jr} for use of node j by a unit amount of flow on route r and determine flows x_s and y_r by a *primal* algorithm such that

$$x_s = \sum_{r \in \mathcal{R}^S(s)} y_r = \frac{w_s}{\min_{r \in \mathcal{R}^S(s)} \sum_{j \in r} \mu_{jr}} \tag{6}$$

for given quantities w_s and with the proviso that y_r is only positive on routes r that attain the minimum in the denominator of the expression on the right-hand side. Thus, the action is to select flows such that the *rate of spending*, $x_s \sum_{j \in r} \mu_{jr}$, is minimal over the choice of routes $r \in \mathcal{R}^S(s)$ and has value w_s per unit time.

The underlying rationale for this primal algorithm is that of *proportional fairness* which adopts a maximization of utility with the specific choice $U(x) = w \log x$ as the utility function [4,7,10].

The prices μ_{jr} are intended to depend on current flows x_s in order to align demand for resources of bandwidth and power with their provision given in term of the quantities C_j and Γ_j. The dependence is through separate prices for bandwidth and power written μ_j^B and μ_j^P, respectively. Specifically, we write

$$\mu_{jr} = \begin{cases} e_{jr}^T \mu_j^P + \mu_j^B & j \text{ is the source for route } r \text{ (so } r \in \mathcal{R}^S(j)) \\ \left(e^R + e_{jr}^T\right) \mu_j^P + 2\mu_j^B & j \text{ is a transit node for route } r \\ & \qquad \text{(so } j \in r \wedge r \notin \mathcal{R}^S(j) \cup \mathcal{R}^D(j)) \\ e^R \mu_j^P + \mu_j^B & j \text{ is the destination for route } r \text{ (so } r \in \mathcal{R}^D(j)). \end{cases} \tag{7}$$

All the flows x_s, y_r and the various prices $\mu_{jr}, \mu_j^B, \mu_j^P$ will further depend on time and we use this dependence to specify the *dual* algorithm in which prices are adjusted over time according to the following equations

$$\frac{d}{dt}\mu_j^B(t) = \frac{\kappa \mu_j^B(t)}{C_j}(c_j - C_j) \tag{8}$$

$$\frac{d}{dt}\mu_j^P(t) = \frac{\kappa \mu_j^P(t)}{\Gamma_j}(\gamma_j - \Gamma_j) \tag{9}$$

where κ is a small positive constant.

2.3   Coalition Games and Shapley Values

We now turn to coalition games and their use in resource allocation problems. We suppose that a game is composed of a collection of N players corresponding to the source nodes in the ad-hoc network and that for each subset or coalition of players $S \subseteq N$ there is a payoff $v(S)$ given by the *characteristic function*. The characteristic function, $v : \mathcal{P}(N) \mapsto \mathbb{R}$, determines the maximum payoff that the coalition S can guarantee themselves by coordinating the actions of its members, whatever the other players decide. See [8,9] for further discussion of coalition games and the Shapley value approach.

We shall assume that $v(\emptyset) = 0$ and that $v(\cdot)$ is *superadditive*, that is

$$v(S \cup T) \geq v(S) + v(T) \tag{10}$$

whenever $S \cap T = \emptyset$.

An important notion for allocating the value $v(N)$ of the full coalition amongst the players is given by the *Shapley value* $\phi_i(v)$ defined by

$$\phi_i(v) = \sum_{S \subseteq N \setminus \{i\}} \frac{|S|! \, (|N| - |S| - 1)!}{|N|!} \left(v(S \cup \{i\}) - v(S) \right) . \tag{11}$$

It may shown that the vector of Shapley values $(\phi_i(v) : i \in N)$ forms an *imputation*. That is, they are an assignment of the value $v(N)$ between the players where the assignment to player i is at least as great as they could obtain independently of the other players so that

$$\sum_{i \in N} \phi_i(v) = v(N) \tag{12}$$

$$\phi_i(v) \geq v(\{i\}), \qquad \forall i \in N . \tag{13}$$

Note that to compute the Shapley values we require the characteristic function $v(S)$ to be determined for each of the 2^N possible coalitions of the full set of players N. In our experiments described later in Sect. 3 we have 10 players and so there are $2^{10} = 1024$ possible coalitions to consider.

The primal and dual algorithms are motivated by the underlying utility maximization framework and there is a large body of work that now supports that approach. A recent survey of this approach is given in [3]. The central notions are economic ones and relate to competitive Walrasian equilibria in exchange markets [8,10]. A connection between the non-cooperative notions of competitive equilibria and the cooperative notion of a Shapley value is through the *value equivalence theorem* of [2]. This work establishes how in a continuum setting of many small players the allocations associated with the Shapley values are the same as the competitive allocations. The approach we take studies these allocations in ad-hoc networks with a finite numbers of players.

The value equivalence theorem builds on earlier work that establishes a *core equivalence theorem* relating in a similar continuum setting the competitive allocations to those in the core of the game [1,8]. In general, the *core* of a cooperative game in characteristic form is the (possibly empty) set of imputations $(x_i : i \in N)$ such that

$$\sum_{i \in S} x_i \geq v(S) \qquad \forall S \subset N. \tag{14}$$

We do not study the core further in this paper but instead concentrate on the use of the Shapley values.

2.4 Stochastic Features

The experiments discussed in Sect. 3 add stochastic features to the model discussed so far. We shall suppose that each source node s is controlled by a Markov process $D_s(t)$ that assigns the destination node for the current flow starting at s. We also allow $D_s(t)$ to take a further state, labelled 0, indicating that there is *no* current flow associated with source node s. Thus, $D_s(t) \in \{0, 1, \ldots, N\} \setminus \{s\}$ and $D_s(t) = d$ $(d \neq 0)$ means that the user s has flow starting at s and terminating at d. If $D_s(t) = 0$ then the user s is currently inactive and necessarily $x_s(t) = 0$. As the Markov processes $(D_s(t) : s \in N)$ change state then so do the sets $\mathcal{R}^S(\cdot)$ and $\mathcal{R}^D(\cdot)$ describing the sets of routes corresponding to the random source-destination pairs.

In our experiments the random holding times in the different states are independent exponentially distributed random variables with a common parameter λ. The permitted transitions of $D_s(t)$ are such that from the inactive state $(D_s(t) = 0)$ the source will select any destination from $\{1, 2, \ldots, N\} \setminus \{s\}$ equally likely. In the active state $(D_s(t) \neq 0)$ the only transition is to the inactive state. Thus a source alternates between inactive and active periods with a destination node chosen uniformly at random for each successive active period.

The stochastic effects of the traffic patterns changing over time will accordingly imply a continual adjustment of flows and prices using the joint primal and dual algorithms. In our experiments we have further adopted the simplification that routes chosen (by the least cost primal algorithm) do not subsequently change during an active period even though prices may fluctuate to an extent that the chosen routes are nolonger least cost ones.

Thus, the primal algoirthm of equation (6) is revised to

$$x_s(t) = \frac{w_s(t)}{\min_{r \in \mathcal{R}^S(s)} \sum_{j \in r} \mu_{jr}(t)} \tag{15}$$

if $D_s(t) \neq 0$ and $x_s(t) = 0$ if $D_s(t) = 0$. The routes r for which $y_r > 0$ are determined by the minimum above when the flow initially becomes active and are then maintained without change throughout the active period.

2.5   Dynamic Schemes

A net balance of transit costs earned over those paid is maintained by the quantity, $b_s(t)$, defined for each $s \in N$ by

$$\frac{d}{dt}b_s(t) = \sum_{r \in \mathcal{R}^S(s)} y_r(t)\mu_{sr}(t) - w_s(t) \qquad (16)$$

with the initial condition $b_s(0) = 1$. The first term measures revenue from transit fees per unit time and the second term, $w_s(t)$, as already noted, is the spending rate of source node s on its transit costs incurred by its flow of $x_s(t) = \sum_{r \in \mathcal{R}^S(s)} y_r(t)$. Here we allow the possibility that spending rates will vary over time as the function $w_s(t)$.

Furthermore, we assume that each source node has an initial endowment, $b_s(0)$, of one unit.

3   Experiments

Having reviewed the basic theory and choice of mechanisms which underly our model we now explore through a number of experiments the joint behaviour of the ad-hoc network.

Figure 1 shows the set of 10 nodes, here labelled $N = \{A, B, \ldots, J\}$, placed uniformly at random in a square of side 100 units as considered in [5]. Edges are shown between pairs of distinct nodes that are a Euclidean distance of no more than 56 units apart. The purpose of these edges is to define the set of routes available for flow. A possible flow must be along a route corresponding to a path in the network. The network shown in Fig. 1 is connected but we shall also consider subnetworks defined by subsets of nodes (and just the edges incident to nodes within the subset). It is possible for such subnetworks to fail to be connected (the choice of distance threshold 56 controls which subnetworks are disconnected). In our experiments we have taken that for a disconnected subnetwork the flows are set to $x(t) = 0$ throughout the subnetwork. Other possibilites could include considering connected components separately.

We shall also consider in our experiments the effect of movement by the nodes to new locations. Figure 2 shows a new network where node B has moved from a location on the extreme of the network to a new location determined by the centroid of the remaining 9 nodes.

In the experiments we have taken the following choice of parameters. The gain parameter in the primal and dual algorithms is $\kappa = 0.05$, the power coefficients are $e^R = 0.001$ and $e^T = 0.0001 \times d$ where d is the Euclidean distance from the transmitting node to the receiver. The bandwidth capacity is $C_j = 10$ and the power constraint is $\Gamma_j = 0.5$. The distributions of the holding times for the active and inactive periods are independent exponential distributions with means $\lambda^{-1} = 0.5$ seconds.

In the next section we shall discuss the choice of the spending rate parameter, $w_s(t)$, for the primal algorithm and discuss two important classes of scheme: a static case and a dynamic case.

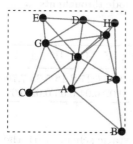

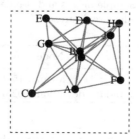

Fig. 1. A network of 10 nodes $N = \{A, B, \ldots, J\}$ located uniformly at random in a square of side 100 units. An edge is shown between each pair of nodes separated by a Euclidean distance of at most 56 units.

Fig. 2. A second network of 10 nodes $N = \{A, B, \ldots, J\}$: compared to the first network B has now moved to the centroid of the remaining 9 nodes

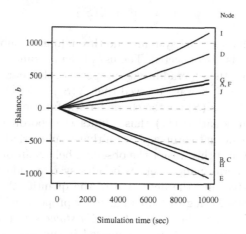

Fig. 3. Node balances over time with a static choice of $w_s(t) = 0.3$. The labels in the right hand margin identify the node(s). The node balances show clear trends: some increasing and some decreasing.

3.1 Static Schemes

We now describe a series of experiments with a static choice of spending rate parameter fixed over time at a value $w_s(t) = 0.3$ the same for each source node.

Figure 3 shows the net balance $b_s(t)$ over time for a simulation of duration 10,000 seconds. All the balances show either clear increasing trends or clear decreasing trends according to whether the spending rate of 0.3 units per second is above or below the rate of earning from transit fees charged to other flows.

Such inbalances between spending and earning reflect a disparity between the market power of the users and the allocation obtained. We demonstrate this by

Table 1. Sample means and standard deviations of node throughputs calculated from 50 independent replicates

Node	A	B	C	D	E	F	G	H	I	J
Mean	4.15	4.16	5.50	6.76	7.56	3.75	6.17	8.16	5.38	7.70
Standard deviation	0.02	0.02	0.03	0.03	0.05	0.02	0.04	0.05	0.02	0.06

considering the observed throughputs under the static scheme with the Shapley values constructed from a characteristic function formed from the system-wide throughput of the subcoalitions $S \subset N$ for each of the $2^{10} = 1024$ possible subcoalitions of the 10 nodes.

The specific details of the Shapley value calculation are described as follows. First, for each subcoalition, S, of users the subnetwork was tested for connectedness. If the subnetwork was disconnected then we set $v(S) = 0$. For the connected subnetworks we set

$$v(S) = \max_{S' \subseteq S} X(S') \tag{17}$$

where $X(S')$ was the observed system-wide throughput of subnetwork S'. Note that $X(S') = 0$ if S' is disconnected. The use of the maximum over all subcoalitions S' was to ensure the superadditivity property of the characteristic function. In some subnetworks, S, it was observed that $X(S') > X(S)$ for $S' \subset S$ which prevents taking $v(S) = X(S)$ for the characteristic function.

The definition in equation (17) thus admits the coordinated action of the players of the subcoalition to drop a player if that would strictly *increase* system-wide throughput even if this wasn't the observed behaviour of the scheme when simulated. Further performance metrics besides system-wide throughput could easily be incorporated into the definition of the quantity $X(\cdot)$. See [6] for an alternative means of ensuring the superadditivity property which has important connections with the game-theoretic notion of a *stable set* of imputations.

The random quantity, $X(S)$, was estimated in our experiments by a long-run average over the randomly varying traffic patterns driven by the Markov processes $(D_s(t) : s \in N)$. Table 1 shows the sample means and sample standard deviations of the node throughputs from 50 independent simulation replicates. The standard deviations show little variability in the estimates of the mean.

Figures 4 and 5 show the correspondence between the observed throughputs and the Shapley values and in each case the proportion of throughput or of the sum of the Shapley values, $v(N) = \sum_{i \in N} \phi_i(v)$ is given. As just noted it is possible for the system-wide throughput to be less than $v(N)$ and so we consider proportions only throughout our comparisons.

Figure 4 shows the Shapley values and observed throughputs according to the locations of the nodes. It is clear that nodes at extreme locations (such as B, C, E and H) receive far larger shares of the system-wide throughput than is allocated by the share of the Shapley value. Conversely, nodes close to the centre

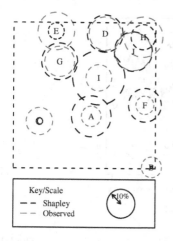

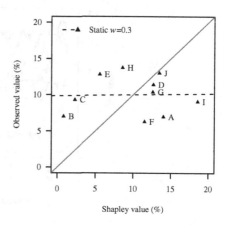

Fig. 4. Shapley values and observed values for the proportion of throughputs by node with static choice of $w_s(t) = 0.3$. The radius of the circles measure either the proportion of the Shapley value or the observed throughput.

Fig. 5. Scatter plot of observed values and Shapley values with the static choice of $w_s(t) = 0.3$. The near horizontal dashed line is a least squares fit to the data points.

of the network (such as A, F and I) receive smaller shares of the system-wide throughput than those allocated according to the Shapley value.

These effects are also apparent from Fig. 5 which shows a scatter plot of the shares of the throughputs and Shapley values. Also shown here is the dashed line given by least squares fit to the data which is far from diagonal.

The dynamic models of the next set of experiments attempt to correct this bias which favours nodes at the extreme of the network with little market power in preference to those near the centre of the network with the largest Shapley values.

3.2 Dynamic Schemes

Here we take the spending rate parameters as the functions of the balance $b_s(t)$ given by

$$w_s(t) = \alpha b_s(t) \tag{18}$$

for a constant $\alpha \in (0, 1)$. For the experiments discussed here we took $\alpha = 0.3$. In this way a larger balance feeds through to a higher spending rate and if the net balance drops the spending rate is reduced accordingly.

Figure 6 shows the net balances under the dynamic scheme. All balances start at $b_s(0) = 1$ given by the initial endowment and then evolve over time to fluctuate about constant levels. The sum of the net balances, $\sum_{s \in N} b_s(t) = N = 10$ remains fixed over time since all spending by one node is earnt by other nodes.

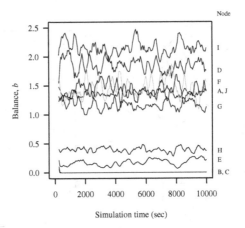

Fig. 6. Node balances over time with a dynamic choice of $w_s(t) = \alpha b_s(t)$. The labels in the right hand margin identify the node(s). Here node balances fluctuate about values without any long-term trend to increase or decrease. The lines shown are smoothed according to a moving average of 250 seconds.

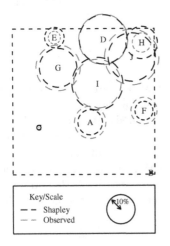

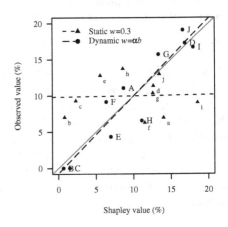

Fig. 7. Shapley values and observed values for proportion of throughputs by node with dynamic choice of $w_s(t) = \alpha b_s(t)$. There is a much closer correspondence between the Shapley values and the observed throughputs in the dynamic case.

Fig. 8. Scatter plot of observed values and Shapley values, including static and dynamic choices of $w_s(t) = \alpha b_s(t)$. The near diagonal dashed line is a least squares fit between the Shapley values and the observed throughputs in the dynamic case.

From Fig. 6 we can see that nodes B and C have net balances $b_s(t)$ that converge to zero while node I has the highest net balance. Figures 7 and 8 show the shares of system-wide throughput and of Shapley value obtained under the

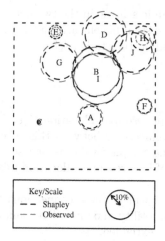

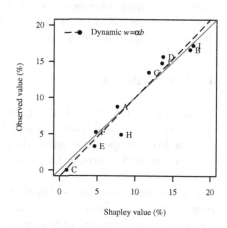

Fig. 9. Shapley values and observed values after node B has moved to the centroid

Fig. 10. Scatter plot of the proportion of Shapley values and observed throughputs after node B has moved to the centroid

Table 2. Estimated node throughputs under the dynamic scheme in the two networks

Node	A	B	C	D	E	F	G	H	I	J
B at extreme	5.81	0.01	0.02	9.06	2.27	4.81	8.25	3.45	8.76	10.02
B at centroid	5.86	11.16	0.02	10.51	2.19	3.52	9.05	3.28	11.56	9.92

dynamic scheme. We can see that there is a much closer correspondence between the observed share of the throughput obtained by each player and that given by the Shapley value approach. Thus market power and outcomes have been more closely aligned than under the static scheme. The line of least squares fit to the dynamic data is very close to the diagonal line and the departures from the diagonal line are more modest than for those in the static scheme.

Figures 9 and 10 show the dynamic scheme in operation in the second network where node B is nolonger on the periphery of the network but placed at the centroid of the remaining 9 nodes. The figures show that the shares of the system-wide throughput and of the Shapley value are quite tightly aligned around the diagonal line in Fig. 10. In [5], the authors considered a trajectory for node B which started at the extreme position and passed through the centroid position and then on towards the upper boundary of the square. The system-wide throughput increases as B moves along this trajectory towards the centroid. In our experiments the throughput increased from 52.45 to 67.07. In both network scenarios the allocations obtained by the dynamic scheme closely align with Shapley values and thus the market power of the node. Table 2 shows the

R.J. Gibbens and P.B. Key

node throughputs obtained under the dynamic scheme in the two networks and reveals that the majority of the additional system benefit arising as B moves to the more favourable position at the centroid accrues to B itself.

4 Conclusions

In this paper we have studied the scheme for resource allocation given in [5] for ad-hoc networks and have explored the connections between this scheme and notions from cooperative game theory. The Shapley value is one such notion that has enabled a broader understanding of how resource allocation takes place with this scheme.

Further work remains to investigate how widely these connections between cooperative game theory and the underlying utility maximization framework for flow control and routing in networks can be extended.

Acknowledgements. RJG acknowledges support from the UK EPSRC grant reference GR/S86266/01 and the International Technology Alliance in Network and Information Science (ITA).

References

1. Aumann, R.J.: Markets with a continuum of traders. Econometrica 32, 39–50 (1964)
2. Aumann, R.J.: Values of markets with a continuum of traders. Econometrica 43(4), 611–646 (1975)
3. Chiang, M., Low, S.H., Calderbank, A.R., Doyle, J.C.: Layering as optimization decomposition: A mathematical theory of network architectures. Proceedings of the IEEE 95(1), 255–312 (2007)
4. Chiang, M.: Balancing transport and physical layers in wireless mulithop networks: Jointly optimal congestion control and power control. IEEE Journal on Selected Areas in Communications 23(1), 104–116 (2005)
5. Crowcroft, J., Gibbens, R.J., Kelly, F.P., Östring, S.: Modelling incentives for collaboration in mobile ad hoc networks. Performance Evaluation 57, 427–439 (2004)
6. Gillies, D.B.: Solutions to general non-zero-sum games. In: Tucker, A.W., Luce, D.R. (eds.) Contributions to the Theory of Games, vol. IV. Princeton University Press, Princeton (1959)
7. Kelly, F.P., Maulloo, A.K., Tan, D.K.H.: Rate control for communication networks: shadow prices, proportional fairness and stability. Journal of the Operational Research Society 49, 237–252 (1998)
8. Mas-Colell, A., Whinston, M.D., Green, J.R.: Microeconomic Theory. Oxford University Press, Oxford (1995)
9. Thomas, L.C.: Games, Theory and Applications. Dover Publications (2003)
10. Varian, H.R.: Microeconomic Analysis, 3rd edn. W.W. Norton and Company (1992)

Bio-Inspired Topology Maintenance Protocols for Secure Wireless Sensor Networks

Andrea Gabrielli and Luigi V. Mancini

Dipartimento di Informatica
Università di Roma "La Sapienza"
00198 Rome, Italy
andrea.gabrielli@di.uniroma1.it

Abstract. We analyze the security vulnerabilities of some well-known topology maintenance protocols (TMPs) for wireless sensor networks. These protocols aim to increase the lifetime of the sensor network by only maintaining a subset of nodes in an active or awake state. The design of these protocols assumes that the sensor nodes will be deployed in a trusted, non-adversarial environment, and does not take into account the impact of attacks launched by malicious insider or outsider nodes. We describe three attacks against these protocols that may be used to reduce the lifetime of the sensor network, or to degrade the functionality of the sensor application by reducing the network connectivity and the sensing coverage that can be achieved. Further, we describe countermeasures, inspired by biological systems and processes, that can be taken to increase the security and fault-tolerance of the protocols.

Keywords: Wireless Networks, Security, Bio-Inspired.

1 Introduction

Topology maintenance protocols (TMPs), such as SPAN [1], ASCENT [2], PEAS [3], and CCP [4] are critical to the operation of wireless sensor networks. These protocols aim to increase the lifetime of the sensor network by only maintaining a subset of nodes in an active or awake state, while turning off redundant nodes. There have to be enough active nodes to maintain the connectivity of the network as well as to obtain sensing coverage in the area where the sensor network is deployed.

The various topology maintenance protocols that have been proposed in the literature differ in their objectives as well as in the approaches that are used to achieve their objectives. For example, SPAN and ASCENT attempt to maintain network connectivity, but do not guarantee sensing coverage. On the other hand, PEAS and CCP are designed to address both connectivity and the application's coverage requirements in a configurable fashion.

All these protocols involve some form of coordination and message exchange between neighboring nodes in order to elect coordinators and determine sleep schedules. These protocols were designed assuming a non-adversarial, trusted

P. Liò et al. (Eds.): BIOWIRE 2007, LNCS 5151, pp. 399–410, 2008.

environment. Consequently, they are vulnerable to security attacks in which malicious nodes send spoofed or false messages to their neighbors in an effort to defeat the objectives of the protocol.

Attacks on the topology maintenance protocols can be carried out either by entities that are external to the network (outsider attacks) or by compromised nodes (insider attacks). Insider attacks are a particularly challenging problem for sensor networks because many sensor applications involve deploying nodes in an unattended environment, thus leaving them vulnerable to capture and compromise by an adversary. Unlike outsider attacks, insider attacks cannot be prevented by authentication mechanisms since the adversary knows all the keying material possessed by the compromised nodes.

In this paper, we analyze the security vulnerabilities of ASCENT, a well-known topology maintenance protocol.

We describe three types of attacks that can be launched against this protocols: *sleep deprivation* attacks that increase the energy expenditure of sensor nodes and thus reduce the lifetime of the sensor network; *snooze* attacks that result in inadequate sensing coverage or network connectivity; and *network substitution* attacks in which multiple attackers collude to take control of part of the sensor network.

Furthermore, we describe countermeasures that can be taken to increase the robustness of the protocols and the resilience to such attacks. The proposed countermeasures are inspired by biological systems and processes.

We previously analyzed TMP vulnerabilities and we proposed countermeasures for a static non-mobile network based on authentication and cryptography mechanisms in [5]. To the best of our knowledge, the only research work that has pointed out the security issues on topology maintenance protocols is [6]. In another related work [7], Stajano and Anderson introduced the problem of the *sleep deprivation* for Wireless Ad-hoc Networks, but not in the context of topology maintenance protocols.

In this paper, we have analyzed the security vulnerabilities of topology maintenance protocols for wireless sensor networks. The main contributions of the paper are:

- The description of how the *sleep deprivation*, the *snooze* and the *network substitution* attacks can be launched against ASCENT, a well-known protocol for sensor networks. Although not discussed in this paper, protocols such as PEAS, CCP, GAF [8], CEC [9], AFECA [10], and SPAN [1] are also vulnerable to these attacks.
- The proposal of biologically inspired countermeasures that can be applied to make the protocols robust against these attacks.

The rest of this paper is organized as follows: Section 2 discusses the threat model and the different kinds of adversaries we expect to encounter in sensor networks. Next, in Section 3 we present a taxonomy of the attacks that can be launched against topology maintenance protocols. Section 4 presents a brief overview of the ASCENT protocol and discusses the specific attacks against the protocol. In Section 5 we discuss the bio-inspired countermeasures for the TMP protocols. Finally, Section 6 concludes this paper and points out several future research directions.

2   Threat Model

In this section, we describe our assumptions with respect to the sensor network and the behavior and capabilities of an adversary.

Due to the wireless nature of communications in sensor networks, we assume that the adversary can eavesdrop on the communications of other nodes and can also inject data packets into the network.

We assume that the nodes are not tamper-proof. Thus, if the adversary captures a node, all the information including cryptographic keys stored in the node is compromised. Furthermore, the adversary can clone the identity of a compromised device, and can store the information obtained from that node in other malicious nodes.

Finally, we assume that the adversary can deploy malicious nodes, and that these nodes can collude together to attack the system.

2.1   Attacker Classification

We may classify the attacker into various categories based on both its hardware capabilities and on its knowledge of the cryptographic keys that are used to provide authenticated and/or confidential communication.

Laptop-Class Vs Node-Class Attackers. A laptop-class attacker uses a relatively powerful device as compared to a sensor node. An attacker with these capabilities has access to greater battery, storage and computational resources than a typical sensor node, e.g. a Berkeley MICA mote [11]. It may also use a high-power radio transmitter and a very sensitive antenna that could allow the attacker to eavesdrop on the entire network and to transmit messages with enough power to be heard by any node.

On the other hand, a node-class attacker uses one or more devices with the same capabilities as legitimate sensor nodes. Therefore, it is only able to listen to or transmit messages within a limited range, and it faces constraints such as limited battery power, small memory and a relatively slow CPU.

Outsider Vs Insider Attackers. An outsider attacker has no more knowledge than the definition of the protocols that are used in the network and the information that is gathered by eavesdropping on network communications. It has no access to cryptographic keys or data that are used to secure the network. For example, it does not possess any credentials that enable it to authenticate itself to the other nodes.

In contrast, an insider is an attacker that has all the information used by a node to be a legitimate member of the network, such as its cryptographic keys. It can be a captured node, but also a device, such a node-class or laptop-class, in which the attacker has stored information retrieved from a compromised node.

3   Attacks on Topology Maintenance Protocols

The use of topology maintenance protocols introduces new vulnerabilities in sensor networks. In particular, an adversary can launch new kinds of attacks by exploiting the ability of these protocols to increase or decrease the number of active nodes. In the following discussion, we present three different attacks on topology maintenance protocols.

Sleep Deprivation Attack. In this type of attack, the adversary tries to induce a node in a specific area to remain active. This attack has two effects. First, by increasing the energy expenditure of sensor nodes, it reduces the estimated lifetime of the network. Second, in the case of a densely populated area, it can lead to increased energy consumption due to congestion and contention at the data link layer.

Snooze Attack. In this type of attack, the adversary forces the nodes to remain in the sleeping state. This kind of attack can be applied to the whole network or to a subset of nodes. In the latter case, the adversary can launch an attack to jeopardize the connectivity of the network or to reduce the sensing coverage in a region. For example, an adversary can selectively turn off nodes that are monitoring an intruder's path through an area in which a sensor field has been deployed for surveillance.

Network Substitution Attack. In this type of attack, the adversary takes control of the entire network or a portion of it by using a set of colluding malicious nodes. The adversary deploys a set of nodes that are included in the set that has been elected by the topology maintenance protocol to maintain network connectivity or the sensing of the area. Once the protocol has chosen the malicious nodes as its working nodes, the portion of the network under attack is totally in the hands of the adversary.

When the adversary controls a portion of the network, it can carry out other attacks such as traffic analysis and selective or complete packet dropping. This attack cannot be easily detected because the adversary can maintain network connectivity and keep it operating as usual. For example, if the application is supposed to receive readings from sensors at a certain frequency, the adversary can send false readings at the same rate and avoid detection.

4   Analysis of ASCENT

4.1   Brief Review of ASCENT

ASCENT [2] adaptively elects a set of active nodes that stay awake all the time and perform multi-hop packet routing, whereas the rest of the nodes remain passive and periodically check whether they should become active. Each node decides to be active or passive by using a local measurement of its connectivity degree (number of active neighbors) and a measurement of its data loss (DL) rate.

The protocol is based on the idea that the DL rate of the network should not be higher than the specified loss threshold (LT), and the number of active nodes in a communication range should not exceed the neighbor threshold (NT). A node is considered an active neighbor if its neighbor link loss is below a specified threshold. Each node adds a unitary monotonically increasing sequence number to each piece of data and control packet that is transmitted. This allows for neighbor link loss detection when a sequence number is skipped. Similarly, a separate sequence number is used to detect lost application data packets. Note that the DL rate is estimated on the basis of application data packets, and that control packets are not taken into consideration in its calculation.

Nodes can be in one of four states: ACTIVE, SLEEP, PASSIVE, or TEST:

- ACTIVE: The node works until its energy is depleted. If it measures a DL rate greater than LT, the active node broadcasts a HELP MESSAGE to its neighbors.
- SLEEP: In this state, the node turns off its radio and sleeps for a time T_s. When T_s expires, the node moves into the PASSIVE state.
- PASSIVE: The intuition behind the PASSIVE state is to collect information regarding the state of the network without causing interference with other nodes. The passive node turns its radio on and sets the network interface in promiscuous mode to overhear all the packets that are transmitted by the neighbors. When a node enters this state, it sets up a timer T_p and sends a NEW PASSIVE NODE ANNOUNCEMENT message. This message is used by active nodes to estimate the density of the nodes in the neighborhood. Active nodes transmit this density estimate to any new passive node. When T_p expires, the passive node enters the SLEEP state. If the number of active neighbors is below the NT before T_p expires, and either the DL rate is higher than the LT, or the DL is below the LT but the node has received a HELP MESSAGE from an active neighbor, then the passive node enters the TEST state.
- TEST STATE: A node in this state probes the network to see whether adding itself may improve connectivity. The test node starts by exchanging data and routing control messages with its neighbors. It sets up a timer T_t and sends a NEIGHBOR ANNOUNCEMENT message. If a node in TEST state receives a NEIGHBOR ANNOUNCEMENT message from a node with higher ID, then it goes back to PASSIVE state. When T_t expires, the node enters the ACTIVE state. If the number of active neighbors is above NT before T_t expires, or if the average DL rate is higher than the average DL rate before the node entered the TEST state, then the node moves into the PASSIVE state.

The timers T_p and T_t are fixed. However, the value of T_s is dynamically adjusted using the estimated node density in the neighborhood. It increases as the node density increases.

4.2   Attacks on ASCENT

We now describe *Snooze, Network Substitution,* and *Sleep Deprivation* attacks on ASCENT. A node that enters the active state in ASCENT does not go back to sleep for any reason, so an attacker is not able to put it in sleeping mode. The aim of the *snooze* attacks we describe is to keep nodes that are not in ACTIVE state either in PASSIVE or in SLEEP state. These attacks only become effective when the initial set of active nodes fail. This is because these nodes will not be replaced by new nodes leaving the area of the network that is under attack without a sufficient number of active nodes. Therefore, if the attacker wants to disable the network at time T, it has to start the attack before T and wait for active nodes in the area to start running out of battery power. Thus, the attacker cannot simply launch the attack when he wants to turn off the network because some legitimate nodes will continue to work. Further, it could be difficult to choose when to start the attack that would disable the network at time T because the attacker may not be able to estimate the remaining battery power of the nodes that are in active state.

Snooze Attack Using Impersonation. The adversary impersonates multiple active nodes so that the legitimate nodes estimate the number of active neighbors to be greater than NT. Thus, all the nodes that are in TEST state enter the PASSIVE state, and all the nodes in PASSIVE state transition to the SLEEP state. Although the adversary cannot affect nodes in ACTIVE state, when these nodes fail they are not replaced by new nodes, and this causes incremental degradation of the connectivity and sensing coverage of the network.

This attack is more effective when launched by a *laptop-class attacker.* In fact, a *laptop-class* adversary only has to impersonate NT identities to launch an attack against all the nodes in the network, while a *node-class* adversary has to impersonate NT nodes to simply launch an attack against its neighbors.

Snooze Attack Using NEIGHBOR ANNOUNCEMENT Messages. The adversary periodically broadcasts a NEIGHBOR ANNOUNCEMENT message with an *ID* that is higher than the identifiers of all the legitimate nodes of the network. This induces all the nodes that are in TEST state to think that they have a neighbor with higher *ID* in TEST mode. Each legitimate node in TEST state will go into SLEEP mode. This way all the nodes that are not already active can be kept in the SLEEP state by the adversary. Thus, when active nodes fail they are not replaced by new nodes, leading to incremental degradation of the connectivity and sensing of the network.

This attack can be launched either by a *laptop-class attacker* or by a *node-class attacker.* In the former case, the adversary can attack the entire network by simply sending the message periodically throughout the network. In the latter case, the malicious node can launch the attack against all its neighbors. In order to minimize its energy consumption, the malicious node can wait for the NEIGHBOR ANNOUNCEMENT message of a neighbor, and then send a NEIGHBOR ANNOUNCEMENT message with higher *ID* to make the node go back to sleep. The adversary can deploy a set of malicious nodes to launch the attack in a selected part of the network and to block communication or sensing coverage in the target area.

Both of the previous *Snooze* attacks can be made more effective by increasing the time interval for which a legitimate node remains in SLEEP state. In AS-CENT, a node adjusts the timer T_s on the basis of the neighbor density estimates it receives from the active nodes, and the time interval increases as node density does. The attacker can send false NEW PASSIVE NODE ANNOUNCEMENT messages with different *ID*s in an effort to lead active nodes to increase their estimates of the node density, or it can announce a high node density directly to the other nodes.

Network Substitution Attack. In this attack, the adversary has to deploy enough nodes to form a connected network and so that any legitimate node in the area being attacked is within the transmission range of at least one malicious node. After their deployment, all the malicious nodes launch a *snooze attack using NEIGHBOR ANNOUNCEMENT messages*. When the currently active nodes run out of power, the adversary controls the part of the network that has been attacked because no new nodes will enter the TEST or ACTIVE states. The only services that are available in the attacked area are the ones provided by the malicious nodes.

Sleep Deprivation Attack. Under normal operating circumstances, the number of active nodes selected by ASCENT suffices to keep the data loss rate below the LT that is specified by the application. However, the adversary can use HELP messages to increase the number of active nodes and reduce the efficiency of ASCENT as follows.

First, the adversary simulates an increase in the data loss rate by sending u several messages which contain sequence numbers that differ by more than 1. This leads node u to compute an erroneous estimate of the data loss rate. The adversary then sends a HELP message, and if a node u has less than NT active neighbors it will then transition to the TEST state. Once u enters the TEST state, the adversary can manipulate the sequence numbers of the messages it sends to convince u that the data loss rate has decreased. Consequently, u will then transition to the active state.

If u has more than NT active neighbors, the adversary has to simulate a decrease in the number of active neighbors of u first. The adversary can then carry out the attack on the above mentioned data loss rate in order to bring u to the ACTIVE state. To simulate a decrease in the number of active neighbors of u, the adversary starts sending u a sequence of messages in which the adversary maliciously skips a few numbers in the sequence every time a message is sent in order to simulate an increase in the neighbor link loss. In particular, the adversary forges the sender identifier of each message to convince u that many links are lossy. At this point u, follows the ASCENT protocol and starts decreasing the number of its active neighbors due to the increase in the measured neighbor link loss. In particular, when the number of active neighbors is less than NT, the adversary can convince u to go into ACTIVE state by repeating the attack explained above.

Thus, the adversary can induce most of its neighbors to transition to the ACTIVE state. This attack can be performed either by a *laptop-class attacker* or by a *node-class attacker*. Even if the adversary has to apply the attack to its neighbors one by one, it does not need to repeat the attack periodically because nodes in ACTIVE state do not go back to sleep. Thus, the adversary only incurs an energy expenditure once for each node that is attacked.

5   Bio-Inspired Solutions

In this section, we propose the countermeasures to the TMP protocol vulnerabilities described in the Section 3. We have studied several biological systems, such as the quorum sensing in bacteria [12] and the immune system [13], and we took inspiration for our solutions from the interesting properties and behaviours of these systems. In particular, the countermeasures are inspired by the robustness and the resilience of the immune system, and by the free cooperation and equipotency of its participants in quorum sensing protocols.

In the immune system the adversary must be recognized and should be forced to interact with cells able to neutralized it, while in the TMP we can say that the adversary must not be able to interact with the nodes of the network; in this way the nodes cannot be attacked. It appears to us that the heterogeneity of the system, typical of the immune system, could be a countermeasure also for the attacks to the TMP. In fact, if an attacker comes within the range of a network with heterogeneous devices then the attacker can interact only with a subset of the nodes but not with the entire network. As a consequence, if the adversary launches an attack then only a subset of nodes are vulnerable to the attack, and the functioning of the network as a whole could be granted by the remaining nodes.

We proposed the following classification of the *heterogeneous networks*:

- **Heterogeneous Hardware Network (HHN):** nodes are manufactured with different hardware devices. They could have communications system with non-compatible technologies, as for example different radio frequency systems, acoustic systems and optical systems; or they can be differentiated based on the interpretation of data, as for example using big-endian or little-endian systems.
- **Heterogeneous Protocols Network (HPN):** nodes use multiple protocols for the same service. For example, the nodes could be loaded with one protocol from a set, or they can be loaded with a set C of protocols to use during the network lifetime.

With nodes loaded with a single TMP protocol from the set of possible choices and with hardware heterogeneity we obtained different networks where nodes take part to only a network during their lifetime. These solutions are logically equivalent to multiple networks with different behaviours and where the nodes of a network do not interact with the nodes of another network.

More interesting solutions are obtained with the HPN methods, where the nodes are loaded with a set of TMPs. For example, applying the above techniques to the TMPs we can image the following scenarios:

- **Uniform Selection:** the node choice of the TMP protocol is uniform distributed:
  - **Uniform TMP Selection (UTS):** a set C of TMP is used in the network and each node is able to take part to each TMP of the set C. Each node divides the time into slot and in each slot S the node randomly chooses, according to a uniform distribution, which TMP to follow during S.
  - **Uniform Results Selection (URS):** each node takes part to all the protocols chosen for the network, but each time a node is in a state-transition test it takes a decision based on the result of a single TMP, randomly selected.
- **Weighted Selection:** the node choice is not uniformly distributed over the set of possible choices but each choice has a different weight. We can further divide this solution in two sub-cases:
  - **Weighted TMP Selection (WTS):** each node divides time into slots and in each slot S selects the TMP to follow during S; each TMP has a different probabilities to be chosen. In this solution, assuming that a TMP has a greater probability to be chosen with respect to the other TMPs then, for each slot S, there is a great number of nodes running the same TMP. The sum of all the probabilities must be equal to 1; note that in the case the probabilities of the TMPs selection are all equal, then we have a UTS. The probabilities can be statically chosen before the nodes deployment or they can change dinamically. In the latter case, it is necessary to design secure techniques to update these probabilities because such update could be exploited by the adversary to attack the network. For example, an adversary could induce a node to favour a specific TMP, in this way when all the nodes are running the same TMP, then the adversary can jeopardized a big part of the network with an attack to a single protocol. Note that if the adversary is successful in the previous technique then the network heterogeneity could result as an advantage for the attacker; in fact the adversary could select, from all the protocols in use in the network, the better one, and exploit it for the attack.
  - **Weighted Results Selection (WRS):** each node takes part to all the TMP but it makes its state-transition decision selecting the results of only one single TMP. This solution employs a function that assigns weights to the results obtained with the TMPs, and each node follows the result with higher weight. If multiple results has the same higher weight, then the node uses the random selection as a tie breaking mechanism.

As application examples we describe some of the above techniques applied using Peas and ASCENT. In the following solutions each node of the network is loaded with the Peas and ASCENT code.

Uniform TMP Selection with Peas and ASCENT (UTS_Peas_ASCENT). Each node u, that is not in *Sleeping* state, divides the time into slot S. At the beginning of each slot the node randomly chooses the TMP to execute during S. During S node u follows Peas with probability $1/2$ and ASCENT with probability $1/2$. When u goes to *Sleeping* it starts a timer T_s and it moves from *Sleeping* only when T_s expires, during the *Sleeping* period the division of time into slot does not take place neither the TMP selection. When T_s expires the node wakes up and flips the coin to select the TMP to execute during the incoming S. To avoid that the nodes become unstable and they move from a TMP to another without completing state-transition tests, we can: fix the length of S such that it completely contains the TMPs state-transition tests or, in case u is in the middle of a test when S finishes, we can force u to select the same TMP of the previous slot, in this way the node can complete the state-transition.

Uniform Result Selection with Peas and ASCENT (URS_Peas_ASCENT). At the same time, each node u takes part both to Peas and to ASCENT. However, after each state-transition test the node chooses with equal probability (that is $1/2$ in our example) the results of a protocol or of the other. On the contrary of the *TMP Selection* solutions, node u executes either Peas or ASCENT and, after it gathers the results, the node chooses which protocol to comply with; in this solution an eavesdropper cannot establish which protocol is executed by the node for the state-transition decision before the transition itself takes place.

Weighted TMP Selection with Peas and ASCENT (WTS_Peas_ASCENT). This solution is similar to UTS_Peas_ASCENT but the TMP choice is not uniformly distributed over the two protocols, the choice of a TMP has a weight different to the other. In each slot S a node u chooses Peas with probability P_p and ASCENT with probability P_a, where $P_p \neq P_a$ and $P_p + P_a = 1$.

Weighted Result Selection with Peas and ASCENT (WRS_Peas_ASCENT). Each node executes either Peas or ASCENT. It is defined a function to assign scores to the results of the protocols state-transition tests. When state-transition tests end, the node changes its state complying with the results with the higher score. In case of a tie it is randomly choose one.

6   Conclusion and Future Works

Based on our analysis of some TMPs, we can make the following general observations with respect to the security considerations in the design of topology maintenance protocols:

- TMPs should be designed so that a node makes its state-transition decisions, e.g., a decision regarding whether to sleep or remain active, based on input from multiple neighbor nodes in order to be resilient to false messages injected by malicious nodes.

– TMPs should be designed so that state-transition decisions are revisited periodically. For example, without a periodic check of a node's eligibility to be in a sleeping or active state, it becomes possible for an adversary to launch a resource-consumption attack that results in a node staying in the active state until its energy is depleted.

Among the bio-inspired solutions we have proposed, the most secure solutions are the *Uniform/Weighted Result Selection*. In fact, in the other proposals, the *Uniform/Weighted TMP Selection* solutions, although both the protocols are executed, an eavesdropper can understand which protocol is used by node u during the state-transition tests. After node u has selected a protocol P, it only executes the tests of P, thus listening to the node communication an adversary can infer the protocol P that u is executing. On the contrary, with *Uniform/Weighted Result Selection* solutions a node u executes the tests of all the protocols and after the tests, node u chooses which protocol to comply with. In these solutions, the adversary cannot infer the TMP used by node u for its state-transition decision before the transition itself takes place.

This work is an initial foray into the design of secure TMP based on Bio-Inspired systems.

In the near future, we plan to build a test-bed to evaluate two new TMPs for Heterogeneous Protocols Networks (HPNs). The goal is to increase the resilience of the network designing the two protocols strictly coupled together so that they maximized the effectiveness of the HPN. For this scope, we are defining a set of functions to measure the effort required to launch the attacks and to measure the efficiency of the attacks; the functions can be used to evaluate and compare different HPNs. In particular we are studying three functions: *AttackCost*, *AttackEffect* and *AttackLength*. The first measures the energy needed to launch the attacks, the second quantify the consequence of the attacks in terms of network performance degradations and the last measures how long it takes for the attack to be effective. The best HPN will have an high value from the *AttackCost* function and low values from the *AttackEffect* and *AttackLength* functions.

References

1. Chen, B., Jamieson, K., Balakrishnan, H., Morris, R.: Span: An energy-efficient coordination algorithm for topology maintenance in ad hoc wireless networks. ACM Wireless Networks Journal 8(5) (2002)
2. Cerpa, A., Estrin, D.: ASCENT: Adaptive Self-Configuring Sensor Networks Topologies. IEEE Transactions on Mobile Computing 3(3) (2004)
3. Ye, F., Zhong, G., Lu, S., Zhang, L.: PEAS: A Robust Energy Conserving Protocol for Long-lived Sensor Networks. In: Proceedings of the 23rd IEEE International Conference on Distributed Computing System (ICDCS 2003), Rhode Island, Providence (2003)

4. Wang, X., Xing, G., Zhang, Y., Lu, C., Pless, R., Gill, C.: Integrated Coverage and Connectivity Configuration in Wireless Sensor Networks. In: Proceedings of the 1st ACM International Conference on Embedded Networked Sensor Systems (SenSys 2003), Los Angeles, California (2003)

5. Gabrielli, A., Mancini, L.V., Setia, S., Jajodia, S.: Securing Topology Maintenance Protocols for Sensor Networks: Attacks and Countermeasures. In: Proceedings of the 1st IEEE/CreateNet International Conference on Security and Privacy for Emerging Areas in Communication Networks (SecureComm 2005), Athens, Greece (2005)

6. Karlof, C., Wagner, D.: Secure Routing in Wireless Sensor Networks: Attacks and Countermeasure. In: Proceedings of the 1st IEEE International Workshop on Sensor Network Protocols and Applications (SNPA 2003), Anchorage, Alaska (2003)

7. Stajano, F., Anderson, R.: The Resurrecting Duckling: Security Issues for Ad-hoc Wireless Networks. Springer, London (1999)

8. Xu, Y., Heidemann, J., Estrin, D.: Geography-informed energy conservation for ad-hoc routing. In: Proceedings of the 7th ACM International Conference on Mobile Computing and Networking (MobiCom 2001), Rome, Italy (2001)

9. Xu, Y., Heidemann, J., Estrin, D.: Energy conservation by adaptive clustering for ad-hoc networks. In: Poster Session of the 3rd ACM International Symposium on Mobile Ad-Hoc Networking and Computing (MobiHoc 2002), Lausanne, Switzerland (2002)

10. Xu, Y., Heidemann, J., Estrin, D.: Adaptive Energy-Conserving Routing for Multihop Ad-Hoc Networks. Research Report527, USC/Information Sciences Institute (2000)

11. Mica sensor node, http://www.xbow.com

12. Miller, M.B., Bassler, B.L.: Quorum Sensing in Bacteria. Annual Review of Microbiology 55, 165–199 (2001)

13. Janeway, C.A.: Immunobiology: The Immune System in Health and Disease. Garland Publishing (2001)

Dynamic Topologies for Robust Scale-Free Networks

Shishir Nagaraja and Ross Anderson

Computer Laboratory
JJ Thomson Avenue, Cambridge CB3 0FD, UK
forename.surname@cl.cam.ac.uk

Abstract. In recent years, the field of anonymity and traffic analysis have attracted much research interest. However, the analysis of subsequent dynamics of attack and defense, between an adversary using such topology information gleaned from traffic analysis to mount an attack, and defenders in a network, has recieved very little attention. Often an attacker tries to disconnect a network by destroying nodes or edges, while the defender counters using various resilience mechanisms. Examples include a music industry body attempting to close down a peer-to-peer file-sharing network; medics attempting to halt the spread of an infectious disease by selective vaccination; and a police agency trying to decapitate a terrorist organisation. Albert, Jeong and Barabási famously analysed the static case, and showed that vertex-order attacks are effective against scale-free networks. We extend this work to the dynamic case by developing a framework to explore the interaction of attack and defence strategies. We show, first, that naive defences don't work against vertex-order attack; second, that defences based on simple redundancy don't work much better, but that defences based on cliques work well; third, that attacks based on centrality work better against clique defences than vertex-order attacks do; and fourth, that defences based on complex strategies such as delegation plus clique resist centrality attacks better than simple clique defences. Our models thus build a bridge between network analysis and traffic analysis, and provide a framework for analysing defence and attack in networks where topology matters. They suggest definitions of efficiency of attack and defence, and may even explain the evolution of insurgent organisations from networks of cells to a more virtual leadership that facilitates operations rather than directing them. Finally, we draw some conclusions and present possible directions for future research.

Keywords: Scale-free networks, robustness, covert groups, topology, security.

1 Introduction

Many modern conflicts turn on connectivity. In conventional war, much effort is expended on disrupting the other side's command, control and communications

P. Liò et al. (Eds.): BIOWIRE 2007, LNCS 5151, pp. 411–426, 2008.

by jamming or destroying her facilities. Counterterrorism operations involve a similar effort but with different tools: traffic analysis to trace communications, coupled with surveillance of the flows of money, material and recruits, followed by the arrest and interrogation of individuals who appear to be significant nodes. Terrorists are aware of this, and take measures to prevent their networks being traced. Usama bin Laden described his strategy on the videotape captured in Afghanistan as 'Those who were trained to fly didn't know the others. One group of people didn't know the other group' (see [14], which describes the hijackers' networks).

Connectivity matters for social dominance too, as a handful of leading individuals do much of the work of holding a society together. Subverting or killing these leaders is likely to be the cheapest way to make an invaded country submit. When the Norman French invaded England in the eleventh century, they killed or impoverished most of the indigenous landowners; when the Turks, and then the Mongols, invaded India, they killed both landowners and priests; when England suppressed the Scottish highlands after the 1745 uprising, landowners were induced to move to Edinburgh or London; and in many of the dreadful events of the last century, rulers targeted the elite (Russian kulaks, Polish officers, Tutsi schoolteachers, . . .).

Moving from politics to commerce, the music industry spends a lot of money attempting to disrupt peer-to-peer file-sharing networks. Techniques range from technical attacks to aggressive litigation against individuals believed to have been running major nodes.

Networks of personal contacts are important in other applications too. In public health, for example, it often happens that a small number of individuals account for much of the transmission of a disease. Thus Senegal has been more effective at tackling the spread of HIV/AIDS than other African countries, as they targeted prostitutes [19]. In fact, interest in social networks has grown greatly over the last 15 years in the humanities and social sciences [20,9].

Recent advances in the theory of networks have provided us with the mathematical and computational tools to understand such phenomena better. One striking result is that a network much of whose connectivity comes from a small number of highly-connected nodes can be very efficient, but at the cost of extreme vulnerability. As a simple example, if everyone in the county communicates using one telephone exchange, and that burns down, then everyone is isolated.

This paper starts to explore the tactical and strategic options open to combatants in such conflicts. What strategies can one adopt, when building a network, to provide good trade-offs between efficiency and resilience? We are particularly interested in complex networks, involving thousands or millions of nodes, which are so complicated (or under such dispersed control) that the resilience rules can only be implemented locally, rather than by a central planner who deliberately designs a network with multiple redundant backbones.

Is it possible, for example, to create a virtual high-degree node, by combining a number of nodes which appear on external inspection to have lower degree? For example, a number of individuals might join together in a ring, and use some covert communications channel to route sensitive information round the ring in a manner shielded from casual external inspection. There is a loose precedent in Chaum's 'dining cryptographers' construction [10], in which a number of cryptographers pass messages round a ring in such a way as to mask, from insiders, the source and destination of encrypted traffic. Can we build a similar construction, but in which the fact of systematic message routing is concealed from outsiders, with the result that the participants appear to be 'ordinary' nodes making a modest contribution in the network, rather than important nodes that should be targeted for close inspection and/or destruction?

2 Previous Work

There has been rapid progress in recent years in understanding how networks can develop organically, how their growth influences their topology, and how the topology in turn affects both their capacity and their robustness. There is now a substantial literature: for a book-length introduction, see Watts [21], while literature surveys are [1,17].

Early work by Erdös and Renyi modelled networks as random graphs [11,7]; this is mathematically interesting but does not model most real-world networks accurately. In real networks, path lengths are generally shorter; it is well known that any two people are linked by a chain of maybe half a dozen others who are pairwise acquainted – known as the 'small-world' phenomenon. This idea was popularised by Milgram in the 60s [16]. An explanation started to emerge in 1998 when Watts and Strogatz produced the alpha model. Alpha is a parameter that expresses the tendency of nodes to introduce their neighbours to each other; with $\alpha = 0$, each node is connected to its neighbours' neighbours, so the network is a set of disconnected cliques, while with $\alpha = \infty$, we have a random graph. They discovered that, for critical values of α, a small-world network resulted. The alpha model is rather complex to analyse, so they next introduced the beta network: this is constructed by arranging nodes in a ring, each node being connected to its r neighbours on either side, then replacing existing links with random links according to a parameter β; for $\beta = 0$ no links are replaced, and for $\beta = 1$ all links have been replaced, so that the network has again become a random graph [22]. The effect is to provide a mix of local and long-distance links that models observed phenomena in social and other networks.

How do networks with short path lengths come about in the real world? The simplest explanation involves preferential attachment. Barabási and Albert showed in 1999 how, if new nodes in a network prefer to attach to nodes that already have many edges, this leads to a power-law distribution of vertex order which in turn gives rise to a *scale-free* network [6], which turns out to be a more common type of network than the alpha or beta types. In a social network, for example, people who already have many friends are useful to know, so their friendship is particularly sought by newcomers. In friendship terms, the rich get richer. There are many economic contexts in which such dynamics are also of interest [13].

The key paper for our purposes was written by Albert, Jeong, and Barabási in 2000. They observed that the connectivity of scale-free networks, which depends on the highly-connected nodes, comes at a price: the destruction of these nodes will disconnect the network. If an attacker removes the best-connected nodes one after another, then past some threshold point the size of the largest component of the graph collapses [2].

Later work by Holme, Kim, Yoon and Han in 2002 extended this from attacks on vertices to attacks on edges; here, the attacker removes edges connecting high-degree nodes, and again, past some critical point, the network becomes disconnected [15]. They also suggested using *centrality* – technically, this is the 'betweenness centrality' of Freeman [12] – as an alternative to degree for attack targeting. (A node's centrality is, roughly speaking, the proportion of paths on which it lies.) Computing centrality is harder work for the attacker than observing vertex degree, but it enables him to attack networks (such as beta networks) where there is little or no variability in vertex order. Finally, in 2004, Zhao, Park and Lai modelled the circumstances in which a scale-free network can suffer cascading breakdown from the successive failure of high-connectivity nodes [23]. These ideas find some resonance in the field of strategic studies: for example, Soviet doctrine called for destroying a third of the enemy's network, jamming a further third, and hoping that the remaining third would collapse under the increased weight of traffic.

3   Naive Defences Don't Work

Given the obvious importance of the subject, and the fact that the Albert-Jeong-Barabási paper appeared in 2000, one obvious question is why there has been no published work since on how a network can defend itself against a decapitation attack. Here is one possible explanation: the two obvious defences don't work.

One of these is simply to replenish destroyed nodes with new nodes, and furnish them with edges according to the same scale-free rule that was used to generate the network initially. One might hope that some equilibrium would be found between attack and defence.

The other obvious defence is to replenish destroyed nodes, but to wire their edges according to a random graph model. In this way, we might hope that, under attack, a network would evolve from an efficient scale-free structure into a less efficient but more resilient random structure. In a real application, this might happen either as a result of nodes learning new behaviour, or by selective pressure on a node population with heterogeneous connectivity preferences: in peacetime the nodes with higher degree would become hubs, while in wartime they would be early casualties.

Nice as these ideas may seem in theory, they do not work at all well in practice. Figure 1 shows first (solid line) how the vertex-order attack of Albert, Jeong and Barabási works against a simulated network with no replenishment, then with random replenishment, then with scalefree replenishment. In the vanilla case the

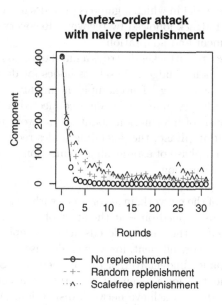

Fig. 1. Naive defences against vertex-order decapitation attack

attack takes two rounds to disconnect the network; with random replenishment it takes three, and with scale-free replenishment it takes four.

It seems that, to defend against these kinds of decapitation attacks on networks, we will need smarter defence strategies. But how should these be evolved, and what sort of framework should we use to evaluate them?

4 A Model of Repeated Attack and Defense

Previous researchers considered disruptive attacks on networks to be a single-round game. Such a model is suitable for applications such as a conventional war, in which the attacker has to expend a certain amount of effort to destroy the defender's command, control and communications, and one wishes to estimate how much; or a single epidemic in which a certain amount of resource must be spent to bring the disease under control.

However, there are many applications in which attack and defense evolve through multiple rounds: terrorism and music-sharing are only two examples. We now develop a framework for considering this more general case. Our ideas are inspired from evolutionary game theory developed by Axelrod and others [3,4]. This theory studies how games of multiple rounds differ from single-round games, and it has turned out to have significant explanatory power in applications from ethology to economics.

We now formalise a model in which a game is played with a number of rounds. Each round consists of attack followed by recovery. Recovery in turn consists of two phases: replenishment and adaptation.

In the **attack phase**, the attacker destroys a number of nodes (or, in a variant, of edges); this number is his budget. He selects nodes for destruction according to some rule, which is his strategy. For example, he might at each round destroy the ten nodes with the largest number of edges connected to them. He executes this strategy on the basis of information about the network topology.

In the **replenishment phase**, the defending nodes recruit a number of new nodes, and go through a phase of establishing connections – again, according to given strategies and information.

In the **adaptation** phase, the defending nodes may rewire links within each connected component of the network, in accordance with some defensive strategy. The adaptation phase is applied once at the start of the game, before the first round of attack; thereafter the game proceeds attack – replenish – adapt.

An attack strategy is more efficient, for a given defense strategy, if an attacker using it requires a smaller budget to disrupt the network. Similarly, a defense strategy is more efficient if, for a given attack strategy, it compels the attacker to expend a higher budget to achieve network disruption. (We will clarify this later once we have presented and discussed a few simulations.)

We assume initially that the attacker has perfect information about the network topology, and that her goal is simply to partition the network – that is, divide it into two or more nontrivial disjoint components. We assume that the defender has only local information, that it, each node shares the information available to those nodes with which it is connected. Thus, for example, if the attacker manages to split the network into two components, there is no way for them to reconnect. We also start off by assuming that the defence strategy affects only the adaptation phase, as only once nodes have connected to a network can they be programmed to follow it; so the replenishment phase is exogenous.

A further initial assumption is that the attack and defence budgets are roughly equal. By this we will mean that for each node destroyed in the attack phase, one node will be replaced in the resource addition phase. Thus the network will neither grow or shrink in absolute size and we can concentrate on connectivity effects. We will discuss other possible assumptions later, but the static budgets and global attack / local defence assumptions will get us started.

5   Defence Evolution – First Round

To analyse the vulnerability of a network, the selection of network elements (nodes or edges) destroyed in each round is the attacker's choice and constitutes her strategy. The attacker wishes to maximize the network damage caused per unit of work.

We will start off by considering a static attacker, using what we know to be a reasonable attack (vertex-order), and examine how the defence strategy can adapt. Then we will see what better attacks can be found against the best defence we found. Then we will look for a defence against the best attack we found in the last round, and so on. There is no guarantee that the process converges – there may be a specialised attack that works well against each defence, and vice versa – but if evolutionary games on networks behave like more traditional evolutionary games, we may expect to find some strategies that do well overall, as 'tit for tat' does in multi-round prisoners' dilemma. We may also expect to gain useful insights in the process.

5.1 Defense Strategy 1 – Random Replenishment

Our first defensive strategy is the simplest of all, and is one of the naive defences introduced in the above section. New nodes are joined to the graph at random. We assume that each attack round removes r nodes, and the replenishment round adds exactly r nodes, each of which is joined to the surviving vertices with probability p. r remains constant for each run of the simulation, while p increases from $k/(N - r)$ to $k/(N - 1)$ as the replenishment proceeds. In this strategy, the defender does nothing in the adaptation phase.

This models the case where new recruits to a subversive network simply contact any other subversives they can find; no attempt is made to reshape the network in response to the capture of leaders but the network is simply allowed to become more amorphous.

5.2 Defense Strategy 2 – Dining Steganographers

Our second defensive strategy is more sophisticated, and is inspired by the theory of anonymous communication as developed by computer scientists, most notably Chaum [10]. A node that acquires a high vertex order, and thus could be threatened by a vertex-order attack, splits itself into n nodes, arranged in a ring. The rings have two functions. First, they provide resilience: a ring broken at one point still supports communications between all its surviving nodes, and it is the simplest such structure. Second, nodes can route covert traffic between appropriate input and output links, and use encryption and other information-hiding mechanisms to conceal the traffic. This model was originally presented in Chaum's seminal 'dining cryptographers' paper cited above, so we might refer to it as the 'dining steganographers'. The collaborating nodes in each ring cannot conceal the existence of communication between them, as the cover traffic is visible to the attacker. However, from the attacker's viewpoint it is not obvious that these n nodes are acting as a virtual supernode.

Our focus here is on the effects of network topology, rather than on the higher-layer mechanisms that actually implement the covertness property and that provide any confidentiality of content or of routing data. We assume a world in which there is sufficient encrypted traffic (SSL, SSH, DRM, . . .) that encrypted traffic is not of itself suspicious so long as it is wrapped in a common ciphertext type. The attacker's input consists of traffic data collected from the backbone or from ISPs, and her output consists of decisions to send police officers to raid the premises associated with particular IP addresses. Her problem is this: given an observed pattern of communications, whom should she investigate first?

The precise mechanism of ring formation in our simulation is as follows. A vulnerable node decides to create a ring and recruits for the purpose a further $n - 1$ nodes from the new nodes introduced in the most recent replenishment round, or, if they are inadequate, from among its immediate neighbours. Existing ring members cannot be recruited, so rings may not overlap. Finally, recruits to a ring relinquish any existing links with the rest of the network, and the ring-forming node shares its external links uniformly among all the members of the ring.

5.3   Defense Strategy 3 – Revolutionary Cells

Our third defensive strategy is inspired by cells of revolutionaries, along the model favoured historically by a number of insurgent organisations. A node that acquires a high vertex order splits itself into n nodes, all linked with each other, with the previous outside connections split uniformly between them. In graph-theoretic language, each supernode is a clique.

As in ring formation, a node that considers itself vulnerable is allowed to split itself into a clique of nodes. The new nodes are drawn either from the pool of new nodes, or, if they are insufficient, from low-vertex-order neighbours of the clique-forming node. As before, this node's external edges are distributed uniformly among members, while other member nodes' former external edges are deleted.

Simulations – First Set. For our first set of simulations, we consider a scalefree network of $N = 400$ nodes. We use a Barabási-Albert network created by the following algorithm:

1. *Growth:* Starting with $m_0 = 40$ nodes, at each round we add $m = 10$ new nodes, each with 3 edges.
2. *Preferential Attachment:* The probability that a new node connects to node i is $\Pi(k_i) = k_i / \sum_j k_j$ where k_i is the degree of node i.

Having created the scalefree network, we then ran each of the above defensive strategies against a vertex-order attack.

Results. The results of the initial three simulations are given in Figure 2.

The red graph in Figure 2 provides a calibration baseline. As seen in the above section, random replenishment without adaptation is ineffective: within three rounds the size of the largest connected component has fallen by a half, from 400 nodes to well under 200.

The green graph shows that rings give only a surprisingly short-term defence benefit. They postpone network collapse from about two rounds to about a dozen rounds. Thereafter, the network is almost completely disconnected. In fact, the outcome is even worse than with random replenishment.

Cliques, on the other hand, work well. A few vertices are disconnected at each attack round, but as the cyan graph shows, the network itself remains robustly connected. This may provide some insight into why, although rings have seemed

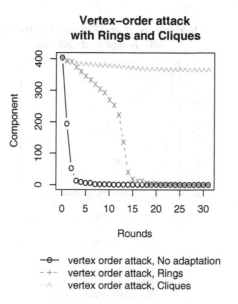

**Vertex–order attack
with Rings and Cliques**

—⊖— vertex order attack, No adaptation
- +- vertex order attack, Rings
··∧·· vertex order attack, Cliques

Fig. 2. Vertex order decapitation attack in rings, cliques and with no adaptation

attractive to theoreticians, those real revolutionary movements that have left some trace in the history books have used a cell structure instead.

6   Attack Evolution – First Round

Having tried a number of defence strategies and found that one of them – cliques – is effective, the next step is to try out a number of attack strategies to see if any of them is effective against our defences, and in particular against cliques.

Of the attack strategies we tried against a clique defence, the best performer is an attack based on centrality. We used the centrality algorithm of Brandes [8] to select the highest-centrality nodes for destruction at each round. As before, our calibration baseline is random replenishment. For this, the red and black graphs show performance against vertex-order and centrality attacks respectively. Both are equally effective; within two or three rounds the size of the largest connected component has been halved.

The green and blue graphs show that the same holds for rings: the network collapses completely after about a dozen rounds. Centrality attacks are very slightly more effective but there is not much in it.

The most interesting results from these simulations come from the magenta and cyan graphs, which show how cliques behave. Cyan shows, as before, a vertex-order attack with severity $m = 10$ being ineffective against a clique defence. Magenta shows the effect on such a network of a centrality attack. Here the largest connected component retains about 400 nodes until the network

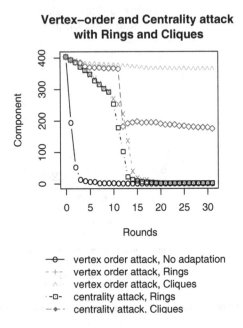

Fig. 3. Rings and Cliques defense under vertex order and centrality attacks

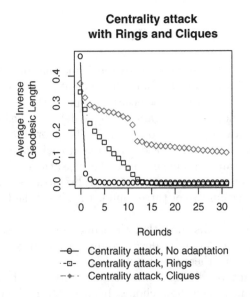

Fig. 4. Average inverse geodesic lengths of rings and clique adaption, under centrality attack

suddenly partitions at 14 rounds, whereafter a largest-component size of about 200 is maintained stably.

Some insight into the internal mechanics can be gleaned from Figure 4. This shows the *average inverse geodesic length.* To calculate this, for each node, we find the length of the shortest path to each other node, and take the inverse (we take the length to be infinite, and thus the inverse to be zero, if the nodes are in disjoint components). We average this value over all $n(n-1)/2$ pairs of nodes. This value falls sharply for defense without adaptation, and falls steadily for defense with rings. These falls reflect increasing difficulty in internode communication. With cliques, the vertex-order attack has little effect, while the centrality attack makes steadily increasing progress on a graph of 400 vertices, until it achieves partition and reduces the largest component to about 200 vertices. But it makes only slow progress thereafter.

6.1   Clique Sizes

We next ran a simulation comparing how well defense works when using different sizes of rings and cliques. Ring size appears to make little difference; rings are just not an effective defence other than in the very short term. However, varying the clique size yields the results displayed in Figure 5.

This shows that under a centrality attack, the performance of the defense increases steadily with the size of the clique. There is still a phase transition after about 14 rounds or so after which the largest connected component becomes

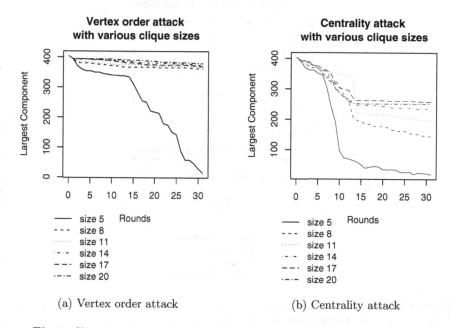

(a) Vertex order attack   (b) Centrality attack

Fig. 5. Clique recovery with different clique sizes under a centrality attack

significantly smaller, but the size of this equilibrium component increases steadily from about 150 with clique size 8 to almost 300 at clique size 20.

7   Defence Evolution – Second Round

Now that we know centrality attacks are powerful, we have tried a number of other possible defences. The most promising at present appears to be a compound defence based on cliques and delegation.

The idea behind delegation is fairly simple. A node that is becoming too well-connected selects one of its neighbours as a 'deputy' and connects it to a second neighbour, with which it then disconnects. This reflects normal human behaviour even in peacetime: busy leaders pass new recruits on to colleagues. In wartime, and with an enemy that might resort to vertex-order attacks, the incentive to delegate is even greater. Thus a terrorist leader who gets an offer from a wealthy businessman to finance an attack might simply introduce him to a young militant who wants to carry one out. The leader need now maintain communications with at most one of the two.

Delegation on its own is rather slow; it takes dozens of rounds for delegation to 'immunise' a network against vertex-order attack. If a vanilla scale-free network is going to be exposed to either a vertex-order or centrality attack from the next

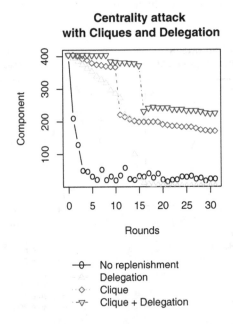

Fig. 6. Component size: clique, immunization by delegation, and combined clique and delegation defenses against centrality attack

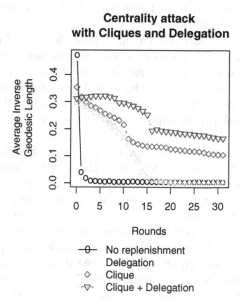

Fig. 7. Clique, immunization by delegation, and combined clique and delegation defenses against centrality attack

round, then drastic action (such as clique formation) is needed at once; else it will be disconnected within two or three rounds. Slower defences like delegation can however play a role, provided they are started from network formation or a reasonable time period (say 20 rounds) before the attack begins.

It turns out that the delegation defence, on its own, is rather like the rings of dining steganographers. Network fragmentation is postponed (about 14 rounds with the parameters used here) though not ultimately averted.

What is interesting, however, is this. If we form a network and immunise it by running the delegation strategy, then run a clique defence as well from the initiation of hostilities, this compound strategy works rather better than ordinary cliques. Figure 6 shows the simulation results.

Figure 7 may give some insight into the mechanisms. Delegation results in shorter path lengths under attack: it postpones and slows down the growth of path length that otherwise results from hub elimination. As a result, equilibrium is achieved later, and with a larger minimum connected component.

8   Conclusions and Future Work

In this paper, we have built a bridge between network science and traffic analysis.

For some years, people have discussed what sort of communications topologies might be ideal for covert communication in the presence of powerful adversaries, and whether network science might be of practical use in covert conflicts – whether to insurgents or to counterinsurgency forces [5,18]. Our work makes a start on dealing with this question systematically.

Albert, Jeong and Barabási showed that although a scalefree network provides better connectivity, this comes at a cost in robustness – an opponent can disconnect a network quickly by concentrating its firepower on well-connected nodes. In this paper, we have asked the logical next questions. What sort of defence should be planned by operators of such a network? And what sort of framework can be developed in which to test successive refinements of attack, defense, counterattack and so on?

First, we have shown that naive defences don't work. Simply replacing dead hubs with new recruits does not slow down the attacker much, regardless of whether link replacement follows a random or scale-free pattern.

Moving from a single-shot game to a repeated game provides a useful framework. It enables concepts of evolutionary game theory to be applied to network problems.

Next, we used the framework to explore two more sophisticated defensive strategies. In one, potentially vulnerable high-order nodes are replaced with rings of nodes, inspired by a standard technique in anonymous communications. In the other, they are replaced by cliques, inspired by the cell structure often used in revolutionary warfare. To our surprise we found that rings were all but useless, while cliques are remarkably effective. This may be part of the reason why cell structures have been widely used by capable insurgent groups.

Next, we searched for attacks that work better against clique defences. We found that the centrality attack of Holme et al does indeed appear to be more powerful, although it can be more difficult to mount as evaluating node centrality involves knowledge of the entire topology of the network. Centrality attacks may reflect the modern reality of counterinsurgency based on pervasive communications intelligence and, in particular, traffic analysis.

Now we are searching for defences that work better against centrality attacks. A promising candidate appears to be the delegation defence, combined with cliques. This combination may in some ways reflect the reported 'virtualisation' strategies of some modern insurgent networks.

Another promising direction would be to consider the role of communities in the robustness of networks. For instance, social networks can be both scale-free and navigable by random walks if they are divided into communities. Hence, it might be more prudent for an attacker to take on the network community by community rather than removing important nodes at a network level, we leave further investigation in this direction to future work.

Above all, this work provides a systematic way to evolve and test security concepts relating to the topology of networks. Clearly the coevolution of attack and defense can be taken much further. Further work includes testing:

1. networks that grow or shrink, maybe with endogenous replenishment (current recruitment a function of past operational success)
2. imperfectly informed attackers, such as policemen who have access to the records of some but not all phone companies or email service providers, or who must use purely local measures of centrality

3. perfectly informed defenders, who can coordinate connectivity globally
4. budget tradeoffs – for example, a defender might be able to hide specific edges but only at some cost to his replenishment budget
5. heterogeneous networks, with subpopulations having different robustness preferences
6. dynamic strategies that detect opponents' strategies and respond
7. different attacker goals. For example, some say that the Iraqi rebel leader Al-Zarqawi is not bin Laden's subordinate but his competitor. So an attack objective might be not just partition, but to divide the opposition into groups of less than a certain size. When attacking an ad-hoc sensor network, the goal might be to reduce the effective bandwidth, and there might be interaction with routing algorithms.

Preliminary though it is, we suggest that this work has broad potential applicability – from making the Internet more resilient against natural disasters and malicious attacks, to the question of how best to disrupt (or design) subversive networks.

Acknowledgements. We have had useful feedback on an early versions of this paper from Albert-László Barabási, Mike Bond, George Danezis, Karen Spärck Jones and Chris Lesniewski-Laas. The first author was supported by a scholarship from the Gates Trust.

References

1. Albert, R., Barabási, A.L.: Statistical Mechanics of Complex Networks. Reviews of Modern Physics 74, 47 (2002)
2. Albert, R., Jeong, H., Barabási, A.L.: Error and attack tolerance of complex networks. Nature 406, 387–482 (2000)
3. Axelrod, R.: The Evolution of Cooperation. Basic Books, New York (1984)
4. Axelrod, R.: The Complexity of Cooperation. Princeton University Press, Princeton (1997)
5. Ballester, C., Calvó-Armengol, A., Zenou, Y.: Who's Who in Crime Networks – Wanted the Key Player, IUI Working Paper Series 617, The Research Institute of Industrial Economics (2004)
6. Barabási, A.L., Albert, R.: Emergence of scaling in random networks. Science 286, 509–512 (1999)
7. Bollobás, B.: Random Graphs. Academic Press, London (1985)
8. Brandes, U.: A Faster Algorithm for Betweenness Centrality. J. Math. Soc. 25(2), 163–177 (2001)
9. Carrington, P.J., Scott, J., Wassermann, S.: Models and Methods in Social Network Analysis. Cambridge University Press, Cambridge (2005)
10. Chaum, D.: The Dining Cryptographers Problem: Unconditional Sender and Recipient Untraceability. Journal of Cryptology 1, 65–75 (1989)
11. Erdös, P., Renyi, A.: On Random Graphs. Publicationes Mathematicae 6, 290–297 (1959)
12. Freeman, L.C.: A set of measuring centrality based on betweenness. Sociometry 40, 35–41 (1977)

426   S. Nagaraja and R. Anderson

13. Jackson, M.O.: A Survey of Models of Network Formation: Stability and Efficiency A. In: Demange, G., Wooders, M. (eds.) Group Formation in Economics: Networks, Clubs, and Coalitions, Cambridge University Press, Cambridge
14. Krebs, V.E.: Mapping Networks of Terrorist Cells. Connections 12(3), http://www.locative.net/tcmreader/index.php?mapping;krebs
15. Holme, P., Kim, B.J., Yoon, C.N., Han, S.K.: Attack Vulnerability of Complex Networks. Phys. Rev. E 65, art. no. 018101 (2002)
16. Milgram, S.: The Small World Problem. Psychology Today 2, 60–87 (1967)
17. Newmann, M.E.J.: Structure and Function of Complex Networks. SIAM Review 45, 167–256 (2003)
18. Sparrow, M.K.: The Application of Network Analysis to Criminal Intelligence: An assessment of the prospects. Social Networks 13, 253–274 (1990)
19. Thompson, N.: The Network Effect – Why Senegal's bold anti-AIDS program is working. The Boston Globe (January 5, 2003), http://www.newamerica.net/index.cfm?pg=article&DocID=1092
20. Wassermann, S., Faust, K.: Social Network Analysis. Cambridge University Press, Cambridge (1994)
21. Watts, D.J.: Six Degrees: The Science of a Connected Age. Norton, New York (2003)
22. Watts, D.J., Strogatz, S.H.: Collective Dynamics of Small-World Networks. Nature 393, 440–442 (1998)
23. Zhao, L.A., Park, K.H., Lai, Y.C.: Attack vulnerability of scale-free networks due to cascading breakdown. Physical review E 70, 035101 (R) (2004)

Author Index

Lecture Notes in Computer Science

Sublibrary 1: Theoretical Computer Science and General Issues

For information about Vols. 1– 5035
please contact your bookseller or Springer